Student's Solutions Manual

Volume 2

to Accompany Swokowski's

Calculus: The Classic Edition

Jeffery A. Cole
Anoka-Ramsey Community College

Gary K. Rockswold
Minnesota State University, Mankato

Brooks/Cole
Thomson Learning

Australia • Canada • Mexico • Singapore • Spain • United Kingdom • United States

Sponsoring Editor: *Gary Ostedt*
Assistant Editor: *Carol Ann Benedict*
Production Editor: *Scott Brearton*
Cover Design: *Vernon Boes*
Marketing Team: *Karin Sandberg, Beth Kroenke*
Print Buyer: *Tracy Brown*
Cover Printing: *Malloy Lithographing, Inc.*
Printing and Binding: *Malloy Lithographing, Inc.*

For more information about this or any other Brooks/Cole products, contact:
BROOKS/COLE
511 Forest Lodge Road
Pacific Grove, CA 93950 USA
www.brookscole.com
1-800-423-0563 (Thomson Learning Academic Resource Center)

Printed in the United States of America

5 4 3 2 1

ISBN 0-534-38281-9

PREFACE

This *Student's Solutions Manual* contains selected solutions and strategies for solving typical exercises in Chapters 11 through 19 of the text, *Calculus: The Classic Edition*, by Earl W. Swokowski. No particular problem pattern has been followed in this manual, other than to include solutions to approximately two-thirds of the odd-numbered problems in each section, with extra attention paid to the applied "word" problems. We have tried to illustrate enough solutions so that the student will be able to obtain an understanding of all types of problems in each section.

This manual is *not* intended to be a substitute for regular class attendance. However, a significant number of today's students are involved in various outside activities, and find it difficult, if not impossible, to attend all class sessions. This manual should help supplement the needs of these students. In addition, it is our hope that this manual's solutions will enhance the understanding of all readers of the material and provide insights to solving other exercises.

We have included extra figures when appropriate to enhance a solution. All figures are new and have been plotted using computer software to provide a high degree of precision.

We would appreciate any feedback concerning errors, solution correctness, solution style, or manual style. These and any other comments may be sent directly to us or in care of the publisher.

We would like to thank: Editor Dave Geggis, for entrusting us with this project and continued support; Sally Lifland and Gail Magin of Lifland et al., Bookmakers, for assembling the final manuscript; and George and Bryan Morris, for preparing the new figures. We dedicate this book to Earl Swokowski, and thank him for his confidence in us and for being a great mentor.

Jeffery A. Cole
Anoka-Ramsey Community College
11200 Mississippi Blvd. NW
Coon Rapids, MN 55433

Gary K. Rockswold
Minnesota State University, Mankato
P.O. Box 41
Mankato, MN 56002

To the Student

This is a text supplement and should be read along *with* the text. Read all exercise solutions in this manual since explanations of concepts are given and then these concepts appear in subsequent solutions. We do not usually review all concepts necessary to solve a particular problem. If you are having difficulty with a previously covered concept, look back to the section where it was covered for more complete help. The writing style we have used in this manual reflects the way we explain concepts to our own students. It is not as mathematically precise as that of the text, but our students have found that these explanations help them understand difficult concepts with ease.

The most common complaint about solutions manuals that we receive from our students is that there are not enough exercise solutions in them. We believe there is a sufficient number of solutions in this manual, with about one-third of the exercises solved in every section—all of them are odd-numbered exercises.

Lengthier explanations and more steps are given for the more difficult problems. We have included additional intuitive information that our students have found helpful—see page 9, Exercise 13. We have included ample references to the numbered items in the text, that is, definitions, theorems, guidelines, and so on.

In the review sections, the solutions are somewhat abbreviated since more detailed solutions were given in previous sections. However, this is not true for the word problems in these sections since they are unique. In easier groups of exercises, representative solutions are shown. Occasionally, alternate solutions are also given. When possible, we tried to make each piece of art with the same scale to show a realistic and consistent graph.

This manual was done using EXP: *The Scientific Word Processor*. There are some limitations to the number of boxes and characters per line, which sometimes causes the appearance of an "incomplete" line. We have used a variety of display formats for the mathematical equations, including centering, vertical alignment, and flushing text to the right. We hope that these make reading and comprehending the material easier for you.

Notations

The following notations are used in the manual.

Note:	Notes to the student pertaining to hints on solutions, common mistakes, or conventions to follow.
{ }	{comments to the reader are in braces}
LHS, RHS	{Left Hand Side, Right Hand Side – used for identities}
$\Rightarrow$	{implies, next equation, logically follows}
$\Leftrightarrow$	{if and only if, is equivalent to}
•	{bullet, used to separate problem statement from solution or explanation}
★	{used to identify the answer to the problem}
§	{<u>section</u> references}
$\forall$	{For all, i.e., $\forall x$ means "for all x".}
$\mathbb{R} - \{a\}$	{The set of all real numbers except a.}
$\therefore$	{therefore}

(continued on next page)

The following notations are defined in the manual, and also listed here for convenience.

DNE	{ Does Not Exist }
L, I, S	{ the original limit, integral, or series }
T, S	{ the result is obtained from using the trapezoidal rule or Simpson's rule }
$\stackrel{\text{A}}{=}$	{ integration by parts has been applied— the parts are defined following the solution }
$\{\frac{\infty}{\infty}\}$, $\{\frac{0}{0}\}$	{ L'Hôpital's rule is applied when this symbol appears }
C, D	{ converges or convergent, diverges or divergent }
AC, CC	{ absolutely convergent, conditionally convergent }
DERIV	{ see notes in §11.8 and §11.9 for this notation }
V, F, l	{ vertex, focus, and directrix of a parabola }
C, V, F, M	{ center, vertices, foci, and end points of the minor axis of an ellipse }
C, V, F, W	{ center, vertices, foci, and end points of the conjugate axis of a hyperbola }
D	{ discriminant value ($B^2 - 4AC$) in §12.4 }
VT, HT	{ vertical tangent, horizontal tangent }
↑, ↓	{ increasing, decreasing }
CN	{ critical number(s) }
PI	{ point(s) of inflection }
CU, CD	{ concave up, concave down }
MAX, MIN	{ absolute maximum or minimum }
SP	{ saddle point }
$[\![n]\!]$	{ equation number n }
$\mathbb{S}$	{ surface area }
J	{ Jacobian }
$\mathbb{A}$	{ the value of $\mathbf{F} \cdot < -f_x,\ -f_y,\ 1 >$ }
IF	{ integrating factor }
LMAX, LMIN	{ local maximum or minimum }
VA, HA, OA	{ vertical, horizontal, or oblique asymptote }
QI, QII, QIII, QIV	{ quadrants I, II, III, IV }
NTH, INT, BCT, LCT, RAT, ROT, AST	{ various series tests: nth-term, integral, basic comparison, limit comparison, ratio, root, the alternating series }

Table of Contents

Chapter 11: Infinite Series

Exercises 11.1

Note: In Exercises 1–16, the first four terms are found by

substituting 1, 2, 3, and 4 for n in the nth term.

[1] $a_n = \dfrac{n}{3n+2} \Rightarrow a_1 = \dfrac{1}{3(1)+2} = \dfrac{1}{5},\ a_2 = \dfrac{2}{3(2)+2} = \dfrac{2}{8} = \dfrac{1}{4},$

$$a_3 = \frac{3}{3(3)+2} = \frac{3}{11},\ a_4 = \frac{4}{3(4)+2} = \frac{4}{14} = \frac{2}{7}.$$

$$\lim_{n\to\infty} a_n = \lim_{n\to\infty} \frac{n}{3n+2} = \lim_{n\to\infty} \left(\frac{n}{3n+2} \cdot \frac{1/n}{1/n}\right) = \lim_{n\to\infty} \frac{1}{3+2/n} = \frac{1}{3+0} = \frac{1}{3}$$

[5] Since $a_n = -5$ for every n, it follows that $a_1 = -5$, $a_2 = -5$, $a_3 = -5$,

and $a_4 = -5$. $\lim\limits_{n\to\infty} a_n = \lim\limits_{n\to\infty} (-5) = -5$

[11] $a_n = (-1)^{n+1} \dfrac{3n}{n^2+4n+5}$. First, consider the factor $(-1)^{n+1}$. $(-1)^{n+1}$ is equal to $+1$ when $n+1$ is an even integer and is equal to -1 when $n+1$ is an odd integer. $(-1)^{n+1}$ has the effect of causing the terms of the sequence to alternate in sign.

Thus, $a_1 = (-1)^{1+1} \dfrac{3(1)}{1^2+4(1)+5} = \dfrac{3}{10},\ a_2 = (-1)^{2+1} \dfrac{3(2)}{2^2+4(2)+5} = -\dfrac{6}{17},$

$$a_3 = (-1)^{3+1} \frac{3(3)}{3^2+4(3)+5} = \frac{9}{26},\ a_4 = (-1)^{4+1} \frac{3(4)}{4^2+4(4)+5} = -\frac{12}{37}.$$

Since the terms alternate in sign, we will examine $\lim\limits_{n\to\infty} |a_n|$.

$$\lim_{n\to\infty} |a_n| = \lim_{n\to\infty} \left|(-1)^{n+1} \frac{3n}{n^2+4n+5}\right| = \lim_{n\to\infty} \frac{3n}{n^2+4n+5} = 0,$$

since the highest power of n is greater in the denominator than in the numerator.

Thus, by (11.8), $\lim\limits_{n\to\infty} a_n = \lim\limits_{n\to\infty} (-1)^{n+1} \dfrac{3n}{n^2+4n+5} = 0.$

[15] $a_n = 1 + (-1)^{n+1} \Rightarrow$ $a_1 = 1 + (-1)^{1+1} = 1 + 1 = 2,$

$a_2 = 1 + (-1)^{2+1} = 1 + (-1) = 0,$

$a_3 = 1 + (-1)^{3+1} = 1 + 1 = 2,$

$a_4 = 1 + (-1)^{4+1} = 1 + (-1) = 0.$

The sequence follows the pattern 2, 0, 2, 0, 2, 0, ..., and a_n will never approach one value. Thus, $\lim\limits_{n\to\infty} a_n = \lim\limits_{n\to\infty} \left[1 + (-1)^{n+1}\right]$ DNE since it does not converge to a

real number L.

Note: Let C denote *converges* and D denote *diverges*.

Also, let L denote the limit of the sequence, if it exists.

[17] $\lim_{n\to\infty} a_n = \lim_{n\to\infty} 6(-\frac{5}{6})^n = 6 \lim_{n\to\infty} (-\frac{5}{6})^n = 6 \cdot 0 = 0$ by (11.6)(i) with $r = -\frac{5}{6}$; C

[19] In order to determine the limit, refer to the graph of $y = \arctan x$ (Figure 8.6, page 429). From the graph, we can see that $\lim_{x\to\infty} \arctan x = \frac{\pi}{2}$. Thus, by (11.5)(i), $\lim_{n\to\infty} a_n = \lim_{n\to\infty} \arctan n = \frac{\pi}{2}$; C.

[23] Using (11.5)(i) and L'Hôpital's rule (10.2), $\lim_{x\to\infty} \frac{\ln x}{x} \{\frac{\infty}{\infty}\} = \lim_{x\to\infty} \frac{1/x}{1} = 0$,

and hence $\lim_{n\to\infty} |a_n| = \lim_{n\to\infty} \left|(-1)^n \frac{\ln n}{n}\right| = \lim_{n\to\infty} \frac{\ln n}{n} = 0$.

Thus, by (11.8), $\lim_{n\to\infty} a_n = \lim_{n\to\infty} (-1)^n \frac{\ln n}{n} = 0$; C.

[25] Using (11.5)(ii), $\lim_{x\to\infty} \frac{4x^4 + 1}{2x^2 - 1} \{\frac{\infty}{\infty}\} = \lim_{x\to\infty} \frac{16x^3}{4x} = \lim_{x\to\infty} 4x^2 = \infty$.

Thus, $\lim_{n\to\infty} a_n = \lim_{n\to\infty} \frac{4n^4 + 1}{2n^2 - 1} = \infty$; D.

[29] By (7.32)(ii), $\lim_{n\to\infty} a_n = \lim_{n\to\infty} (1 + \frac{1}{n})^n = e$; C.

[31] This exercise is similar to Example 8.

Since $\lim_{n\to\infty} \sin n$ DNE, we will apply (11.7) to determine $\lim_{n\to\infty} a_n = \lim_{n\to\infty} \frac{\sin n}{2^n}$.

Now, $-1 \le \sin n \le 1$ for all $n \Rightarrow -\frac{1}{2^n} \le \frac{\sin n}{2^n} \le \frac{1}{2^n}$.

Since $\lim_{n\to\infty} -\frac{1}{2^n} = 0$ and $\lim_{n\to\infty} \frac{1}{2^n} = 0$, it follows that $\lim_{n\to\infty} \frac{\sin n}{2^n} = 0$ by (11.7); C.

[33] $\lim_{n\to\infty} a_n = \lim_{n\to\infty} \left(\frac{n^2}{2n - 1} - \frac{n^2}{2n + 1}\right) = \lim_{n\to\infty} \frac{n^2(2n + 1) - n^2(2n - 1)}{(2n - 1)(2n + 1)} =$

$\lim_{n\to\infty} \frac{2n^2}{4n^2 - 1} = \lim_{n\to\infty} \frac{2n^2/n^2}{4n^2/n^2 - 1/n^2} = \frac{1}{2}$; C.

[35] $a_n = \cos \pi n \Rightarrow a_1 = \cos \pi = -1$, $a_2 = \cos 2\pi = 1$, $a_3 = \cos 3\pi = -1$, $a_4 = \cos 4\pi = 1$. The sequence follows the pattern $-1, 1, -1, 1, -1, 1, \ldots$, and does not approach one value for large values of n. Thus, $\lim_{n\to\infty} a_n = \lim_{n\to\infty} \cos \pi n$ DNE; D.

[37] By Exercise 17, §10.2, $\lim_{x\to\infty} x^{1/x} = 1$.

Thus, by (11.5)(i), $\lim_{n\to\infty} a_n = \lim_{n\to\infty} n^{1/n} = 1$; C.

[41] $\lim_{n\to\infty} a_n = \lim_{n\to\infty} (\sqrt{n + 1} - \sqrt{n}) = \lim_{n\to\infty} \left(\frac{\sqrt{n + 1} - \sqrt{n}}{1} \cdot \frac{\sqrt{n + 1} + \sqrt{n}}{\sqrt{n + 1} + \sqrt{n}}\right) =$

$\lim_{n\to\infty} \frac{(n + 1) - (n)}{\sqrt{n + 1} + \sqrt{n}} = \lim_{n\to\infty} \frac{1}{\sqrt{n + 1} + \sqrt{n}} = 0$; C.

[43] (a) Next year's bird population on island A is determined by the number of birds that stay (90% of its present population) and the number of birds that migrate (5% of island C's present population). Thus, $A_{n+1} = 0.90A_n + 0.05C_n$. In a similar manner, $B_{n+1} = 0.80B_n + 0.10A_n$ and $C_{n+1} = 0.95C_n + 0.20B_n$.

(b) Let $\lim_{n\to\infty} A_n = a$, $\lim_{n\to\infty} B_n = b$, and $\lim_{n\to\infty} C_n = c$. Then, as $n \to \infty$,

$$\lim_{n\to\infty} A_{n+1} = \lim_{n\to\infty} 0.90A_n + \lim_{n\to\infty} 0.05C_n \Rightarrow$$

$$a = 0.90a + 0.05c \Rightarrow 0.1a = 0.05c \Rightarrow c = 2a,$$

$$\lim_{n\to\infty} B_{n+1} = \lim_{n\to\infty} 0.80B_n + \lim_{n\to\infty} 0.10A_n \Rightarrow$$

$$b = 0.80b + 0.10a \Rightarrow 0.2b = 0.1a \Rightarrow a = 2b,$$

and $$\lim_{n\to\infty} C_{n+1} = \lim_{n\to\infty} 0.95C_n + \lim_{n\to\infty} 0.20B_n \Rightarrow$$

$$c = 0.95c + 0.20b \Rightarrow 0.05c = 0.2b \Rightarrow c = 4b.$$

Now, $35{,}000 = a + b + c = 2b + b + 4b = 7b \Rightarrow b = 5000$.

Thus, there will be 5000 birds on B, 10,000 birds on A, and 20,000 birds on C.

[47] (a) $a_1 = 1$, $a_2 = \cos a_1 = \cos 1 \approx 0.540302$, $a_3 = \cos a_2 \approx 0.857553$, $a_4 = \cos a_3 \approx 0.654290$, $a_5 = \cos a_4 \approx 0.793480$. As this process is repeated *many* times, the sequence appears to converge to approximately 0.7390851.

(b) $a_{k+1} = \cos a_k \Rightarrow \lim_{k\to\infty} a_{k+1} = \lim_{k\to\infty} \cos a_k \Rightarrow \lim_{k\to\infty} a_{k+1} = \cos\left(\lim_{k\to\infty} a_k\right) \Rightarrow$

$L = \cos L$.

[51] (a) $f(x) = \frac{1}{4}\sin x \cos x + 1 = \frac{1}{8}(2\sin x \cos x) + 1 = \frac{1}{8}\sin 2x + 1 \Rightarrow$
$|f'(x)| = \frac{1}{4}|\cos 2x| \le \frac{1}{4} < 1$, since $|\cos 2x| \le 1$ for any x. Let $B = \frac{1}{4}$.

Thus, the sequence converges for any a_1.

(b) Letting $a_{k+1} = f(a_k) = \frac{1}{4}\sin a_k \cos a_k + 1 = \frac{1}{8}\sin 2a_k + 1$ with $a_1 = 1$ and then $a_1 = -100$ gives the following results.

$a_1 = 1$	$a_1 = -100$
$a_2 = 1.113662$	$a_2 = 1.109162$
$a_3 = 1.099015$	$a_3 = 1.099697$
$a_4 = 1.101207$	$a_4 = 1.101107$
$a_5 = 1.100884$	$a_5 = 1.100899$

It appears that $\lim_{n\to\infty} a_n \approx 1.10$.

Exercises 11.2

Note: It is important that you understand the difference between a sequence and a series. Carefully compare (11.1) and (11.11). In a series, the terms are added together, whereas in a sequence, they are not.

1 (a) $a_n = \dfrac{-2}{(2n+5)(2n+3)} \Rightarrow S_1 = a_1 = \dfrac{-2}{(2\cdot 1+5)(2\cdot 1+3)} = \dfrac{-2}{7\cdot 5} = -\dfrac{2}{35}$;

$S_2 = (a_1) + a_2 = S_1 + a_2 = -\frac{2}{35} + \frac{-2}{63} = -\frac{4}{45}$;

$S_3 = (a_1 + a_2) + a_3 = S_2 + a_3 = -\frac{4}{45} + \frac{-2}{99} = -\frac{6}{55}$.

(b) Using partial fractions, $a_n = \dfrac{-2}{(2n+5)(2n+3)} = \dfrac{1}{2n+5} - \dfrac{1}{2n+3}$.

$$\begin{aligned} S_n &= a_1 + a_2 + a_3 + \cdots + a_n \\ &= \left(\tfrac{1}{7} - \tfrac{1}{5}\right) + \left(\tfrac{1}{9} - \tfrac{1}{7}\right) + \left(\tfrac{1}{11} - \tfrac{1}{9}\right) + \cdots + \left(\frac{1}{2n+5} - \frac{1}{2n+3}\right) \\ &= -\tfrac{1}{5} + \left(\tfrac{1}{7} - \tfrac{1}{7}\right) + \left(\tfrac{1}{9} - \tfrac{1}{9}\right) + \cdots + \left(\frac{1}{2n+3} - \frac{1}{2n+3}\right) + \frac{1}{2n+5} \\ &= \frac{1}{2n+5} - \frac{1}{5} = -\frac{2n}{5(2n+5)}. \end{aligned}$$

Note: It may be easier to compute the values in part (a) *after* finding the general formula in part (b).

(c) $$\begin{aligned} S &= \lim_{n\to\infty} S_n \\ &= \lim_{n\to\infty}\left[-\frac{2n}{5(2n+5)}\right] \quad \{\text{consider only the terms with the highest powers of } n\} \\ &= -\frac{2}{5\cdot 2} = -\frac{1}{5}. \end{aligned}$$

5 (a) $S_1 = a_1 = \ln\frac{1}{2} = \ln 1 - \ln 2 = 0 - \ln 2 = -\ln 2$;

$S_2 = S_1 + a_2 = -\ln 2 + \ln\frac{2}{3} = -\ln 2 + (\ln 2 - \ln 3) = -\ln 3$;

$S_3 = S_2 + a_3 = -\ln 3 + \ln\frac{3}{4} = -\ln 3 + (\ln 3 - \ln 4) = -\ln 4$.

(b) $a_n = \ln\dfrac{n}{n+1} = \ln n - \ln(n+1)$.

$$\begin{aligned} S_n &= (\ln 1 - \ln 2) + (\ln 2 - \ln 3) + (\ln 3 - \ln 4) + \cdots + \left[\ln n - \ln(n+1)\right] \\ &= \ln 1 + (\ln 2 - \ln 2) + (\ln 3 - \ln 3) + \cdots + (\ln n - \ln n) - \ln(n+1) \\ &= \ln 1 - \ln(n+1) = -\ln(n+1). \end{aligned}$$

(c) $S = \lim_{n\to\infty} S_n = \lim_{n\to\infty}\left[-\ln(n+1)\right] = -\lim_{n\to\infty}\ln(n+1) = -\infty$.

Since $\{S_n\}$ diverges, the series $\sum a_n$ diverges and has no sum.

Note: From (11.15), the series converges if $|r| < 1$ and diverges is $|r| \geq 1$.

[7] $3 + \frac{3}{4} + \cdots + \frac{3}{4^{n-1}} + \cdots = 3 + 3(\frac{1}{4}) + 3(\frac{1}{4})^2 + \cdots + 3(\frac{1}{4})^{n-1} + \cdots$.

From this series, we see that $a = 3$ and $r = \frac{1}{4}$. Thus, $S = \frac{a}{1 - r} = \frac{3}{1 - \frac{1}{4}} = 4$; C.

[11] $0.37 + 0.0037 + \cdots + \frac{37}{(100)^n} + \cdots =$

$$0.37 + 0.37(\tfrac{1}{100}) + 0.37(\tfrac{1}{100})^2 + \cdots + 0.37(\tfrac{1}{100})^{n-1} + \cdots.$$

From this series, we see that $a = 0.37$ and $r = \frac{1}{100}$.

Thus, $S = \frac{a}{1 - r} = \frac{0.37}{1 - \frac{1}{100}} = \frac{37}{99}$; C.

[13] $\sum_{n=1}^{\infty} 2^{-n}3^{n-1} = \sum_{n=1}^{\infty} \frac{3^{n-1}}{2^n} = \sum_{n=1}^{\infty} \frac{1}{2}(\frac{3}{2})^{n-1} =$

$$\tfrac{1}{2} + \tfrac{1}{2}(\tfrac{3}{2}) + \tfrac{1}{2}(\tfrac{3}{2})^2 + \cdots + \tfrac{1}{2}(\tfrac{3}{2})^{n-1} + \cdots \Rightarrow a = \tfrac{1}{2} \text{ and } r = \tfrac{3}{2} > 1 \Rightarrow \text{D}$$

[17] $1 - x + x^2 - x^3 + \cdots + (-1)^n x^n + \cdots =$

$$1 + 1(-x) + 1(-x)^2 + 1(-x)^3 + \cdots.$$

This is a geometric series with $a = 1$ and $r = -x$. S converges if $|r| < 1$.

Thus, S converges if $|-x| < 1$, or, equivalently, $-1 < x < 1$.

$$a = 1,\ r = -x \Rightarrow S = \frac{a}{1 - r} = \frac{1}{1 - (-x)} = \frac{1}{1 + x}.$$

[19] $\frac{1}{2} + \frac{(x - 3)}{4} + \frac{(x - 3)^2}{8} + \cdots + \frac{(x - 3)^n}{2^{n+1}} + \cdots =$

$$\frac{1}{2} + \frac{1}{2}\left(\frac{x - 3}{2}\right) + \frac{1}{2}\left(\frac{x - 3}{2}\right)^2 + \cdots + \frac{1}{2}\left(\frac{x - 3}{2}\right)^n + \cdots.$$

This is a geometric series with $a = \frac{1}{2}$ and $r = \frac{x - 3}{2}$.

$|r| < 1 \Rightarrow \left|\frac{x - 3}{2}\right| < 1 \Rightarrow |x - 3| < 2 \Rightarrow -2 < x - 3 < 2 \Rightarrow 1 < x < 5.$

$$S = \frac{a}{1 - r} = \frac{1/2}{1 - \left(\frac{x - 3}{2}\right)} = \frac{1}{5 - x}.$$

[21] $0.\overline{23} = 0.23 + 0.0023 + 0.000023 + \cdots$

$= \frac{23}{100} + \frac{23}{(100)^2} + \frac{23}{(100)^3} + \cdots$

$= \frac{23}{100} + \frac{23}{100}(\frac{1}{100}) + \frac{23}{100}(\frac{1}{100})^2 + \cdots$

$= \frac{\frac{23}{100}}{1 - \frac{1}{100}} = \frac{23}{99}$ since $a = \frac{23}{100}$ and $r = \frac{1}{100}$

[23] $3.2\overline{394} = 3.2 + 0.0394 + 0.0000394 + 0.0000000394 + \cdots$

$$= 3.2 + \frac{394}{10^4} + \frac{394}{10^7} + \frac{394}{10^{10}} + \cdots$$

$$= 3.2 + \frac{394}{10{,}000} + \frac{394}{10{,}000}\left(\frac{1}{1000}\right) + \frac{394}{10{,}000}\left(\frac{1}{1000}\right)^2 + \cdots$$

$$= 3.2 + \frac{\frac{394}{10{,}000}}{1 - \frac{1}{1000}} = 3.2 + \frac{394}{9990} = \frac{16{,}181}{4995}$$

[25] $\frac{1}{4 \cdot 5} + \frac{1}{5 \cdot 6} + \cdots + \frac{1}{(n+3)(n+4)} + \cdots = \sum_{n=1}^{\infty} \frac{1}{(n+3)(n+4)} = \sum_{n=4}^{\infty} \frac{1}{n(n+1)}$.

By Example 1, $\sum_{n=1}^{\infty} \frac{1}{n(n+1)} = 1$.

By (11.19) with $k = 3$, $\sum_{n=4}^{\infty} \frac{1}{n(n+1)}$ also converges.

[29] $\frac{1}{4} + \frac{1}{5} + \cdots + \frac{1}{n+3} + \cdots = \sum_{n=1}^{\infty} \frac{1}{n+3} = \sum_{n=4}^{\infty} \frac{1}{n}$.

Since $\sum_{n=1}^{\infty} \frac{1}{n}$ diverges by (11.14), so does $\sum_{n=4}^{\infty} \frac{1}{n}$ by (11.19).

[33] Since $\lim_{n\to\infty} a_n = \lim_{n\to\infty} \frac{3n}{5n-1} = \frac{3}{5} \neq 0$, the series diverges.

[35] Since $\lim_{n\to\infty} a_n = \lim_{n\to\infty} \frac{1}{n^2+3} = 0$, further investigation is necessary.

Note: The nth-term test *never* indicates that a series is convergent.

[39] Since $\lim_{n\to\infty} a_n = \lim_{n\to\infty} \frac{n}{\ln(n+1)} \left\{\frac{\infty}{\infty}\right\} = \lim_{n\to\infty} \frac{1}{1/(n+1)} = \infty \neq 0$,

the series diverges.

[41] Since $\sum_{n=3}^{\infty} \left(\frac{1}{4}\right)^n = \sum_{n=1}^{\infty} \left(\frac{1}{4}\right)^3\left(\frac{1}{4}\right)^{n-1}$ and $\sum_{n=3}^{\infty} \left(\frac{3}{4}\right)^n = \sum_{n=1}^{\infty} \left(\frac{3}{4}\right)^3\left(\frac{3}{4}\right)^{n-1}$ are both convergent

geometric series, $\sum_{n=3}^{\infty} \left(\frac{1}{4}\right)^n + \sum_{n=3}^{\infty} \left(\frac{3}{4}\right)^n = \sum_{n=3}^{\infty} \left[\left(\frac{1}{4}\right)^n + \left(\frac{3}{4}\right)^n\right]$ is also convergent.

$$\sum_{n=3}^{\infty} \left[\left(\tfrac{1}{4}\right)^n + \left(\tfrac{3}{4}\right)^n\right] = \sum_{n=3}^{\infty} \left(\tfrac{1}{4}\right)^n \left\{a = \left(\tfrac{1}{4}\right)^3,\ r = \tfrac{1}{4}\right\} + \sum_{n=3}^{\infty} \left(\tfrac{3}{4}\right)^n \left\{a = \left(\tfrac{3}{4}\right)^3,\ r = \tfrac{3}{4}\right\}$$

$$= \frac{1/4^3}{1 - \frac{1}{4}} + \frac{3^3/4^3}{1 - \frac{3}{4}} = \frac{1}{48} + \frac{27}{16} = \frac{41}{24};\ \mathrm{C}$$

[45] Since $\sum_{n=1}^{\infty} \frac{1}{8^n} = \sum_{n=1}^{\infty} \left(\frac{1}{8}\right)\left(\frac{1}{8}\right)^{n-1}$ {of the form $\sum_{n=1}^{\infty} ar^{n-1}$ with $a = \frac{1}{8}$ and $r = \frac{1}{8}$}

$= \frac{\frac{1}{8}}{1 - \frac{1}{8}} = \frac{1}{7}$ and by Example 1, $\sum_{n=1}^{\infty} \frac{1}{n(n+1)} = 1$,

the series $\sum_{n=1}^{\infty} \left[\frac{1}{8^n} + \frac{1}{n(n+1)}\right]$ converges to $\frac{1}{7} + 1 = \frac{8}{7}$.

[47] $\sum_{n=1}^{\infty}\left(\frac{5}{n+2}-\frac{5}{n+3}\right) = \sum_{n=3}^{\infty}\left(\frac{5}{n}-\frac{5}{n+1}\right)$ {replace $n+2$ with n, or, equivalently, n with $n-2$}

$= 5\sum_{n=3}^{\infty}\left(\frac{1}{n}-\frac{1}{n+1}\right) = 5\sum_{n=3}^{\infty}\frac{1}{n(n+1)}$

$= 5\left[\sum_{n=1}^{\infty}\frac{1}{n(n+1)} - \frac{1}{1\cdot 2} - \frac{1}{2\cdot 3}\right]$ {subtract the first two terms from the sum of the series in Example 1}

$= 5\left[1-\frac{1}{2}-\frac{1}{6}\right] = 5(\frac{1}{3}) = \frac{5}{3}$; C

[51] We must find an m such that $S_m = 1 + \frac{1}{2} + \frac{1}{3} + \cdots + \frac{1}{m} \geq 3$.

$S_m = S_{2^k} > (k+1)(\frac{1}{2})$ {from Example 3} $= 3 \Rightarrow k+1 = 6 \Rightarrow k = 5$ and

$m = 2^k = 2^5 = 32$. $S_{32} = 1 + \frac{1}{2} + \frac{1}{3} + \cdots + \frac{1}{32} \approx 4.058495$.

[55] The ball initially falls 10 m, after which, in each up and down cycle it travels: 5 m up, 5 m down; $\frac{5}{2}$ m up, $\frac{5}{2}$ m down; $\frac{5}{4}$ m up, $\frac{5}{4}$ m down; etc. This total distance traveled is $d = 10 + 2(5) + 2(\frac{5}{2}) + 2(\frac{5}{4}) + \cdots$

$= 10 + 10 + 10(\frac{1}{2}) + 10(\frac{1}{4}) + \cdots$

$= 10 + \sum_{n=0}^{\infty} 10(\frac{1}{2})^n = 10 + \frac{10}{1-\frac{1}{2}} = 10 + 20 = 30$ m.

[57] (a) Immediately after the first dose, there are Q units of the drug in the bloodstream. $A(1) = Q$. After the second dose, there is a new Q units plus $A(1)e^{-cT} = Qe^{-cT}$ units from the first dose. Thus, $A(2) = Q + Qe^{-cT}$. Similarly,

$A(3) = Q + A(2)e^{-cT} = Q + (Q + Qe^{-cT})e^{-cT} = Q + Qe^{-cT} + Qe^{-2cT}$,

$A(4) = Q + A(3)e^{-cT} = Q + (Q + Qe^{-cT} + Qe^{-2cT})e^{-cT} =$

$Q + Qe^{-cT} + Qe^{-2cT} + Qe^{-3cT}, \ldots \Rightarrow A(k) = \sum_{n=0}^{k-1} Qe^{-ncT}$.

(b) Since all terms are positive, $A(k)$ is an increasing sequence with the form $a + ar + ar^2 + \cdots + ar^{k-1}$, where $a = Q$ and $r = e^{-cT}$.

Since $c > 0$, $|r| = |e^{-cT}| < 1$. Thus, by (11.15)(ii),

the upper bound is $\lim_{k\to\infty} A(k) = \sum_{n=0}^{\infty} Q(e^{-cT})^n = \frac{Q}{1-e^{-cT}}$.

(c) We must find T such that the upper bound found in (b) is less than M. Thus,

$\frac{Q}{1-e^{-cT}} < M \Rightarrow Q < M(1-e^{-cT}) \Rightarrow Q - M < -Me^{-cT} \Rightarrow$

$Me^{-cT} < M - Q \Rightarrow e^{-cT} < \frac{M-Q}{M} \Rightarrow -cT < \ln\frac{M-Q}{M} \Rightarrow$

$T > -\frac{1}{c}\ln\frac{M-Q}{M}$ $\{-c < 0\}$.

61 (a) From the second figure we see that $(\frac{1}{4}a_k)^2 + (\frac{3}{4}a_k)^2 = (a_{k+1})^2 \Rightarrow$

$$\frac{10}{16}a_k^2 = a_{k+1}^2 \Rightarrow a_{k+1} = \frac{1}{4}\sqrt{10}\,a_k.$$

(b) From part (a), $a_n = (\frac{1}{4}\sqrt{10})a_{n-1}$

$$= \frac{1}{4}\sqrt{10}(\frac{1}{4}\sqrt{10})a_{n-2} = (\frac{1}{4}\sqrt{10})^2 a_{n-2}$$

$$= (\frac{1}{4}\sqrt{10})^2(\frac{1}{4}\sqrt{10})a_{n-3} = (\frac{1}{4}\sqrt{10})^3 a_{n-3} = \cdots = (\frac{1}{4}\sqrt{10})^{n-1}\,a_1.$$

$A_{k+1} = a_{k+1}^2 = \frac{10}{16}a_k^2 = \frac{5}{8}A_k$ since $A_k = a_k^2$. In a manner similar to finding an expression for a_n, $A_n = \frac{5}{8}A_{n-1} = (\frac{5}{8})^2 A_{n-2} = (\frac{5}{8})^3 A_{n-3} = \cdots$,

and hence $A_n = (\frac{5}{8})^{n-1}A_1$.

$P_{k+1} = 4a_{k+1} = 4\cdot\frac{1}{4}\sqrt{10}\,a_k = \sqrt{10}\,a_k = \sqrt{10}\,(\frac{1}{4}P_k)$.

As with a_n, it follows that $P_n = (\frac{1}{4}\sqrt{10})^{n-1}P_1$.

(c) $\sum_{n=1}^{\infty} P_n = \sum_{n=1}^{\infty} P_1(\frac{1}{4}\sqrt{10})^{n-1}$ is an infinite geometric series with first term $a = P_1$ and $r = \frac{1}{4}\sqrt{10} < 1$. $\sum_{n=1}^{\infty} P_n = \frac{a}{1-r} = \frac{P_1}{1-\frac{1}{4}\sqrt{10}} = \frac{4P_1}{4-\sqrt{10}} = \frac{16a_1}{4-\sqrt{10}}$.

$$\sum_{n=1}^{\infty} A_n = \sum_{n=1}^{\infty} (\tfrac{5}{8})^{n-1}A_1 = \sum_{n=1}^{\infty} (\tfrac{5}{8})^{n-1}a_1^2 = \frac{a_1^2}{1-\frac{5}{8}} = \tfrac{8}{3}a_1^2.$$

Exercises 11.3

1 (a) In (11.23), let $f(n) = a_n = \dfrac{1}{(3+2n)^2}$ and $f(x) = \dfrac{1}{(3+2x)^2}$.

(i) Since $(3+2x)^2 > 0$ if $x \geq 1$, f is a positive-valued function.

(ii) f is continuous on $\mathbb{R} - \{-\frac{3}{2}\}$ and hence is continuous on $[1, \infty)$.

(iii) By the reciprocal rule, $f'(x) = -\dfrac{2(3+2x)\cdot 2}{\left[(3+2x)^2\right]^2} = -\dfrac{4}{(2x+3)^3}$. If $x \geq 1$, then $(2x+3)^3 > 0$ and f' is negative. Thus, f is decreasing on $[1, \infty)$.

It now follows that f satisfies the hypotheses of (11.23).

(b) $\displaystyle\int_1^{\infty} f(x)\,dx = \int_1^{\infty}(3+2x)^{-2}\,dx = \lim_{t\to\infty}\int_1^t (3+2x)^{-2}\,dx$

$$= \lim_{t\to\infty}\left[\frac{1}{2}\cdot\frac{(3+2x)^{-1}}{-1}\right]_1^t = \lim_{t\to\infty}\left[-\frac{1}{2(3+2x)}\right]_1^t$$

$$= \left[\lim_{t\to\infty} -\frac{1}{2(3+2t)}\right] - \left[-\frac{1}{2(5)}\right] = 0 - \left(-\frac{1}{10}\right);\ \text{C}$$

It is important to realize that $S = \sum_{n=1}^{\infty}\dfrac{1}{(3+2n)^2} \neq \dfrac{1}{10}$, but $\displaystyle\int_1^{\infty} f(x)\,dx = \frac{1}{10}$ and, therefore, S converges. We did not find out what S actually converges to.

[3] (a) In (11.23), let $f(n) = \dfrac{1}{4n + 7}$ and $f(x) = \dfrac{1}{4x + 7}$.

(i) Since $4x + 7 > 0$ if $x \geq 1$, f is a positive-valued function.

(ii) f is continuous on $\mathbb{R} - \{-\frac{7}{4}\}$ and hence is continuous on $[1, \infty)$.

(iii) $f(x) = \dfrac{1}{4x + 7} \Rightarrow f'(x) = -\dfrac{4}{(4x + 7)^2} < 0$ if $x \geq 1$.

Thus, f is decreasing on $[1, \infty)$.

It now follows that f satisfies the hypotheses of (11.23).

(b) $\displaystyle\int_1^\infty f(x)\,dx = \int_1^\infty (4x + 7)^{-1}\,dx = \lim_{t \to \infty} \int_1^t (4x + 7)^{-1}\,dx =$

$\displaystyle\lim_{t \to \infty} \left[\frac{1}{4}\ln|4x + 7|\right]_1^t = \left[\lim_{t \to \infty} \frac{1}{4}\ln|4t + 7|\right] - \left[\frac{1}{4}\ln|11|\right] = \infty$; D

[11] (a) In (11.23), let $f(n) = \dfrac{\arctan n}{1 + n^2}$ and $f(x) = \dfrac{\arctan x}{1 + x^2}$.

(i) (Refer to Figure 8.6 for the graph of $y = \arctan x$.)

Since $\arctan x > 0$ and $1 + x^2 > 0$ for $x \geq 1$, f is a positive-valued function.

(ii) Since $1 + x^2 \neq 0$ and both $\arctan x$ and $1 + x^2$ are continuous,

f is continuous on $[1, \infty)$.

(iii) $f(x) = \dfrac{\arctan x}{1 + x^2} \Rightarrow$

$$f'(x) = \frac{(1 + x^2)\cdot 1/(1 + x^2) - (\arctan x)(2x)}{(1 + x^2)^2} = \frac{1 - 2x\arctan x}{(1 + x^2)^2} < 0 \text{ if}$$

$x \geq 1$ since $(\arctan x) \geq \frac{\pi}{4}$ on $[1, \infty)$. Thus, f is decreasing on $[1, \infty)$.

(b) $\displaystyle\int_1^\infty f(x)\,dx = \lim_{t \to \infty} \int_1^t \arctan x\left(\frac{1}{1 + x^2}\right)dx = \lim_{t \to \infty} \left[\tfrac{1}{2}(\arctan x)^2\right]_1^t =$

$\displaystyle\tfrac{1}{2}\lim_{t \to \infty} (\arctan t)^2 - \tfrac{1}{2}(\arctan 1)^2 = \tfrac{1}{2}(\tfrac{\pi}{2})^2 - \tfrac{1}{2}(\tfrac{\pi}{4})^2 = \tfrac{3\pi^2}{32}$; C

Note: Exer. 13–20: S (the given series) converges by (11.26)(i) or diverges by (11.26)(ii).

[13] Generally, if the highest power of n in the denominator is greater than the highest power of n in the numerator by *more* than one, the series will converge. It is often a good idea to choose a p-series (11.25) to use in the basic comparison test.

In this problem, we believe the given series converges because the degree of the numerator is zero and the degree of the denominator is four, so we will pick a p-series that converges and is greater (term-by-term) than the given series. In (11.26)(i), let $a_n = \dfrac{1}{n^4 + n^2 + 1}$ and $b_n = \dfrac{1}{n^4}$. Then, S converges since $\dfrac{1}{n^4 + n^2 + 1} < \dfrac{1}{n^4}$ for $n \geq 1$, and $\displaystyle\sum_{n=1}^{\infty} \frac{1}{n^4}$ converges by (11.25) with $p = 4$.

[17] For $n \geq 1$, $\arctan n$ increases from a value of $\frac{\pi}{4}$. We can see that for large n, the terms of $\frac{\arctan n}{n}$ become larger than those of the divergent harmonic series (11.14). Therefore, we suspect that the given series diverges. Let $a_n = \frac{\arctan n}{n}$ and $b_n = \frac{\pi/4}{n}$ in (11.26)(ii). Then, S diverges since $\frac{\arctan n}{n} \geq \frac{\pi/4}{n}$ and $\frac{\pi}{4} \sum_{n=1}^{\infty} \frac{1}{n}$ diverges.

[19] Since n^n increases much faster than n^2, $\frac{1}{n^n}$ decreases much faster than $\frac{1}{n^2}$, and we suspect that the given series converges. Let $a_n = \frac{1}{n^n}$ and $b_n = \frac{1}{n^2}$ in (11.26)(i). Then, S converges since $\frac{1}{n^n} \leq \frac{1}{n^2}$ for $n \geq 1$ and $\sum_{n=1}^{\infty} \frac{1}{n^2}$ converges by (11.25). *Note:* We could have picked n^3 or n^4 instead of n^2—the choice was really arbitrary—we just wanted a known convergent series.

[21] Let $a_n = \frac{\sqrt{n}}{n + 4}$. By deleting all terms in the numerator and denominator of a_n except those that have the greatest effect on the magnitude, we obtain $b_n = \frac{\sqrt{n}}{n} = \frac{1}{\sqrt{n}}$, which is a divergent p-series with $p = \frac{1}{2}$. Using the limit comparison test (11.27), we have $\lim_{n \to \infty} \frac{a_n}{b_n} = \lim_{n \to \infty} \frac{\sqrt{n}}{n + 4} \cdot \frac{\sqrt{n}}{1} = \lim_{n \to \infty} \frac{n}{n + 4} = 1 > 0$. Thus, S diverges by (11.27).

[23] Since the numerator is a constant and the term having the greatest effect in the denominator is $\sqrt{n^3}$ (ignore the coefficient 4), let $a_n = \frac{1}{\sqrt{4n^3 - 5n}}$ and $b_n = \frac{1}{\sqrt{n^3}} = \frac{1}{n^{3/2}}$. $\lim_{n \to \infty} \frac{a_n}{b_n} = \lim_{n \to \infty} \frac{n^{3/2}}{\sqrt{4n^3 - 5n}}$. This limit can be evaluated easily since the highest power of n in both the numerator and denominator are equal. The limit will simply be the ratio of the leading coefficients, which is $\frac{1}{\sqrt{4}} = \frac{1}{2} > 0$. S converges by (11.27) since $\sum b_n$ converges by (11.25) with $p = \frac{3}{2}$.

[25] The term having the greatest effect in the numerator is n^2 and in the denominator is $e^n n^2$. This ratio is $\frac{n^2}{e^n n^2} = \frac{1}{e^n}$. Thus, let $a_n = \frac{8n^2 - 7}{e^n(n + 1)^2}$ and $b_n = \frac{1}{e^n}$. $\lim_{n \to \infty} \frac{a_n}{b_n} = \lim_{n \to \infty} \frac{8n^2 - 7}{e^n(n + 1)^2} \cdot \frac{e^n}{1} = \lim_{n \to \infty} \frac{8n^2 - 7}{(n + 1)^2} = \frac{8}{1} = 8 > 0$. S converges by (11.27) since $\sum b_n$ is a geometric series with $r = \frac{1}{e} < 1$ and converges by (11.15).

[31] $\dfrac{1 + 2^n}{1 + 3^n} < \dfrac{1 + 2^n}{3^n} = \dfrac{1}{3^n} + \dfrac{2^n}{3^n} = \left(\dfrac{1}{3}\right)^n + \left(\dfrac{2}{3}\right)^n$ and

$\sum\limits_{n=1}^{\infty} \left[(\frac{1}{3})^n + (\frac{2}{3})^n\right] = \sum\limits_{n=1}^{\infty} (\frac{1}{3})^n + \sum\limits_{n=1}^{\infty} (\frac{2}{3})^n$ converges by (11.15) and (11.20)(i).

Let $a_n = \dfrac{1 + 2^n}{1 + 3^n}$ and $b_n = \left(\dfrac{1}{3}\right)^n + \left(\dfrac{2}{3}\right)^n$. Then, S converges by (11.26).

[33] The term having the greatest effect in the numerator is 1 and in the denominator is

$\sqrt[3]{n^2} = n^{2/3}$. Thus, let $a_n = \dfrac{1}{\sqrt[3]{5n^2 + 1}}$ and $b_n = \dfrac{1}{n^{2/3}}$.

$$\lim_{n \to \infty} \frac{a_n}{b_n} = \lim_{n \to \infty} \frac{1}{\sqrt[3]{5n^2 + 1}} \cdot \frac{n^{2/3}}{1} = \lim_{n \to \infty} \frac{n^{2/3}}{\sqrt[3]{5n^2 + 1}} = \frac{1}{\sqrt[3]{5}} > 0.$$

S diverges by (11.27) since $\sum b_n$ diverges by (11.25) with $p = \frac{2}{3}$.

[37] If we let $f(x) = xe^{-x}$, then $f'(x) = (1 - x)e^{-x} < 0$ for $x > 1$, and f is a positive-valued, continuous, decreasing function on $[1, \infty)$. Since we can readily integrate f using integration by parts or Formula 96, p. A25, we will apply the integral test.

$$\int_1^{\infty} f(x)\,dx = \lim_{t \to \infty} \int_1^t xe^{-x}\,dx = \lim_{t \to \infty} \Big[-xe^{-x} - e^{-x}\Big]_1^t = 0 - (-2e^{-1}) = \tfrac{2}{e};\ \text{C}$$

[39] Let $a_n = \sin \dfrac{1}{n^2}$ and $b_n = \dfrac{1}{n^2}$.

$$\lim_{n \to \infty} \frac{a_n}{b_n} = \lim_{n \to \infty} \frac{\sin(1/n^2)}{1/n^2}\ \left\{\tfrac{0}{0}\right\} = \lim_{n \to \infty} \frac{(-2/n^3)\cos(1/n^2)}{-2/n^3} = \lim_{n \to \infty} \cos(1/n^2) =$$

$\cos 0 = 1 > 0$. S converges by (11.27) since $\sum b_n$ converges by (11.25) with $p = 2$.

[41] After deleting terms of least magnitude, we have $\dfrac{(2n)^3}{(n^3)^2} = \dfrac{8n^3}{n^6} = \dfrac{8}{n^3}$.

Let $a_n = \dfrac{(2n + 1)^3}{(n^3 + 1)^2}$ and $b_n = \dfrac{1}{n^3}$. $\lim\limits_{n \to \infty} \dfrac{a_n}{b_n} = \lim\limits_{n \to \infty} \dfrac{(2n + 1)^3 \cdot n^3}{(n^3 + 1)^2} = 8 > 0$.

S converges by (11.27) since $\sum b_n$ converges by (11.25) with $p = 3$.

[45] When series contain terms with ln and polynomial terms, we often use the fact that $\ln n < n$ for $n \geq 1$, as illustrated in this solution. Since $\dfrac{\ln n}{n^3} < \dfrac{n}{n^3} = \dfrac{1}{n^2}$ and $\sum\limits_{n=1}^{\infty} \dfrac{1}{n^2}$ converges, S converges by (11.26) with $a_n = \dfrac{\ln n}{n^3}$ and $b_n = \dfrac{1}{n^2}$.

47 Since the value of k will affect how we determine the convergence of the given series, we will consider four cases for values of k.

(i) $k > 1$. Let $a_n = \dfrac{1}{n^k \ln n}$ and $b_n = \dfrac{1}{n^k}$. Then since

$$\frac{1}{n^k \ln n} < \frac{1}{n^k} \ \{\ln n > 1 \text{ if } n \geq 3\} \text{ and } \sum_{n=2}^{\infty} \frac{1}{n^k} \text{ converges by (11.25), S converges.}$$

(ii) $k = 1$. Using the integral test, $\displaystyle\int_2^{\infty} \frac{1}{x \ln x}\, dx =$

$$\lim_{t \to \infty} \int_{\ln 2}^{t} \frac{1}{u}\, du \ \{u = \ln x;\ du = \tfrac{1}{x}\, dx\} = \lim_{t \to \infty} \Big[\ln |u|\Big]_{\ln 2}^{t} = \infty; \text{ D.}$$

(iii) $0 \leq k < 1$. $u = \ln x$, $x = e^u$, and $dx = e^u\, du \Rightarrow \displaystyle\int_2^{\infty} \frac{1}{x^k \ln x}\, dx =$

$$\int_{\ln 2}^{\infty} \frac{e^u}{e^{ku}\, u}\, du = \int_{\ln 2}^{\infty} \frac{e^{(1-k)u}}{u}\, du. \text{ Since } 1 - k > 0, \text{ the integrand approaches } \infty$$

as $u \to \infty$. Thus, the integral must diverge and S also diverges.

(iv) $k < 0$. Since $\dfrac{1}{n^k \ln n} = \dfrac{n^{-k}}{\ln n} > \dfrac{1}{\ln n} > \dfrac{1}{n}$ $(n \geq 2)$ and $\displaystyle\sum_{n=2}^{\infty} \frac{1}{n}$ diverges,

S diverges.

Thus, S converges iff $k > 1$.

49 (a) First use the results of the proof of (11.23) with $f(x) = \dfrac{1}{x+1}$.

(i)

$$\sum_{k=2}^{n} f(k) \leq \int_1^n f(x)\, dx \leq \sum_{k=1}^{n-1} f(k) \Rightarrow$$

$$\sum_{k=2}^{n} \frac{1}{k+1} \leq \int_1^n \frac{1}{x+1}\, dx \leq \sum_{k=1}^{n-1} \frac{1}{k+1} \Rightarrow$$

$$\frac{1}{3} + \frac{1}{4} + \cdots + \frac{1}{n+1} \leq \Big[\ln|x+1|\Big]_1^n \leq \frac{1}{2} + \frac{1}{3} + \cdots + \frac{1}{n} \Rightarrow$$

$$\frac{1}{3} + \frac{1}{4} + \cdots + \frac{1}{n+1} \leq \ln(n+1) - \ln 2 \leq \frac{1}{2} + \frac{1}{3} + \cdots + \frac{1}{n} \Rightarrow$$

$$\ln(n+1) \leq \ln 2 + \frac{1}{2} + \frac{1}{3} + \cdots + \frac{1}{n}.$$

$$\{\text{since } \ln 2 < 1\}\ \ln(n+1) < 1 + \frac{1}{2} + \frac{1}{3} + \cdots + \frac{1}{n}.$$

(ii) Now repeat (i) with $f(x) = \dfrac{1}{x}$.

$$\sum_{k=2}^{n} f(k) \leq \int_1^n f(x)\, dx \leq \sum_{k=1}^{n-1} f(k) \Rightarrow$$

$$\sum_{k=2}^{n} \frac{1}{k} \leq \int_1^n \frac{1}{x}\, dx \leq \sum_{k=1}^{n-1} \frac{1}{k} \Rightarrow$$

$$\frac{1}{2} + \frac{1}{3} + \cdots + \frac{1}{n} \leq \ln n \leq 1 + \frac{1}{2} + \frac{1}{3} + \cdots + \frac{1}{n-1} \Rightarrow$$

$$1 + \frac{1}{2} + \frac{1}{3} + \cdots + \frac{1}{n} \leq 1 + \ln n.$$

By (i) and (ii), $\ln(n+1) < 1 + \frac{1}{2} + \frac{1}{3} + \cdots + \frac{1}{n} < 1 + \ln n$ $(n > 1)$.

(b) From part (a), $S_n = 1 + \frac{1}{2} + \frac{1}{3} + \cdots + \frac{1}{n} > 100$ if $\ln(n + 1) > 100$.

Thus, $n + 1 > e^{100} \Rightarrow n > e^{100} - 1 \approx 2.688 \times 10^{43}$.

[51] Since $\lim_{n \to \infty} \frac{a_n}{b_n} = 0$, then by (11.3), there is an $N \geq 1$ such that if $n > N$, then $\frac{a_n}{b_n} < 1$, or $a_n < b_n$. Since $\sum b_n$ converges and $a_n < b_n$ for all but at most a finite number of terms, $\sum a_n$ must also converge by (11.26).

[53] $\sum_{n=1}^{\infty} a_n = \sum_{n=1}^{N} a_n + \sum_{n=N+1}^{\infty} a_n$. The error in approximating $\sum_{n=1}^{\infty} a_n$ by $\sum_{n=1}^{N} a_n$ is $\sum_{n=N+1}^{\infty} a_n$ since all the a_n are positive.

Thus, we must show that $\sum_{n=N+1}^{\infty} a_n < \int_N^{\infty} f(x)\,dx$.

Since $a_{n+1} < \int_n^{n+1} f(x)\,dx$ for every n and f is decreasing, it follows that the error

$$E = a_{N+1} + a_{N+2} + a_{N+3} + \cdots$$
$$= \sum_{n=N+1}^{\infty} a_n < \int_N^{N+1} f(x)\,dx + \int_{N+1}^{N+2} f(x)\,dx + \int_{N+2}^{N+3} f(x)\,dx + \cdots = \int_N^{\infty} f(x)\,dx.$$

(See Figure 11.8.)

[55] We must determine an N such that $\int_N^{\infty} \frac{1}{x^3}\,dx < 0.01$.

Error $< \int_N^{\infty} \frac{1}{x^3}\,dx = \lim_{t \to \infty} \left[-\frac{1}{2x^2}\right]_N^t = \frac{1}{2N^2} \leq 0.01 \Rightarrow N^2 \geq 50 \Rightarrow N \geq \sqrt{50}$; 8 terms.

[57] Since $\sum a_n$ converges, $\lim_{n \to \infty} a_n = 0$ by (11.26) and therefore, $\lim_{n \to \infty} \frac{1}{a_n} = \infty$.

By (11.17), $\sum \frac{1}{a_n}$ diverges.

[61] From the graphs, it is apparent that $x \geq \ln(x^k)$ for $k = 1, 2, 3$ and $x \geq 5$. Thus, $\frac{1}{n} \leq \frac{1}{\ln(n^k)}$ for $k = 1, 2, 3$ and $n \geq 5$. By the basic comparison test, since $\sum_{n=1}^{\infty} \frac{1}{n}$ diverges, $\sum_{n=1}^{\infty} \frac{1}{\ln(n^k)}$ also diverges for $k = 1, 2$, and 3.

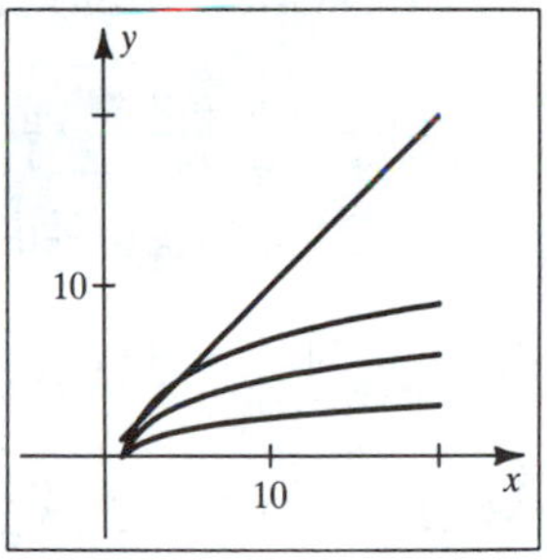

Figure 61

Exercises 11.4

[1] $\lim_{n\to\infty} \frac{a_{n+1}}{a_n} = \lim_{n\to\infty} \left(a_{n+1} \cdot \frac{1}{a_n}\right) = \lim_{n\to\infty} \frac{3(n+1)+1}{2^{n+1}} \cdot \frac{2^n}{3n+1} =$

$\lim_{n\to\infty} \frac{2^n(3n+4)}{2^{n+1}(3n+1)} = \lim_{n\to\infty} \frac{3n+4}{2(3n+1)} = \frac{1}{2} < 1.$

Since $\lim_{n\to\infty} \frac{a_{n+1}}{a_n} < 1$, the series is convergent.

Note: For limits involving only powers of n, we will not divide by the highest power of n, but will merely compare leading terms to determine the limit. For example, in this exercise, the highest power of n in the numerator and denominator of $\frac{3n+4}{2(3n+1)}$ is 1. The ratio of their coefficients is $\frac{3}{2(3)} = \frac{1}{2}$ and this ratio is equal to the limit.

[3] $\lim_{n\to\infty} \frac{a_{n+1}}{a_n} = \lim_{n\to\infty} \frac{5^{n+1}}{(n+1)3^{n+2}} \cdot \frac{n3^{n+1}}{5^n}$

$= \lim_{n\to\infty} \frac{5^{n+1} \cdot 3^{n+1} \cdot n}{5^n \cdot 3^{n+2} \cdot (n+1)} = \lim_{n\to\infty} \frac{5n}{3(n+1)} = \frac{5}{3} > 1$; D

Since $\lim_{n\to\infty} \frac{a_{n+1}}{a_n} > 1$, the series is divergent.

[7] $\lim_{n\to\infty} \frac{a_{n+1}}{a_n} = \lim_{n\to\infty} \frac{n+4}{(n+1)^2 + 2(n+1) + 5} \cdot \frac{n^2 + 2n + 5}{n+3} = 1$. The highest power of n in the numerator and denominator is 3. The ratio of their coefficients is $\frac{1}{1}$. Note that you do *not* need to multiply the expressions out to determine this. It can be done by inspecting only the leading terms. Since $\lim_{n\to\infty} \frac{a_{n+1}}{a_n} = 1$, the ratio test is inconclusive. Using the limit comparison test with $b_n = \frac{1}{n}$, we see that $\lim_{n\to\infty} \frac{a_n}{b_n} =$ $\lim_{n\to\infty} \frac{n+3}{n^2+2n+5} \cdot \frac{n}{1} = 1$ and $\sum a_n$ diverges since $\sum \frac{1}{n}$ diverges. However, the purpose here is to call attention to the condition under which the ratio test fails.

[11] $\lim_{n\to\infty} \sqrt[n]{a_n} = \lim_{n\to\infty} \sqrt[n]{\frac{1}{n^n}} = \lim_{n\to\infty} \frac{1}{n} = 0 < 1$; C

[13] $\lim_{n\to\infty} \sqrt[n]{a_n} = \lim_{n\to\infty} \sqrt[n]{\frac{2^n}{n^2}} = \lim_{n\to\infty} \frac{2}{n^{2/n}}.$

Since $\lim_{n\to\infty} n^{2/n} = \lim_{n\to\infty} (n^{1/n})^2 = (1)^2$ {see Exercise 11.1.37},

$\lim_{n\to\infty} \frac{2}{n^{2/n}} = \frac{2}{(1)^2} = 2 > 1$, and $\sum a_n$ diverges.

[17] $\lim_{n\to\infty} \sqrt[n]{a_n} = \lim_{n\to\infty} \sqrt[n]{\left(\frac{n}{2n+1}\right)^n} = \lim_{n\to\infty} \frac{n}{2n+1} = \frac{1}{2} < 1$; C

Note: The following may be solved using several methods. For a convenience, we list an abbreviation of the test used at the beginning of the problem.

NTH — nth-term test (11.17)
INT — integral test (11.23)
BCT — basic comparison test (11.26)
LCT — limit comparison test (11.27)
RAT — ratio test (11.28)
ROT — root test (11.29)

[19] LCT Deleting terms of least magnitude gives $\dfrac{\sqrt{n}}{n^2} = \dfrac{1}{n^{3/2}}$. Let $a_n = \dfrac{\sqrt{n}}{n^2 + 1}$ and $b_n = \dfrac{1}{n^{3/2}}$. $\displaystyle\lim_{n\to\infty} \frac{a_n}{b_n} = \lim_{n\to\infty} \frac{n^{1/2}}{n^2 + 1} \cdot \frac{n^{3/2}}{1} = \lim_{n\to\infty} \frac{n^2}{n^2 + 1} = 1 > 0.$

S converges since $\sum b_n$ converges.

[23] BCT Since $\dfrac{2}{n^3 + e^n} < \dfrac{2}{n^3}$ and $2 \displaystyle\sum_{n=1}^{\infty} \frac{1}{n^3}$ converges since it is a p-series with $p = 3 > 1$, S converges.

[25] RAT
$$\lim_{n\to\infty} \frac{a_{n+1}}{a_n} = \lim_{n\to\infty} \frac{2^{n+1}(n+1)!}{(n+1)^{n+1}} \cdot \frac{n^n}{2^n\, n!}$$
$$= \lim_{n\to\infty} \frac{2^{n+1}(n+1)\, n!\, n^n}{2^n\, n!\,(n+1)^{n+1}} = \lim_{n\to\infty} \frac{2(n+1)\, n^n}{(n+1)^n (n+1)^1}$$
$$= 2 \lim_{n\to\infty} \frac{n^n}{(n+1)^n} = 2 \lim_{n\to\infty} \left(\frac{n}{n+1}\right)^n = 2 \lim_{n\to\infty} \left(\frac{1}{\frac{n+1}{n}}\right)^n$$
$$= 2 \lim_{n\to\infty} \frac{1^n}{\left(1 + \frac{1}{n}\right)^n} = 2 \cdot \tfrac{1}{e} < 1;\ \text{C.}$$
Note: Using Example 2, $\displaystyle\lim_{n\to\infty} \left(\frac{n}{n+1}\right)^n$ is the reciprocal of $\displaystyle\lim_{n\to\infty} \left(\frac{n+1}{n}\right)^n$, which is e.

[27] ROT
$$\lim_{n\to\infty} \sqrt[n]{a_n} = \lim_{n\to\infty} \sqrt[n]{\frac{n^n}{10^{n+1}}} = \lim_{n\to\infty} \frac{(n^n)^{1/n}}{(10^{n+1})^{1/n}}$$
$$= \lim_{n\to\infty} \frac{n}{10^{1+(1/n)}} = \lim_{n\to\infty} \frac{n}{10(10^{1/n})} = \infty \text{ since } \lim_{n\to\infty} n = \infty \text{ and}$$
$$\lim_{n\to\infty} 10(10^{1/n}) = 10(10^0) = 10(1) = 10;\ \text{D}$$

[31] INT
$$\int_2^{\infty} \frac{1}{x\sqrt[3]{\ln x}}\, dx$$
$$= \int_2^{\infty} (\ln x)^{-1/3} \cdot \tfrac{1}{x}\, dx \ \{u = \ln x,\ du = \tfrac{1}{x}\, dx;\ x = 2, \infty \Rightarrow u = \ln 2, \infty\}$$
$$= \lim_{t\to\infty} \int_{\ln 2}^{t} u^{-1/3}\, du = \lim_{t\to\infty} \left[\tfrac{3}{2} u^{2/3}\right]_{\ln 2}^{t} = \infty;\ \text{D}$$

[35] NTH
$$\lim_{n\to\infty} n \tan\tfrac{1}{n}\ \{\infty \cdot 0\} = \lim_{n\to\infty} \frac{\tan(1/n)}{1/n}\ \{\tfrac{0}{0}\} = \lim_{n\to\infty} \frac{(-1/n^2)\sec^2(1/n)}{-1/n^2} =$$
$$\lim_{n\to\infty} \sec^2(1/n) = \sec^2 0 = 1 \neq 0.$$ Thus, S diverges.

[39] RAT $\lim_{n\to\infty} \frac{a_{n+1}}{a_n} = \lim_{n\to\infty} \frac{1\cdot 3\cdot 5\cdot \cdots \cdot(2n-1)(2n+1)}{(n+1)!} \cdot \frac{n!}{1\cdot 3\cdot 5\cdot \cdots \cdot(2n-1)} =$

$$\lim_{n\to\infty} \frac{2n+1}{n+1} = 2 > 1;\ \text{D}$$

Exercises 11.5

[1] (a) $a_k = \frac{1}{k^2+7}$. $a_{k+1} = \frac{1}{(k+1)^2+7} = \frac{1}{(k^2+7)+(2k+1)}$.

Since the denominator of a_{k+1} is larger than the denominator of a_k, $a_k > a_{k+1}$.

Also, $a_{k+1} > 0$ and hence, condition (i) of (11.30) is satisfied.

Condition (ii) is satisfied since $\lim_{n\to\infty} a_n = \lim_{n\to\infty} \frac{1}{n^2+7} = 0$.

(b) The series converges by (11.30) since conditions (i) and (ii) are satisfied.

[3] (a) As in the beginning of Example 1, we will show that $f(x) = (1 + e^{-x})$ is decreasing. $f(x) = 1 + e^{-x} \Rightarrow f'(x) = -e^{-x} < 0$ and f is decreasing. Since $e^{-n} > 0$ for all n, $a_n = 1 + e^{-n} > 0$ for all n. Hence, $a_k > a_{k+1} > 0$ and condition (i) is satisfied. $\lim_{n\to\infty} a_n = \lim_{n\to\infty} (1 + e^{-n}) = 1 + 0 = 1 \neq 0$ and condition (ii) is not satisfied.

(b) The series diverges by the nth-term test.

Note: Let AC denote Absolutely Convergent; CC, Conditionally Convergent; D, Divergent; AST, the Alternating Series Test; and S, the given series.

[5] $a_n = \frac{1}{\sqrt{2n+1}} > 0$. To show that $a_n \geq a_{n+1}$, we will show that $\frac{a_{n+1}}{a_n} \leq 1$.

$\frac{a_{n+1}}{a_n} = \frac{1}{\sqrt{2(n+1)+1}} \div \frac{1}{\sqrt{2n+1}} = \frac{\sqrt{2n+1}}{\sqrt{2n+3}} < 1$ and $\lim_{n\to\infty} a_n = 0 \Rightarrow$

S converges by AST. To determine whether or not this series converges absolutely, we will use the LCT. Let $b_n = \frac{1}{\sqrt{n}}$. Then $\lim_{n\to\infty} \frac{a_n}{b_n} = \lim_{n\to\infty} \frac{1}{\sqrt{2n+1}} \cdot \frac{\sqrt{n}}{1} = \frac{1}{\sqrt{2}} > 0$ and $\sum a_n$ diverges since $\sum \frac{1}{\sqrt{n}}$ is a p-series with $p = \frac{1}{2}$ and it diverges. Thus, S is not AC, but S is CC.

[9] Since $\lim_{n\to\infty} \frac{n}{\ln n} \{\frac{\infty}{\infty}\} = \lim_{n\to\infty} \frac{1}{1/n} = \lim_{n\to\infty} n = \infty$, S is D by the nth-term test.

[11] Since $\frac{5}{n^3+1} < \frac{5}{n^3}$ and $5 \sum_{n=1}^{\infty} \frac{1}{n^3}$ converges by (11.25), $\sum_{n=1}^{\infty} \frac{5}{n^3+1}$ converges by a basic comparison test and hence, S is AC.

Note: Absolute convergence implies *both* convergence and conditional convergence.

[13] $\lim_{n\to\infty}\left|\frac{a_{n+1}}{a_n}\right| = \lim_{n\to\infty}\frac{10^{n+1}}{(n+1)!}\cdot\frac{n!}{10^n} = \lim_{n\to\infty}\frac{10^{n+1}\,n!}{10^n\,(n+1)\,n!} =$

$\lim_{n\to\infty}\frac{10}{n+1} = 0 \Rightarrow$ S is AC by (11.35)(i).

[17] $a_n = \frac{n^{1/3}}{n+1} > 0.$ $f(x) = \frac{x^{1/3}}{x+1} \Rightarrow f'(x) = \frac{1-2x}{3x^{2/3}(x+1)^2} < 0$ for $x > \frac{1}{2}$ and

$a_n \geq a_{n+1}$. Since $\lim_{n\to\infty} a_n = 0$, it follows that S converges by AST.

To determine the absolute convergence of S, we will use the LCT.

Deleting terms of least magnitude results in $\frac{\sqrt[3]{n}}{n} = \frac{1}{n^{2/3}}$. Let $b_n = \frac{1}{n^{2/3}}$.

Then $\lim_{n\to\infty}\frac{a_n}{b_n} = \lim_{n\to\infty}\frac{n^{1/3}}{n+1}\cdot\frac{n^{2/3}}{1} = \lim_{n\to\infty}\frac{n}{n+1} = 1 > 0$ and

$\sum a_n$ diverges since $\sum 1/n^{2/3}$ diverges by (11.25). Thus, S is CC.

[19] Since $\left|\frac{\cos\frac{\pi}{6}n}{n^2}\right| \leq \frac{1}{n^2}$ and $\sum_{n=1}^{\infty}\frac{1}{n^2}$ converges by (11.25), S is AC.

Note that $\left|\cos\frac{\pi}{6}n\right| \leq 1$ for all n.

[21] Since $\lim_{n\to\infty} n\sin\frac{1}{n}\ \{\infty\cdot 0\} = \lim_{n\to\infty}\frac{\sin(1/n)}{1/n}\ \{\frac{0}{0}\} = \lim_{n\to\infty}\frac{(-1/n^2)\cos(1/n)}{-1/n^2} =$

$\lim_{n\to\infty}\cos(1/n) = \cos 0 = 1 \neq 0$, S is D by the nth-term test.

[25] Since both the numerator and denominator are raised to the nth power, we will apply

the root test. $\lim_{n\to\infty}\sqrt[n]{|a_n|} = \lim_{n\to\infty}\sqrt[n]{\left|\frac{n^n}{(-5)^n}\right|} = \lim_{n\to\infty}\frac{n}{5} = \infty \Rightarrow$ S is D.

[29] Since $\cos\pi n = -1$ when n is odd and $\cos\pi n = 1$ when n is even, it follows that

$\sum_{n=1}^{\infty}(-1)^n\frac{\cos\pi n}{n} = \sum_{n=1}^{\infty}(-1)^n\cdot\frac{(-1)^n}{n} = \sum_{n=1}^{\infty}\frac{(-1)^{2n}}{n} = \sum_{n=1}^{\infty}\frac{\left[(-1)^2\right]^n}{n} =$

$\sum_{n=1}^{\infty}\frac{(1)^n}{n} = \sum_{n=1}^{\infty}\frac{1}{n} = 1 + \frac{1}{2} + \frac{1}{3} + \cdots$. This is the harmonic series. S is D.

Note: The value of n in Exercises 33–42 was found by trial and error when the inequality could not be solved using basic algebraic operations.

Note: If the sum S_n is to be approximated to three decimal places, then $a_{n+1} < 0.5\times 10^{-3}$. Equivalently, we use a_n and S_{n-1} to simplify the computations.

[33] In order to apply (11.31), we must determine a value of n such that $a_n < 0.5\times 10^{-3}$.

$a_n = \frac{1}{n!} < 0.0005 \Rightarrow n! > 2000 \Rightarrow n \geq 7$. It now follows that

$a_0 + a_1 + a_2 + a_3 + a_4 + a_5 + a_6$ will be an approximation of S to within three decimal places. Thus, $S \approx S_6 = 1 - \frac{1}{1!} + \frac{1}{2!} - \frac{1}{3!} + \frac{1}{4!} - \frac{1}{5!} + \frac{1}{6!} \approx 0.368$.

[37] $a_n = \dfrac{n+1}{5^n} < 0.0005 \Rightarrow 2000(n+1) < 5^n \Rightarrow n \geq 6.$

Thus, $S \approx S_5 = \frac{2}{5} - \frac{3}{25} + \frac{4}{125} - \frac{5}{625} + \frac{6}{3125} \approx 0.306.$

[39] This exercise is similar to Exercise 33. We must determine an n such that $a_{n+1} < 0.5 \times 10^{-4}$. However, we do not need to estimate S.

$a_{n+1} = \dfrac{1}{(n+1)^2} < 0.00005 \Rightarrow (n+1)^2 > 20{,}000 \Rightarrow n > \sqrt{20{,}000} - 1 \Rightarrow n \geq 141.$

[43] (i) $a_n = \dfrac{(\ln n)^k}{n} > 0 \ (n \geq 3)$. $f(x) = \dfrac{(\ln x)^k}{x} \Rightarrow f'(x) = \dfrac{(\ln x)^{k-1}(k - \ln x)}{x^2} < 0$ for $x > e^k \Rightarrow a_{n+1} < a_n$ for all but a finite number of terms. Remember that a finite number of terms *does not* affect the convergence or divergence of a series.

(ii) We will show that $\lim\limits_{n\to\infty} a_n = 0$ by repeatedly using L'Hôpital's rule (10.2).

$\lim\limits_{x\to\infty} \dfrac{(\ln x)^k}{x} \left\{\frac{\infty}{\infty}\right\} = \lim\limits_{x\to\infty} \dfrac{k(\ln x)^{k-1}\cdot(1/x)}{1} = \lim\limits_{x\to\infty} \dfrac{k(\ln x)^{k-1}}{x} \left\{\frac{\infty}{\infty}\right\} =$

$\lim\limits_{x\to\infty} \dfrac{k(k-1)(\ln x)^{k-2}}{x} \left\{\frac{\infty}{\infty}\right\} = \cdots = \lim\limits_{x\to\infty} \frac{k!}{x} = 0.$ Thus, by AST, S converges.

[45] No. If $a_n = b_n = (-1)^n/\sqrt{n}$, then both $\sum a_n$ and $\sum b_n$ converge by the alternating series test. However, $\sum a_n b_n = \sum 1/n$, which diverges.

Exercises 11.6

Note: Let u_n denote the nth term of the power series. Let AC denote Absolutely Convergent; C, Convergent; D, Divergent; and AST, the Alternating Series Test.

[1] $u_n = \dfrac{x^n}{n+4} \Rightarrow \lim\limits_{n\to\infty}\left|\dfrac{u_{n+1}}{u_n}\right| = \lim\limits_{n\to\infty}\left|\dfrac{x^{n+1}}{n+5}\cdot\dfrac{n+4}{x^n}\right|$

$= \lim\limits_{n\to\infty}\left|\dfrac{n+4}{n+5}\cdot\dfrac{x^{n+1}}{x^n}\right|$

$= \lim\limits_{n\to\infty}\left(\dfrac{n+4}{n+5}\right)|x| = 1\cdot|x| = |x|.$

The series will converge whenever $|x| < 1$. $|x| < 1 \Leftrightarrow -1 < x < 1$.

The end points of the interval of convergence are $x = \pm 1$. To determine the convergence at these points, we must substitute $x = 1$ and $x = -1$ into the series.

If $x = 1$, $\sum\limits_{n=0}^{\infty} \dfrac{1}{n+4}(1)^n = \sum\limits_{n=0}^{\infty} \dfrac{1}{n+4}$ is D {use LCT with $b_n = \frac{1}{n}$}.

If $x = -1$, $\sum\limits_{n=0}^{\infty} (-1)^n \dfrac{1}{n+4}$ is C by AST. ★ $[-1, 1)$

[3] $u_n = \frac{n^2 x^n}{2^n} \Rightarrow \lim_{n\to\infty}\left|\frac{u_{n+1}}{u_n}\right| = \lim_{n\to\infty}\left|\frac{(n+1)^2 x^{n+1}}{2^{n+1}} \cdot \frac{2^n}{n^2 x^n}\right|$

$$= \lim_{n\to\infty}\left|\frac{(n+1)^2}{n^2} \cdot \frac{2^n}{2^{n+1}} \cdot \frac{x^{n+1}}{x^n}\right| = 1 \cdot \tfrac{1}{2} \cdot |x| = \tfrac{1}{2}|x|.$$

The series will converge whenever $\frac{1}{2}|x| < 1$. $\frac{1}{2}|x| < 1 \Leftrightarrow -2 < x < 2$.

The end points of the interval of convergence are $x = \pm 2$.

If $x = 2$, $\sum_{n=0}^{\infty} \frac{n^2}{2^n}(2)^n = \sum_{n=0}^{\infty} n^2$ is D { nth-term test }.

If $x = -2$, $\sum_{n=0}^{\infty} \frac{n^2}{2^n}(-2)^n = \sum_{n=0}^{\infty} \frac{n^2}{2^n}(-1)^n (2)^n = \sum_{n=0}^{\infty} (-1)^n n^2$ is D. ★ $(-2, 2)$

[9] $u_n = \frac{(\ln n)\, x^n}{n^3} \Rightarrow \lim_{n\to\infty}\left|\frac{u_{n+1}}{u_n}\right| = \lim_{n\to\infty}\left|\frac{[\ln(n+1)]\, x^{n+1}}{(n+1)^3} \cdot \frac{n^3}{(\ln n)\, x^n}\right|$

$$= \lim_{n\to\infty}\left|\frac{\ln(n+1)}{\ln n} \cdot \frac{n^3}{(n+1)^3} \cdot \frac{x^{n+1}}{x^n}\right|$$

$$= 1 \cdot 1 \cdot |x| = |x|. \quad |x| < 1 \Leftrightarrow -1 < x < 1.$$

If $x = 1$, $\sum_{n=2}^{\infty} \frac{\ln n}{n^3} \le \sum_{n=2}^{\infty} \frac{n}{n^3} = \sum_{n=2}^{\infty} \frac{1}{n^2}$ is C by BCT.

If $x = -1$, $\sum_{n=2}^{\infty} (-1)^n \frac{\ln n}{n^3}$ is C by AST. ★ $[-1, 1]$

[11] $u_n = \frac{(n+1)(x-4)^n}{10^n} \Rightarrow \lim_{n\to\infty}\left|\frac{u_{n+1}}{u_n}\right| = \lim_{n\to\infty}\left|\frac{(n+2)(x-4)^{n+1}}{10^{n+1}} \cdot \frac{10^n}{(n+1)(x-4)^n}\right|$

$$= \lim_{n\to\infty}\left|\frac{n+2}{n+1} \cdot \frac{10^n}{10^{n+1}} \cdot \frac{(x-4)^{n+1}}{(x-4)^n}\right|$$

$$= 1 \cdot \tfrac{1}{10} \cdot |x-4| = \tfrac{1}{10}|x-4|.$$

$\frac{1}{10}|x-4| < 1 \Leftrightarrow |x-4| < 10 \Leftrightarrow -10 < x-4 < 10 \Leftrightarrow -6 < x < 14$.

If $x = -6$, $\sum_{n=0}^{\infty} \frac{n+1}{10^n}(-10)^n = \sum_{n=0}^{\infty} (-1)^n (n+1)$ is D { nth-term test }.

If $x = 14$, $\sum_{n=0}^{\infty} \frac{n+1}{10^n} 10^n = \sum_{n=0}^{\infty} (n+1)$ is D { nth-term test }. ★ $(-6, 14)$

[13] $u_n = \frac{n!\, x^n}{100^n} \Rightarrow \lim_{n\to\infty}\left|\frac{u_{n+1}}{u_n}\right| = \lim_{n\to\infty}\left|\frac{(n+1)!\, x^{n+1}}{100^{n+1}} \cdot \frac{100^n}{n!\, x^n}\right|$

$$= \lim_{n\to\infty}\left|\frac{(n+1)\, n!}{n!} \cdot \frac{100^n}{100^{n+1}} \cdot \frac{x^{n+1}}{x^n}\right|$$

$$= \lim_{n\to\infty}\left(\frac{n+1}{100}\right)|x| = \infty.$$

Remember, the limit variable is n, not x, and hence the last limit increases without bound. Since the limit is always greater than 1, the series converges only for $x = 0$.

[15] $u_n = \dfrac{x^{2n+1}}{(-4)^n} \Rightarrow \lim_{n\to\infty}\left|\dfrac{u_{n+1}}{u_n}\right| = \lim_{n\to\infty}\left|\dfrac{x^{2n+3}}{(-4)^{n+1}} \cdot \dfrac{(-4)^n}{x^{2n+1}}\right| = \frac{1}{4}x^2.$

$\frac{1}{4}x^2 < 1 \Leftrightarrow x^2 < 4 \Leftrightarrow |x| < 2 \Leftrightarrow -2 < x < 2.$

If $x = \pm 2$, $|u_n| = \left|\dfrac{(\pm 2)^{2n+1}}{(-4)^n}\right| = \dfrac{2^{2n+1}}{4^n} = \dfrac{2^1 \cdot 2^{2n}}{4^n} = \dfrac{2 \cdot (2^2)^n}{4^n} = \dfrac{2 \cdot 4^n}{4^n} = 2\ \forall n,$

and both series are D { nth-term test }. ★ $(-2, 2)$

[17] $u_n = \dfrac{2^n x^{2n}}{(2n)!} \Rightarrow \lim_{n\to\infty}\left|\dfrac{u_{n+1}}{u_n}\right| = \lim_{n\to\infty}\left|\dfrac{2^{n+1} x^{2n+2}}{(2n+2)!} \cdot \dfrac{(2n)!}{2^n x^{2n}}\right|$

$$= \lim_{n\to\infty}\left|\frac{2^{n+1}}{2^n} \cdot \frac{(2n)!}{(2n+2)(2n+1)(2n)!} \cdot \frac{x^{2n}(x^2)}{x^{2n}}\right|$$

$$= \lim_{n\to\infty} \frac{2x^2}{(2n+2)(2n+1)} = 0.$$

The last limit equals 0 since the degree of the denominator is two and the degree of the numerator (with respect to n) is zero. Since the limit is always less than 1, the series converges $\forall x$. ★ $(-\infty, \infty)$

[23] $u_n = (-1)^n \dfrac{n^n (x-3)^n}{n+1} \Rightarrow$

$$\lim_{n\to\infty}\left|\frac{u_{n+1}}{u_n}\right| = \lim_{n\to\infty}\left|\frac{(n+1)^{n+1}(x-3)^{n+1}}{n+2} \cdot \frac{n+1}{n^n (x-3)^n}\right|$$

$$= \lim_{n\to\infty}\left|\frac{n+1}{n+2} \cdot \frac{(n+1)^{n+1}}{n^n} \cdot \frac{(x-3)^{n+1}}{(x-3)^n}\right|$$

$$= \left[\lim_{n\to\infty} \frac{n+1}{n+2}\right] \cdot \left[\lim_{n\to\infty} \frac{(n+1)^n (n+1)}{n^n} \cdot |x-3|\right]$$

$$= 1 \cdot \lim_{n\to\infty} \left(\frac{n+1}{n}\right)^n (n+1)|x-3| = \lim_{n\to\infty} e(n+1)|x-3| = \infty.$$

$\left\{\lim_{n\to\infty}\left(\dfrac{n+1}{n}\right)^n = \lim_{n\to\infty}\left(1 + \dfrac{1}{n}\right)^n = e \text{ by (7.32)(ii)}\right\}$

Converges only when $|x - 3| = 0$, i.e., for $x = 3$.

[27] $u_n = (-1)^n \dfrac{(2x-1)^n}{n6^n} \Rightarrow \lim_{n\to\infty}\left|\dfrac{u_{n+1}}{u_n}\right| = \lim_{n\to\infty}\left|\dfrac{(2x-1)^{n+1}}{(n+1)6^{n+1}} \cdot \dfrac{n6^n}{(2x-1)^n}\right|$

$$= \lim_{n\to\infty}\left|\frac{n}{n+1} \cdot \frac{6^n}{6^{n+1}} \cdot \frac{(2x-1)^{n+1}}{(2x-1)^n}\right|$$

$$= \tfrac{1}{6}|2x - 1|.$$

$\frac{1}{6}|2x-1| < 1 \Leftrightarrow |2x-1| < 6 \Leftrightarrow -6 < 2x - 1 < 6 \Leftrightarrow -\frac{5}{2} < x < \frac{7}{2}.$

If $x = -\frac{5}{2}$, $\sum_{n=1}^{\infty} (-1)^n \dfrac{1}{n6^n} \cdot (-6)^n = \sum_{n=1}^{\infty} (-1)^n \dfrac{1}{n6^n}(-1)^n 6^n = \sum_{n=1}^{\infty} \dfrac{1}{n}$ is D.

If $x = \frac{7}{2}$, $\sum_{n=1}^{\infty} (-1)^n \dfrac{1}{n6^n} \cdot 6^n = \sum_{n=1}^{\infty} (-1)^n \dfrac{1}{n}$ is C by AST. ★ $(-\frac{5}{2}, \frac{7}{2}]$

[29] $u_n = (-1)^n \dfrac{3^n(x-4)^n}{n!} \Rightarrow \lim\limits_{n\to\infty}\left|\dfrac{u_{n+1}}{u_n}\right| = \lim\limits_{n\to\infty}\left|\dfrac{3^{n+1}(x-4)^{n+1}}{(n+1)!}\cdot\dfrac{n!}{3^n(x-4)^n}\right|$

$$= \lim_{n\to\infty}\left|\frac{n!}{(n+1)\,n!}\cdot\frac{3^{n+1}}{3^n}\cdot\frac{(x-4)^{n+1}}{(x-4)^n}\right|$$

$$= \lim_{n\to\infty}\frac{3}{n+1}|x-4| = 0.$$

Since the limit is always less than 1, the series converges $\forall\, x$. ★ $(-\infty, \infty)$

[31] $\lim\limits_{n\to\infty}\left|\dfrac{u_{n+1}}{u_n}\right| =$

$$\lim_{n\to\infty}\left|\frac{1\cdot 3\cdot 5\cdot\cdots\cdot(2n-1)(2n+1)\,x^{n+1}}{3\cdot 6\cdot 9\cdot\cdots\cdot(3n)(3n+3)}\cdot\frac{3\cdot 6\cdot 9\cdot\cdots\cdot(3n)}{1\cdot 3\cdot 5\cdot\cdots\cdot(2n-1)\,x^n}\right| =$$

$\lim\limits_{n\to\infty}\left(\dfrac{2n+1}{3n+3}\right)|x| = \frac{2}{3}|x|$. $\frac{2}{3}|x| < 1 \Leftrightarrow |x| < \frac{3}{2}$. ★ $r = \frac{3}{2}$.

Note: The convergence/divergence at the end points does not affect the radius of convergence.

[33] $\lim\limits_{n\to\infty}\left|\dfrac{u_{n+1}}{u_n}\right| = \lim\limits_{n\to\infty}\left|\dfrac{(n+1)^{n+1}x^{n+1}}{(n+1)!}\cdot\dfrac{n!}{n^n x^n}\right|$

$$= \lim_{n\to\infty}\left|\frac{(n+1)(n+1)^n\,x}{(n+1)n^n}\right|$$

$$= \lim_{n\to\infty}\left(\frac{n+1}{n}\right)^n|x|$$

$= \lim\limits_{n\to\infty}\left(1+\dfrac{1}{n}\right)^n|x| = e|x|$. $e|x| < 1 \Leftrightarrow |x| < \frac{1}{e}$. ★ $r = \frac{1}{e}$.

[35] $\lim\limits_{n\to\infty}\left|\dfrac{u_{n+1}}{u_n}\right| = \lim\limits_{n\to\infty}\left|\dfrac{(n+c+1)!\,x^{n+1}}{(n+1)!\,(n+d+1)!}\cdot\dfrac{n!\,(n+d)!}{(n+c)!\,x^n}\right|$

$$= \lim_{n\to\infty}\left|\frac{(n+c+1)(n+c)!}{(n+c)!}\cdot\frac{(n+d)!}{(n+d+1)(n+d)!}\cdot\frac{n!}{(n+1)\,n!}\cdot\frac{x^{n+1}}{x^n}\right|$$

$= \lim\limits_{n\to\infty}\dfrac{n+c+1}{(n+1)(n+d+1)}|x| = 0$ {deg(num) = 1 < 2 = deg(den)}.

The limit is always less than 1. ★ $r = \infty$.

[41] $\lim\limits_{n\to\infty}\left|\dfrac{u_{n+1}}{u_n}\right| = \lim\limits_{n\to\infty}\left|\dfrac{a_{n+1}\,x^{n+1}}{a_n\,x^n}\right| = \lim\limits_{n\to\infty}\left|\dfrac{a_{n+1}}{a_n}\right||x| = k|x|$.

$k|x| < 1 \Leftrightarrow |x| < \dfrac{1}{k}$ $(k \neq 0)$. ★ $r = \dfrac{1}{k}$

[45] We will prove this by contradiction. Assume that $\sum a_n x^n$ is absolutely convergent at $x = r$. Then $\sum |a_n r^n|$ is convergent by (11.32). Let $x = -r$. Then $\sum |a_n(-r)^n| = \sum |(-1)^n\, a_n r^n| = \sum |a_n r^n|$ is absolutely convergent, which implies that $\sum a_n(-r)^n$ is convergent by (11.34). This is a contradiction.

Exercises 11.7

1 (a) Since $|x| < \frac{1}{3}$, it follows that $|3x| < 1$. Moreover, $\dfrac{1}{1 - 3x}$ can be written as $\dfrac{a}{1 - r}$, where $|r| < 1$. By (11.15)(i) with $a = 1$ and $r = 3x$,

$$f(x) = \frac{1}{1 - 3x} = 1 + (3x) + (3x)^2 + (3x)^3 + \cdots + (3x)^n + \cdots = \sum_{n=0}^{\infty} 3^n x^n.$$

(b) To find the derivative of a power series, we differentiate each term of the series. By (11.40)(i), $f'(x) = \sum_{n=1}^{\infty} n a^n x^{n-1} = \sum_{n=1}^{\infty} n 3^n x^{n-1}$. To find $\int_0^x f(t)\,dt$, where $f(t)$ is a power series, we integrate each term of the series. By (11.40)(ii), $\int_0^x f(t)\,dt = \sum_{n=0}^{\infty} \frac{a^n}{n+1} x^{n+1} = \sum_{n=0}^{\infty} \frac{3^n}{n+1} x^{n+1}$.

3 (a) Since $|x| < \frac{2}{7}$, it follows that $\left|\frac{7}{2}x\right| < 1$. We must rewrite $\dfrac{1}{2 + 7x}$ in the form of $\dfrac{a}{1 - r}$. $f(x) = \dfrac{1}{2 + 7x} = \dfrac{1}{2 - (-7x)} = \dfrac{1}{2} \cdot \dfrac{1}{1 - (-\frac{7}{2}x)}$ {the denominator must be of the form $1 - r$}.

By (11.15)(i), with $a = \frac{1}{2}$ and $r = -\frac{7}{2}x$,

$$f(x) = \tfrac{1}{2}\left[1 + (-\tfrac{7}{2}x) + (-\tfrac{7}{2}x)^2 + (-\tfrac{7}{2}x)^3 + \cdots + (-1)^n(\tfrac{7}{2}x)^n + \cdots\right]$$

$$= \tfrac{1}{2}\sum_{n=0}^{\infty} (-1)^n (\tfrac{7}{2})^n x^n.$$

(b) By (11.40)(i), $f'(x) = \frac{1}{2}\sum_{n=1}^{\infty} (-1)^n (\frac{7}{2})^n n x^{n-1} = \frac{1}{2}\sum_{n=1}^{\infty} (-1)^n \frac{n7^n}{2^n} x^{n-1}$. By (11.40)(ii), $\int_0^x f(t)\,dt = \frac{1}{2}\sum_{n=0}^{\infty} (-1)^n (\frac{7}{2})^n \frac{x^{n+1}}{(n+1)} = \frac{1}{2}\sum_{n=0}^{\infty} (-1)^n \frac{7^n}{(n+1)\,2^n} x^{n+1}$.

5 By (11.15), $\dfrac{1}{1 - r} = 1 + r + r^2 + r^3 + \cdots$, where $|r| < 1$.

If we let $r = x^2$, then $\dfrac{1}{1 - x^2} = 1 + x^2 + (x^2)^2 + (x^2)^3 + \cdots$. Thus,

$$\frac{x^2}{1 - x^2} = x^2 \cdot \frac{1}{1 - x^2}$$

$$= x^2\left[1 + (x^2) + (x^2)^2 + (x^2)^3 + \cdots\right]$$

$$= x^2 \sum_{n=0}^{\infty} (x^2)^n = \sum_{n=0}^{\infty} x^{2n+2}.$$ $\left|x^2\right| < 1 \Rightarrow |x| < 1 \Rightarrow r = 1$.

[9] Using long division, $\dfrac{x^2+1}{x-1} = x + 1 + \dfrac{2}{x-1}$. We will rewrite $\dfrac{2}{x-1}$ in the form $\dfrac{a}{1-r}$ before we represent it as a power series. $\dfrac{2}{x-1} = -\dfrac{2}{1-x} = -2\left(\dfrac{1}{1-x}\right) = -2(1 + x + x^2 + x^3 + \cdots)$ by (11.15) with $a = 1$ and $r = x$. Thus,

$$\begin{aligned}\frac{x^2+1}{x-1} &= x + 1 + \frac{2}{x-1}\\ &= 1 + x - \frac{2}{1-x}\\ &= 1 + x - 2(1 + x + x^2 + x^3 + \cdots)\\ &= 1 + x - 2 - 2x - 2(x^2 + x^3 + x^4 + x^5 + \cdots)\\ &= -1 - x - 2\sum_{n=2}^{\infty} x^n.\end{aligned}$$

$|x| < 1 \Rightarrow r = 1.$

[11] (a) $f(x) = \ln(1-x) \Rightarrow f'(x) = -\dfrac{1}{1-x}$. We do not know the power series for $\ln(1-x)$. However, we do know the power series for its derivative. We can integrate this power series term-by-term to find the power series for $\ln(1-x)$.

$$f'(x) = -\frac{1}{1-x} = -(1 + x + x^2 + \cdots + x^n + \cdots) = -\sum_{n=0}^{\infty} x^n \Rightarrow$$

$$f(x) = \int_0^x f'(t)\,dt = -\sum_{n=0}^{\infty}\frac{x^{n+1}}{n+1} = -\sum_{n=1}^{\infty}\frac{x^n}{n}.$$

(b) To find $\ln(1.2)$ using this series, we must let $x = -0.2$ in the expression $\ln(1-x)$. In this case, the series in part (a) becomes an alternating series that satisfies (11.30). Using (11.31), $\ln(1.2) = \ln\left[1 - (-0.2)\right] \approx$

$$-(-0.2) - \frac{(-0.2)^2}{2} - \frac{(-0.2)^3}{3} \approx 0.183, \text{ with } |\text{error}| < \frac{(-0.2)^4}{4} = 0.4 \times 10^{-3}.$$

Calculator value ≈ 0.182321557.

[15] Using (11.41), we substitute $3x$ for x. Then,

$e^{3x} = 1 + (3x) + \dfrac{(3x)^2}{2!} + \cdots + \dfrac{(3x)^n}{n!} + \cdots$. Multiplying both sides by x gives us

$$f(x) = xe^{3x} = x\left[1 + (3x) + \frac{(3x)^2}{2!} + \frac{(3x)^3}{3!} + \cdots\right] = x\sum_{n=0}^{\infty}\frac{(3x)^n}{n!} = \sum_{n=0}^{\infty}\frac{3^n}{n!}x^{n+1}.$$

[19] From Example 3, $\ln(1+x) = x - \dfrac{x^2}{2} + \dfrac{x^3}{3} - \cdots + (-1)^n\dfrac{x^{n+1}}{n+1} + \cdots$.

Substituting x^2 for x gives us

$\ln(1+x^2) = (x^2) - \dfrac{(x^2)^2}{2} + \dfrac{(x^2)^3}{3} - \dfrac{(x^2)^4}{4} + \cdots + (-1)^n\dfrac{(x^2)^{n+1}}{n+1} + \cdots$. Thus,

$$\begin{aligned} f(x) &= x^2\ln(1+x^2)\\ &= x^2\left[(x^2) - \frac{(x^2)^2}{2} + \frac{(x^2)^3}{3} - \frac{(x^2)^4}{4} + \cdots\right]\\ &= x^2\sum_{n=0}^{\infty}(-1)^n\frac{(x^2)^{n+1}}{n+1} = \sum_{n=0}^{\infty}(-1)^n\frac{1}{n+1}x^{2n+4}.\end{aligned}$$

[21] From Example 5, $\arctan x = x - \dfrac{x^3}{3} + \dfrac{x^5}{5} - \cdots + (-1)^n \dfrac{x^{2n+1}}{2n+1} + \cdots$.

Substituting $\sqrt{x}$ for x gives us

$$\begin{aligned} f(x) &= \arctan \sqrt{x} \\ &= \left[(x^{1/2}) - \frac{(x^{1/2})^3}{3} + \frac{(x^{1/2})^5}{5} - \frac{(x^{1/2})^7}{7} + \cdots \right] \\ &= \sum_{n=0}^{\infty} (-1)^n \frac{(x^{1/2})^{2n+1}}{2n+1} = \sum_{n=0}^{\infty} (-1)^n \frac{1}{2n+1} x^{(2n+1)/2}. \end{aligned}$$

[25] Using the series for $\cosh x$ (just before Example 6), and substituting x^3 for x gives us

$1 + \dfrac{(x^3)^2}{2!} + \dfrac{(x^3)^4}{4!} + \dfrac{(x^3)^6}{6!} + \cdots$. Thus,

$$\begin{aligned} f(x) &= x^2 \cosh(x^3) \\ &= x^2 \left[1 + \frac{(x^3)^2}{2!} + \frac{(x^3)^4}{4!} + \frac{(x^3)^6}{6!} + \cdots \right] \\ &= x^2 \sum_{n=0}^{\infty} \frac{(x^3)^{2n}}{(2n)!} = \sum_{n=0}^{\infty} \frac{1}{(2n)!} x^{6n+2}. \end{aligned}$$

[27] Since x is integrated over the interval $[0, \frac{1}{3}]$, $|x^6| < 1$.

We will use (11.15) and write $\dfrac{1}{1+x^6}$ as $\dfrac{a}{1-r}$ with $a = 1$ and $r = -x^6$.

Then, $\dfrac{1}{1+x^6} = \dfrac{1}{1-(-x^6)} = 1 + (-x^6) + (-x^6)^2 + (-x^6)^3 + \cdots$.

$$\begin{aligned} \int_0^{1/3} \frac{1}{1+x^6}\,dx &= \int_0^{1/3} (1 - x^6 + x^{12} - x^{18} + \cdots)\,dx \\ &= \left[x - \tfrac{1}{7}x^7 + \tfrac{1}{13}x^{13} - \tfrac{1}{19}x^{19} + \cdots \right]_0^{1/3}. \end{aligned}$$

The series inside the brackets is an alternating series that satisfies (11.30) and hence (11.31) applies. Using the first two terms, $I \approx \left[x - \frac{1}{7}x^7 \right]_0^{1/3} = \frac{1}{3} - \frac{1}{7}(\frac{1}{3})^7 \approx 0.3333$,

with $|\text{error}| < \frac{1}{13}(\frac{1}{3})^{13} < 0.5 \times 10^{-7}$.

[29] Dividing the series for $\arctan x$ (found in Example 5) by x, we have $\dfrac{\arctan x}{x} =$

$\dfrac{1}{x}\left(x - \dfrac{x^3}{3} + \dfrac{x^5}{5} - \dfrac{x^7}{7} + \cdots \right) = \left(1 - \dfrac{x^2}{3} + \dfrac{x^4}{5} - \dfrac{x^6}{7} + \cdots \right)$. Thus,

$$\begin{aligned} \int_{0.1}^{0.2} \frac{\arctan x}{x}\,dx &= \int_{0.1}^{0.2} (1 - \tfrac{1}{3}x^2 + \tfrac{1}{5}x^4 - \tfrac{1}{7}x^6 + \cdots)\,dx \\ &= \left[x - \tfrac{1}{9}x^3 + \tfrac{1}{25}x^5 - \tfrac{1}{49}x^7 + \cdots \right]_{0.1}^{0.2}. \end{aligned}$$ Using the first two terms,

$I \approx \left[x - \frac{1}{9}x^3 \right]_{0.1}^{0.2} \approx 0.0992$. Since the series satisfies (11.30), (11.31) applies.

Our estimate of $\arctan 0.2$ is $\left[0.2 - \frac{1}{9}(0.2)^3 \right]$ and has an error of at most $\frac{1}{25}(0.2)^5$.

(cont.)

Our estimate of $\arctan 0.1$ is $\left[0.1 - \frac{1}{9}(0.1)^3\right]$ and has an error of at most $\frac{1}{25}(0.1)^5$.

If we *subtract* these two values, the error will not exceed the *sum* of the two errors. (If the first estimate is high and the second estimate is low, then the total error would be the sum of the two individual errors.)

Thus, $|\text{error}| < \frac{1}{25}(0.2)^5 + \frac{1}{25}(0.1)^5 = 0.0000132 < 0.5 \times 10^{-4}$.

[33] First, notice that $\frac{d}{dx}\left(\frac{1}{1-x^2}\right) = \frac{2x}{(1-x^2)^2}$. If we can write the power series for $\frac{1}{1-x^2}$, we can differentiate it term-by-term to obtain the power series for $\frac{2x}{(1-x^2)^2}$.

By (11.15) with $a = 1$ and $r = x^2$, $\frac{1}{1-x^2} = 1 + x^2 + x^4 + x^6 + \cdots = \sum_{n=0}^{\infty} x^{2n}$.

Differentiating, we have $\frac{2x}{(1-x^2)^2} = \frac{d}{dx}\left(\frac{1}{1-x^2}\right) = \frac{d}{dx}\left(\sum_{n=0}^{\infty} x^{2n}\right) = \sum_{n=1}^{\infty} (2n)x^{2n-1}$

by (11.40)(i).

[35] (a) $J_\alpha(x) = \sum_{n=0}^{\infty} \frac{(-1)^n}{n!\,(n+\alpha)!}\left(\frac{x}{2}\right)^{2n+\alpha} \Rightarrow$

$$\begin{aligned}
D_x\left[J_0(x)\right] &= D_x \sum_{n=0}^{\infty} \frac{(-1)^n}{n!\,n!}\left(\frac{x}{2}\right)^{2n} \\
&= \sum_{n=1}^{\infty} \frac{(-1)^n}{n!\,n!}(2n)\left(\frac{x}{2}\right)^{2n-1}\left(\frac{1}{2}\right) \\
&= \sum_{n=0}^{\infty} \frac{(-1)^{n+1}}{(n+1)!\,(n+1)!}(n+1)\left(\frac{x}{2}\right)^{2n+1} \\
&= \sum_{n=0}^{\infty} \frac{(-1)^n(-1)}{n!\,(n+1)!}\left(\frac{x}{2}\right)^{2n+1} = -J_1(x).
\end{aligned}$$

(b)

$$\begin{aligned}
\int x^3 J_2(x)\,dx &= \int\left[\sum_{n=0}^{\infty} \frac{(-1)^n}{n!\,(n+2)!}\left(\frac{x}{2}\right)^{2n+2} x^3\right] dx \\
&= \int\left[\sum_{n=0}^{\infty} \frac{(-1)^n}{n!\,(n+2)!}\left(\frac{x}{2}\right)^{2n+2}\cdot\left(\frac{x}{2}\right)^3\cdot 8\right] dx \\
&= \int\left[\sum_{n=0}^{\infty} \frac{(-1)^n}{n!\,(n+2)!}(8)\left(\frac{x}{2}\right)^{2n+5}\right] dx \\
&= \left[\sum_{n=0}^{\infty} \frac{(-1)^n}{n!\,(n+2)!}\cdot\frac{8\cdot 2}{2(n+3)}\cdot\left(\frac{x}{2}\right)^{2n+6}\right] + C \\
&= \left[\sum_{n=0}^{\infty} \frac{(-1)^n}{n!\,(n+3)!}\left(\frac{x}{2}\right)^{2n+3} x^3\right] + C = x^3 J_3(x) + C.
\end{aligned}$$

37 Using Exercise 11, $\ln(1-x) = -\sum_{n=1}^{\infty} \frac{x^n}{n} = -\left(x + \frac{x^2}{2} + \frac{x^3}{3} + \frac{x^4}{4} + \cdots\right) \Rightarrow$

$$\frac{\ln(1-t)}{t} = -\frac{1}{t}\left(t + \frac{t^2}{2} + \frac{t^3}{3} + \frac{t^4}{4} + \cdots\right)$$

$$= -\left(1 + \frac{t}{2} + \frac{t^2}{3} + \frac{t^3}{4} + \cdots\right) = -\sum_{n=1}^{\infty} \frac{t^{n-1}}{n}.$$

$$f(x) = \int_0^x \left(-\sum_{n=1}^{\infty} \frac{t^{n-1}}{n}\right) dt = \left[-\sum_{n=1}^{\infty} \frac{t^n}{n^2}\right]_0^x = -\sum_{n=1}^{\infty} \frac{1}{n^2} x^n,\ |x| < 1.$$

Exercises 11.8

Note: Let *DERIV* denote the beginning of the sequence:

$f(x),\ f(c)$‡ $f'(x),\ f'(c)$‡ $f''(x),\ f''(c)$‡ $\ldots$. For a Maclaurin series, $c = 0$.

We have used the double dagger symbol (‡) to separate the terms.

1 In order to determine $a_n = \frac{f^{(n)}(0)}{n!}$, we must find a general formula for $f^{(n)}(0)$.

To do this, we will begin by calculating the first three derivatives of f and try to identify a pattern.

$$f(x) = e^{3x}, \quad f(0) = 1$$
$$f'(x) = 3e^{3x}, \quad f'(0) = 3$$
$$f''(x) = 3^2 e^{3x}, \quad f''(0) = 3^2$$
$$f'''(x) = 3^3 e^{3x}, \quad f'''(0) = 3^3$$

As mentioned in the previous note, we will use the following notation to denote the function, its derivatives, and their values at c.

DERIV: $e^{3x},\ 1$‡ $3e^{3x},\ 3$‡ $3^2e^{3x},\ 3^2$‡ $3^3e^{3x},\ 3^3$‡ $\ldots$.

From this pattern, we can see that $f^{(n)}(x) = 3^n e^{3x}$ and $f^{(n)}(0) = 3^n$.

Thus, $a_n = \frac{f^{(n)}(0)}{n!} = \frac{3^n}{n!}$.

3 Let $f(x) = \sin 2x$. *DERIV*: $\sin 2x,\ 0$‡ $2\cos 2x,\ 2$‡ $-2^2 \sin 2x,\ 0$‡ $-2^3 \cos 2x,\ -2^3$‡ $\ldots$. From this pattern, we see that the derivatives of even order, evaluated at $x = 0$, are equal to zero. The derivatives of odd order, evaluated at $x = 0$, are equal to 2^n, with alternating signs. To obtain an expression that represents these alternating signs, we introduce another variable, k, and let $k = 0, 1, 2, \ldots$. Then $f^{(2k)}(0) = 0$ and $f^{(2k+1)}(0) = (-1)^k 2^{2k+1}$. Thus, if $n = 2k$, then $a_n = 0$, and if $n = 2k + 1$, then

$$a_n = \frac{f^{(n)}(0)}{n!} = (-1)^k \frac{2^{2k+1}}{(2k+1)!}.$$

[7] In order to show that $\lim_{n\to\infty} R_n(x) = 0$, we will try to find an $M \geq 0$ such that $\left|f^{(n+1)}(z)\right| \leq M$ for all z. Let $f(x) = \cos x$. *DERIV*: $\cos x$, 1‡ $-\sin x$, 0‡ $-\cos x$, -1‡ $\sin x$, 0‡ $\cos x$, 1‡ ... . From this pattern, we see that the higher derivatives are always either $\pm\sin x$ or $\pm\cos x$. Since $|\pm\sin x| \leq 1$ and $|\pm\cos x| \leq 1$, let $M = 1$.

(a) $\left|f^{(n+1)}(z)\right| \leq 1 \Rightarrow \left|R_n(x)\right| = \left|\dfrac{f^{(n+1)}(z)}{(n+1)!}x^{n+1}\right| \leq \left|\dfrac{x^{n+1}}{(n+1)!}\right| \to 0$ as $n \to \infty$

by (11.47). Thus, $f(x) = \cos x$ is represented by its Maclaurin series.

(b) By (11.42), $f(x) = f(0) + f'(0)x + \dfrac{f''(0)}{2!}x^2 + \cdots + \dfrac{f^{(n)}(0)}{n!}x^n + \cdots \Rightarrow$

$$\cos x = 1 + 0\cdot x + \frac{-1}{2!}x^2 + \frac{0}{3!}x^3 + \frac{1}{4!}x^4 + \frac{0}{5!}x^5 + \frac{-1}{6!}x^6 + \cdots$$

$$= 1 - \frac{x^2}{2!} + \frac{x^4}{4!} - \frac{x^6}{6!} + \cdots = \sum_{n=0}^{\infty}(-1)^n\frac{1}{(2n)!}x^{2n}.$$

[9] Using (11.48)(a), we substitute $3x$ for x. Then,

$$\sin 3x = 3x - \frac{(3x)^3}{3!} + \frac{(3x)^5}{5!} - \frac{(3x)^7}{7!} + \cdots + (-1)^n\frac{(3x)^{2n+1}}{(2n+1)!} + \cdots$$

$= \sum_{n=0}^{\infty}(-1)^n\dfrac{(3x)^{2n+1}}{(2n+1)!}$. Multiplying by x gives us

$$f(x) = x\sin 3x = x\sum_{n=0}^{\infty}(-1)^n\frac{(3x)^{2n+1}}{(2n+1)!} = \sum_{n=0}^{\infty}(-1)^n\frac{3^{2n+1}}{(2n+1)!}x^{2n+2}.$$

[13] Using a half-angle formula and then (11.48)(b),

$$f(x) = \cos^2 x = \tfrac{1}{2} + \tfrac{1}{2}\cos 2x = \tfrac{1}{2} + \tfrac{1}{2}\sum_{n=0}^{\infty}(-1)^n\frac{(2x)^{2n}}{(2n)!}$$

$$= \tfrac{1}{2} + \sum_{n=0}^{\infty}(-1)^n\frac{2^{2n-1}}{(2n)!}x^{2n}$$

$$= \tfrac{1}{2} + (-1)^0\frac{2^{-1}}{0!}x^0 + \sum_{n=1}^{\infty}(-1)^n\frac{2^{2n-1}}{(2n)!}x^{2n}$$

$$= 1 + \sum_{n=1}^{\infty}(-1)^n\frac{2^{2n-1}}{(2n)!}x^{2n}.$$

[15] *DERIV*: 10^x, 1‡ $10^x\ln 10$, $\ln 10$‡ $10^x(\ln 10)^2$, $(\ln 10)^2$‡ $10^x(\ln 10)^3$, $(\ln 10)^3$‡ ... .

$f^{(n)}(x) = 10^x(\ln 10)^n$ and $f^{(n)}(0) = (\ln 10)^n$. Thus, $10^x =$

$$\sum_{n=0}^{\infty}\frac{f^{(n)}(0)}{n!}x^n = 1 + (\ln 10)x + \frac{(\ln 10)^2}{2!}x^2 + \frac{(\ln 10)^3}{3!}x^3 + \cdots = \sum_{n=0}^{\infty}\frac{(\ln 10)^n}{n!}x^n.$$

[17] *DERIV*: $\sin x$, $\frac{1}{\sqrt{2}}$‡ $\cos x$, $\frac{1}{\sqrt{2}}$‡ $-\sin x$, $-\frac{1}{\sqrt{2}}$‡ $-\cos x$, $-\frac{1}{\sqrt{2}}$‡ $\ldots$.

$$\sin x = f(\tfrac{\pi}{4}) + f'(\tfrac{\pi}{4})(x - \tfrac{\pi}{4}) + \frac{f''(\frac{\pi}{4})}{2!}(x - \tfrac{\pi}{4})^2 + \frac{f'''(\frac{\pi}{4})}{3!}(x - \tfrac{\pi}{4})^3 + \cdots$$

$$= \frac{1}{\sqrt{2}} + \frac{1}{\sqrt{2}}(x - \tfrac{\pi}{4}) - \frac{1}{\sqrt{2}\cdot 2!}(x - \tfrac{\pi}{4})^2 - \frac{1}{\sqrt{2}\cdot 3!}(x - \tfrac{\pi}{4})^3 + \cdots$$

$$= \left[\frac{1}{\sqrt{2}}(x - \tfrac{\pi}{4}) - \frac{1}{\sqrt{2}}\frac{(x - \frac{\pi}{4})^3}{3!} + \frac{1}{\sqrt{2}}\frac{(x - \frac{\pi}{4})^5}{5!} - \cdots\right]\{\text{odd powered terms}\} +$$

$$\left[\frac{1}{\sqrt{2}} - \frac{1}{\sqrt{2}}\frac{(x - \frac{\pi}{4})^2}{2!} + \frac{1}{\sqrt{2}}\frac{(x - \frac{\pi}{4})^4}{4!} - \cdots\right]\{\text{even powered terms}\}$$

$$= \sum_{n=0}^{\infty}(-1)^n\frac{1}{\sqrt{2}\,(2n+1)!}(x - \tfrac{\pi}{4})^{2n+1} + \sum_{n=0}^{\infty}(-1)^n\frac{1}{\sqrt{2}\,(2n)!}(x - \tfrac{\pi}{4})^{2n}.$$

Note: Because the series is complicated, it's easier to represent it in summation notation by breaking it into two parts. However, it is not necessary to do this.

[19] *DERIV*: x^{-1}, $\frac{1}{2}$‡ $-x^{-2}$, $-\frac{1}{2^2}$‡ $2x^{-3}$, $\frac{2}{2^3}$‡ $-6x^{-4}$, $-\frac{6}{2^4}$‡ $\ldots$.

$f^{(n)}(x) = (-1)^n\,n!\,x^{-n-1}$ and $f^{(n)}(2) = (-1)^n\,n!/2^{n+1}$. Thus,

$$\frac{1}{x} = f(2) + f'(2)(x - 2) + \frac{f''(2)}{2!}(x - 2)^2 + \frac{f'''(2)}{3!}(x - 2)^3 + \cdots$$

$$= \frac{1}{2} - \frac{1}{2^2}(x - 2) + \frac{2!}{2^3\cdot 2!}(x - 2)^2 - \frac{3!}{2^4\cdot 3!}(x - 2)^3 + \cdots$$

$$= \frac{1}{2} - \frac{1}{2^2}(x - 2) + \frac{1}{2^3}(x - 2)^2 - \frac{1}{2^4}(x - 2)^3 + \cdots$$

$$= \sum_{n=0}^{\infty}(-1)^n\frac{1}{2^{n+1}}(x - 2)^n.$$

[21] If we are finding a series representation in powers of $x + 1$, then $x - c = x + 1 = x - (-1) \Rightarrow c = -1$. *DERIV*: e^{2x}, e^{-2}‡ $2e^{2x}$, $2e^{-2}$‡ 2^2e^{2x}, 2^2e^{-2}‡ $\ldots$.

$f^{(n)}(x) = 2^n e^{2x}$ and $f^{(n)}(-1) = 2^n e^{-2}$. Thus,

$$e^{2x} = f(-1) + f'(-1)(x + 1) + \frac{f''(-1)}{2!}(x + 1)^2 + \frac{f'''(-1)}{3!}(x + 1)^2 + \cdots$$

$$= e^{-2} + 2e^{-2}(x + 1) + \frac{2^2e^{-2}}{2!}(x + 1)^2 + \frac{2^3e^{-2}}{3!}(x + 1)^3 + \cdots$$

$$= \sum_{n=0}^{\infty}\frac{2^n}{e^2\,n!}(x + 1)^n.$$

[23] *DERIV*: $\sec x$, $2\ddagger$ $\sec x \tan x$, $2\sqrt{3}\ddagger$ $\sec^3 x + \sec x \tan^2 x$, 14.

$$\sec x = f(\tfrac{\pi}{3}) + f'(\tfrac{\pi}{3})(x - \tfrac{\pi}{3}) + \frac{f''(\frac{\pi}{3})}{2!}(x - \tfrac{\pi}{3})^2 + \cdots$$
$$= 2 + 2\sqrt{3}(x - \tfrac{\pi}{3}) + 7(x - \tfrac{\pi}{3})^2 + \cdots.$$

The *first three* terms of the Taylor series for $f(x) = \sec x$ at $x = \frac{\pi}{3}$ are

$$2 + 2\sqrt{3}(x - \tfrac{\pi}{3}) + 7(x - \tfrac{\pi}{3})^2.$$

[25] *DERIV*: $\sin^{-1} x$, $\frac{\pi}{6}\ddagger$ $\dfrac{1}{(1-x^2)^{1/2}}$, $\frac{2}{\sqrt{3}}\ddagger$ $\dfrac{x}{(1-x^2)^{3/2}}$, $\dfrac{4}{3\sqrt{3}}$.

$$\sin^{-1} x = f(\tfrac{1}{2}) + f'(\tfrac{1}{2})(x - \tfrac{1}{2}) + \frac{f''(\frac{1}{2})}{2!}(x - \tfrac{1}{2})^2 + \cdots$$
$$= \frac{\pi}{6} + \frac{2}{\sqrt{3}}(x - \tfrac{1}{2}) + \frac{2}{3\sqrt{3}}(x - \tfrac{1}{2})^2 + \cdots.$$

Note: In Exercises 29–38, all series are alternating, so (11.31) applies. S denotes the sum of the first two nonzero terms and E the absolute value of the maximum error.

[29] By (11.48)(iii),

$$e^{-x} = 1 + (-x) + \frac{(-x)^2}{2!} + \frac{(-x)^3}{3!} + \frac{(-x)^4}{4!} + \cdots$$
$$= 1 - x + \frac{x^2}{2!} - \frac{x^3}{3!} + \frac{x^4}{4!} - \cdots$$
$$= \sum_{n=0}^{\infty} (-1)^n \frac{x^n}{n!}.$$

$\dfrac{1}{\sqrt{e}} = e^{-1/2} \approx 1 - (\tfrac{1}{2}) + \dfrac{(\frac{1}{2})^2}{2!}$ $(x = \tfrac{1}{2})$. $S = 1 - \tfrac{1}{2} = 0.5$. $E = \dfrac{(\frac{1}{2})^2}{2!} = \tfrac{1}{8} = 0.125$.

[33] By (11.48)(e), $\tan^{-1} x = \sum_{n=0}^{\infty} (-1)^n \dfrac{x^{2n+1}}{2n+1}$. $\tan^{-1} 0.1 \approx (0.1) - \dfrac{(0.1)^3}{3} + \dfrac{(0.1)^5}{5}$.

$$S = (0.1) - \frac{(0.1)^3}{3} \approx 0.0997. \quad E = \frac{(0.1)^5}{5} = 2 \times 10^{-6}.$$

[35] Substituting $-x^2$ into (11.48)(c) gives us $e^{-x^2} = \sum_{n=0}^{\infty} \dfrac{(-x^2)^n}{n!} = \sum_{n=0}^{\infty} (-1)^n \dfrac{x^{2n}}{n!} \Rightarrow$

$$\int_0^1 e^{-x^2}\,dx = \int_0^1 \left[\sum_{n=0}^{\infty} (-1)^n \frac{x^{2n}}{n!}\right] dx$$
$$= \left[\sum_{n=0}^{\infty} (-1)^n \frac{x^{2n+1}}{(2n+1)\,n!}\right]_0^1 \quad \{\text{using (11.40)(ii)}\}$$
$$= \sum_{n=0}^{\infty} (-1)^n \frac{1}{(2n+1)\,n!}.$$

$\mathrm{I} \approx 1 - \tfrac{1}{3} + \tfrac{1}{10}$. $S = 1 - \tfrac{1}{3} = \tfrac{2}{3} \approx 0.6667$. $E = \tfrac{1}{10} = 0.1$.

Note: In Exercises 39–42, all series are alternating and satisfy (11.30), so (11.31) applies.

39 $\dfrac{1-\cos x}{x^2} = \dfrac{1-\left(1-\frac{x^2}{2!}+\frac{x^4}{4!}-\frac{x^6}{6!}+\cdots\right)}{x^2}$

$$= \frac{1}{2!} - \frac{x^2}{4!} + \frac{x^4}{6!} - \cdots = \sum_{n=0}^{\infty} (-1)^n \frac{x^{2n}}{(2n+2)!}.$$

$$\int_0^1 \frac{1-\cos x}{x^2}\,dx = \int_0^1 \left[\sum_{n=0}^{\infty} (-1)^n \frac{x^{2n}}{(2n+2)!}\right] dx$$

$$= \left[\sum_{n=0}^{\infty} (-1)^n \frac{x^{2n+1}}{(2n+1)(2n+2)!}\right]_0^1$$

$$= \sum_{n=0}^{\infty} (-1)^n \frac{1}{(2n+1)(2n+2)!} = \sum_{n=0}^{\infty} (-1)^n a_n.$$

We must determine an n so that $a_n < 0.5 \times 10^{-4}$. By substituting 0, 1, 2, and 3 for n, we find that $|\text{Error}| < \dfrac{1}{(2n+1)(2n+2)!} < 0.5 \times 10^{-4}$ for $n \geq 3$.

Thus, $I \approx a_0 - a_1 + a_2 = \dfrac{1}{(1)(2!)} - \dfrac{1}{(3)(4!)} + \dfrac{1}{(5)(6!)} \approx 0.4864.$

41 By (11.48)(d), $\dfrac{\ln(1+x)}{x} = \dfrac{1}{x}\displaystyle\sum_{n=0}^{\infty} (-1)^n \frac{x^{n+1}}{n+1} = \sum_{n=0}^{\infty} (-1)^n \frac{x^n}{n+1}.$

$$\int_0^{1/2} \frac{\ln(1+x)}{x}\,dx = \int_0^{1/2} \left[\sum_{n=0}^{\infty} (-1)^n \frac{x^n}{n+1}\right] dx$$

$$= \left[\sum_{n=0}^{\infty} (-1)^n \frac{x^{n+1}}{(n+1)^2}\right]_0^{1/2}$$

$$= \sum_{n=0}^{\infty} (-1)^n \frac{(\frac{1}{2})^{n+1}}{(n+1)^2} = \sum_{n=0}^{\infty} (-1)^n a_n.$$

We must determine an n so that $a_n < 0.5 \times 10^{-4}$. By substituting successive integer values for n, we find that $|\text{Error}| < \dfrac{(\frac{1}{2})^{n+1}}{(n+1)^2} < 0.5 \times 10^{-4}$ for $n \geq 8$. Thus,

$$I \approx a_0 - a_1 + a_2 - a_3 + a_4 - a_5 + a_6 - a_7$$

$$= \frac{1}{2} - \frac{(1/2)^2}{4} + \frac{(1/2)^3}{9} - \frac{(1/2)^4}{16} + \frac{(1/2)^5}{25} - \frac{(1/2)^6}{36} + \frac{(1/2)^7}{49} - \frac{(1/2)^8}{64} \approx 0.4484.$$

43 (a) $\displaystyle\int_0^1 \sin(x^2)\,dx \approx \int_0^1 (x^2 - \tfrac{1}{6}x^6)\,dx =$

$\left[\tfrac{1}{3}x^3 - \tfrac{1}{42}x^7\right]_0^1 = \tfrac{13}{42} \approx 0.309524.$

$\displaystyle\int_1^2 \sin(x^2)\,dx \approx \int_1^2 (x^2 - \tfrac{1}{6}x^6)\,dx =$

$\left[\tfrac{1}{3}x^3 - \tfrac{1}{42}x^7\right]_1^2 = -\tfrac{29}{42} \approx -0.690476.$

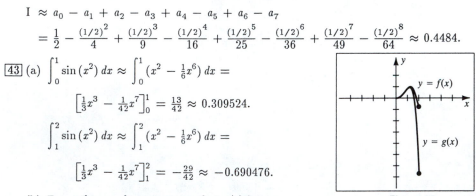

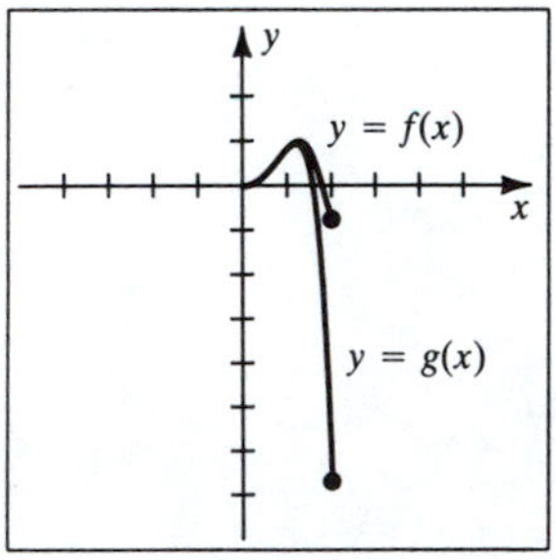

Figure 43

(b) From the graph, we can see that $g(x)$ becomes a worse approximation for $f(x)$ the further x is from zero. The first approximation is more accurate.

[45] $\ln\left(\frac{1+x}{1-x}\right) = \ln(1+x) - \ln(1-x)$.

$$\begin{aligned}\ln(1-x) &= \ln\left[1+(-x)\right] \\ &= \sum_{n=0}^{\infty}(-1)^n \frac{(-x)^{n+1}}{n+1} \\ &= \sum_{n=0}^{\infty}(-1)^n \frac{(-1)^{n+1}x^{n+1}}{n+1} \\ &= \sum_{n=0}^{\infty}(-1)^{2n+1}\frac{x^{n+1}}{n+1} \\ &= -\sum_{n=0}^{\infty}\frac{x^{n+1}}{n+1} \text{ since } 2n+1 \text{ is always an odd integer. Thus,}\end{aligned}$$

$$\begin{aligned}\ln(1+x) - \ln(1-x) &= \sum_{n=0}^{\infty}(-1)^n\frac{x^{n+1}}{n+1} + \sum_{n=0}^{\infty}\frac{x^{n+1}}{n+1} \\ &= \left(x - \frac{x^2}{2} + \frac{x^3}{3} + \cdots\right) + \left(x + \frac{x^2}{2} + \frac{x^3}{3} + \cdots\right) \\ &= 2(x) + 2\left(\frac{x^3}{3}\right) + 2\left(\frac{x^5}{5}\right) + \cdots = 2\sum_{n=0}^{\infty}\frac{x^{2n+1}}{2n+1}.\end{aligned}$$

Since the first series is valid for $-1 < x \le 1$ and the second series is valid for $-1 < -x \le 1 \Leftrightarrow -1 \le x < 1$, the final series is valid for $|x| < 1$.

[49] (a) The central angle θ of a circle of radius R subtended by an arc of length s is $\theta = \frac{s}{R}$. From the figure in the text, $\cos\theta = \frac{R}{R+C} \Rightarrow \sec\theta = \frac{R+C}{R} \Rightarrow$ $R\sec\theta = R + C \Rightarrow C = R(\sec\theta - 1) = R\left[\sec(s/R) - 1\right]$.

(b) *DERIV*: $\sec x$, 1‡ $\sec x\tan x$, 0‡ $\sec^3 x + \sec x\tan^2 x$, 1‡ $5\sec^3 x\tan x + \sec x\tan^3 x$, 0‡ $18\sec^3 x\tan^2 x + 5\sec^5 x + \sec x\tan^4 x$, 5.

$$f(x) = \sec x = f(0) + f'(0) + \frac{f''(0)}{2!}x^2 + \frac{f'''(0)}{3!}x^3 + \cdots \approx 1 + \frac{1}{2!}x^2 + \frac{5}{4!}x^4.$$

If we let $x = \frac{s}{R}$, then $C = R\left[\sec(s/R) - 1\right]$

$$\approx R\left[\left(1 + \frac{s^2}{2R^2} + \frac{5s^4}{24R^4}\right) - 1\right] = \frac{s^2}{2R} + \frac{5s^4}{24R^3}.$$

(c) $R = 3959$, $s = 5 \Rightarrow C \approx \dfrac{5^2}{2(3959)} + \dfrac{5(5)^4}{24(3959)^3} \approx 0.003157 \text{ mi} \approx 16.7 \text{ ft}$.

Exercises 11.9

Note: In the following exercises, $\sin x$ and $\cos x$ have been bound by 1.

In certain intervals, it may be possible to bound them by a smaller value.

1 (a) Use (11.48)(a) to find $P_n(x)$, which consists of terms through x^n. Thus, $P_1(x) = P_2(x) = x$, and $P_3(x) = x - \frac{1}{6}x^3$.

(c) $f(x) = \sin x$ and $f(0.05) \approx P_3(0.05) = 0.05 - \frac{1}{6}(0.05)^3 \approx 0.0500$. $f^{(4)}(x) = \sin x$ and $\left|f^{(4)}(z)\right| \le 1$ for all z. Error $\le \left|R_3(0.05)\right| = \left|\frac{f^{(4)}(z)}{4!}(0.05)^4\right| = \left|\frac{\sin z}{4!}(0.05)^4\right| \le \frac{(1)}{4!}(0.05)^4 \approx 2.6 \times 10^{-7}$.

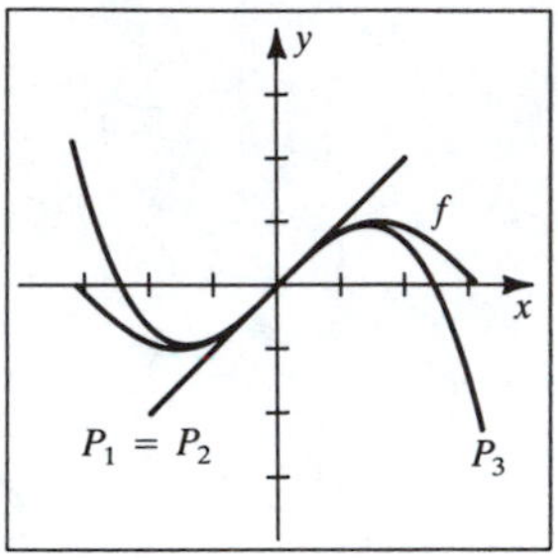

Figure 1

Note: Let *DERIV* denote the beginning of the sequence:

$f(x),\ f(c)\ddagger\ f'(x),\ f'(c)\ddagger\ \ldots\ddagger\ f^{(n)}(x),\ f^{(n)}(c)\ddagger\ f^{(n+1)}(z)$.

7 *DERIV*: $\sin x,\ 1\ddagger\ \cos x,\ 0\ddagger\ -\sin x,\ -1\ddagger\ -\cos x,\ 0\ddagger\ \sin z$.

$$\underline{\sin x} = f(\tfrac{\pi}{2}) + f'(\tfrac{\pi}{2})(x - \tfrac{\pi}{2}) + \frac{f''(\frac{\pi}{2})}{2!}(x - \tfrac{\pi}{2})^2 + \frac{f'''(\frac{\pi}{2})}{3!}(x - \tfrac{\pi}{2})^3 + \frac{f^{(4)}(z)}{4!}(x - \tfrac{\pi}{2})^4$$

$$= 1 + 0\cdot(x - \tfrac{\pi}{2}) + \frac{-1}{2!}(x - \tfrac{\pi}{2})^2 + \frac{0}{3!}(x - \tfrac{\pi}{2})^3 + \frac{\sin z}{4!}(x - \tfrac{\pi}{2})^4$$

$$= 1 - \tfrac{1}{2}(x - \tfrac{\pi}{2})^2 + \frac{\sin z}{24}(x - \tfrac{\pi}{2})^4,\ z \text{ is between } x \text{ and } \tfrac{\pi}{2}.$$

9 *DERIV*: $x^{1/2},\ 2\ddagger\ \frac{1}{2}x^{-1/2},\ \frac{1}{4}\ddagger\ -\frac{1}{4}x^{-3/2},\ -\frac{1}{32}\ddagger\ \frac{3}{8}x^{-5/2},\ \frac{3}{256}\ddagger\ -\frac{15}{16}z^{-7/2}$.

$$\underline{\sqrt{x}} = f(4) + f'(4)(x - 4) + \frac{f''(4)}{2!}(x - 4)^2 + \frac{f'''(4)}{3!}(x - 4)^3 + \frac{f^{(4)}(z)}{4!}(x - 4)^4$$

$$= 2 + \tfrac{1}{4}(x - 4) - \tfrac{1}{64}(x - 4)^2 + \tfrac{1}{512}(x - 4)^3 - \tfrac{5}{128}z^{-7/2}(x - 4)^4,$$

z is between x and 4.

13 *DERIV*: $x^{-1},\ -\frac{1}{2}\ddagger\ -x^{-2},\ -\frac{1}{4}\ddagger\ 2x^{-3},\ -\frac{1}{4}\ddagger\ -6x^{-4},\ -\frac{3}{8}\ddagger$

$24x^{-5},\ -\frac{3}{4}\ddagger\ -120x^{-6},\ -\frac{15}{8}\ddagger\ 720z^{-7}$.

$$\underline{\tfrac{1}{x}} = f(-2) + f'(-2)(x + 2) + \frac{f''(-2)}{2!}(x + 2)^2 + \frac{f'''(-2)}{3!}(x + 2)^3 +$$

$$\frac{f^{(4)}(-2)}{4!}(x + 2)^4 + \frac{f^{(5)}(-2)}{5!}(x + 2)^5 + \frac{f^{(6)}(z)}{6!}(x + 2)^6$$

$$= -\tfrac{1}{2} - \tfrac{1}{4}(x + 2) - \tfrac{1}{8}(x + 2)^2 - \tfrac{1}{16}(x + 2)^3 - \tfrac{1}{32}(x + 2)^4 - \tfrac{1}{64}(x + 2)^5 +$$

$$z^{-7}(x + 2)^6,\ z \text{ is between } x \text{ and } -2.$$

[15] *DERIV*: $\tan^{-1}x$, $\frac{\pi}{4}$‡ $(1+x^2)^{-1}$, $\frac{1}{2}$‡ $-2x(1+x^2)^{-2}$, $-\frac{1}{2}$‡ $(6z^2-2)(1+z^2)^{-3}$.

$$\underline{\tan^{-1}x} = f(1) + f'(1)(x-1) + \frac{f''(1)}{2!}(x-1)^2 + \frac{f'''(z)}{3!}(x-1)^3$$

$$= \tfrac{\pi}{4} + \tfrac{1}{2}(x-1) - \tfrac{1}{4}(x-1)^2 + \frac{3z^2-1}{3(1+z^2)^3}(x-1)^3,\ z \text{ is between } x \text{ and } 1.$$

Note: Exer. 19-30: Since $c = 0$, z is between x and 0.

[19] *DERIV*: $\ln(x+1)$, 0‡ $(x+1)^{-1}$, 1‡ $-1(x+1)^{-2}$, -1‡ $2(x+1)^{-3}$, 2‡ $-6(x+1)^{-4}$, -6‡ $24(z+1)^{-5}$.

$$\underline{\ln(x+1)} = f(0) + f'(0)\,x + \frac{f''(0)}{2!}x^2 + \frac{f'''(0)}{3!}x^3 + \frac{f^{(4)}(0)}{4!}x^4 + \frac{f^{(5)}(z)}{5!}x^5$$

$$= 0 + 1\cdot x + \frac{-1}{2}x^2 + \frac{2}{6}x^3 + \frac{-6}{24}x^4 + \frac{24}{120(z+1)^5}x^5$$

$$= x - \tfrac{1}{2}x^2 + \tfrac{1}{3}x^3 - \tfrac{1}{4}x^4 + \frac{x^5}{5(z+1)^5}$$

[21] *DERIV*: $\cos x$, 1‡ $-\sin x$, 0‡ $-\cos x$, -1‡ $\sin x$, 0‡ $\cos x$, 1‡ $-\sin x$, 0‡ $-\cos x$, -1‡ $\sin x$, 0‡ $\cos x$, 1‡ $-\sin z$. {Note that odd-numbered derivatives will equal 0.}

$$\underline{\cos x} = f(0) + f'(0)\,x + \frac{f''(0)}{2!}x^2 + \frac{f'''(0)}{3!}x^3 + \frac{f^{(4)}(0)}{4!}x^4 + \frac{f^{(5)}(0)}{5!}x^5 + \frac{f^{(6)}(0)}{6!}x^6 + \frac{f^{(7)}(0)}{7!}x^7 + \frac{f^{(8)}(0)}{8!}x^8 + \frac{f^{(9)}(z)}{9!}x^9$$

$$= 1 + \frac{-1}{2!}x^2 + \frac{1}{4!}x^4 + \frac{-1}{6!}x^6 + \frac{1}{8!}x^8 + \frac{-\sin z}{9!}x^9$$

$$= 1 - \frac{x^2}{2!} + \frac{x^4}{4!} - \frac{x^6}{6!} + \frac{x^8}{8!} - \frac{\sin z}{9!}x^9$$

[23] Since $f'(x) = 2e^{2x}$, $f''(x) = 2^2e^{2x}$, $f'''(x) = 2^3e^{2x}$,

we see that $f^{(k)}(x) = 2^ke^{2x}$, $f^{(k)}(0) = 2^k$, and $f^{(6)}(z) = 64e^{2z}$.

$$\underline{e^{2x}} = f(0) + f'(0)\,x + \frac{f''(0)}{2!}x^2 + \frac{f'''(0)}{3!}x^3 + \frac{f^{(4)}(0)}{4!}x^4 + \frac{f^{(5)}(0)}{5!}x^5 + \frac{f^{(6)}(z)}{6!}x^6$$

$$= 1 + 2x + \frac{4}{2!}x^2 + \frac{8}{3!}x^3 + \frac{16}{4!}x^4 + \frac{32}{5!}x^5 + \frac{64e^{2z}}{6!}x^6$$

$$= 1 + 2x + 2x^2 + \tfrac{4}{3}x^3 + \tfrac{2}{3}x^4 + \tfrac{4}{15}x^5 + \tfrac{4}{45}e^{2z}x^6$$

[25] *DERIV*: $(x-1)^{-2}$, 1‡ $-2(x-1)^{-3}$, 2‡ $6(x-1)^{-4}$, 6‡ $-24(x-1)^{-5}$, 24‡ $120(x-1)^{-6}$, 120‡ $-720(x-1)^{-7}$, 720‡ $5040(z-1)^{-8}$.

$$\frac{1}{(x-1)^2} = 1 + 2x + 3x^2 + 4x^3 + 5x^4 + 6x^5 + \frac{7x^6}{(z-1)^8}$$

[29] *DERIV*: $2x^4 - 5x^3$, 0‡ $8x^3 - 15x^2$, 0‡ $24x^2 - 30x$, 0‡ $48x - 30$, -30‡ 48, 48‡ 0.

For both values of n,

$$\underline{f(x)} = f(0) + f'(0)\,x + \frac{f''(0)}{2!}x^2 + \frac{f'''(0)}{3!}x^3 + \frac{f^{(4)}(0)}{4!}x^4 + \frac{f^{(5)}(z)}{5!}x^5$$

$$= 0 + 0 \cdot x + \frac{0}{2!}x^2 + \frac{-30}{3!}x^3 + \frac{48}{4!}x^4 + \frac{0}{5!}x^5$$

$$= -5x^3 + 2x^4.$$

Note: If f is a polynomial of degree n, then $P_n(x) = f(x)$ and $R_n(x) = 0$.

[31] First, we must write 89° in terms of radian measure.

Since $\frac{\pi}{2}$ corresponds to 90° and $\frac{\pi}{180}$ to 1°, let $x = \frac{\pi}{2} - \frac{\pi}{180}$.

Then, $\sin x \approx 1 - \frac{1}{2}(x - \frac{\pi}{2})^2 \Rightarrow \sin 89^\circ \approx 1 - \frac{1}{2}(-\frac{\pi}{180})^2 \approx 0.9998$.

Since $|\sin z| \leq 1$, $|R_3(x)| \leq \left|\frac{1}{24}(-\frac{\pi}{180})^4\right| \approx 4 \times 10^{-9}$.

[33] Let $x = 4 + 0.03$. Then, $\sqrt{x} \approx 2 + \frac{1}{4}(x-4) - \frac{1}{64}(x-4)^2 + \frac{1}{512}(x-4)^3 \Rightarrow$

$\sqrt{4.03} \approx 2 + \frac{1}{4}(0.03) - \frac{1}{64}(0.03)^2 + \frac{1}{512}(0.03)^3 \approx 2.0075$. We must determine where $\left|f^{(4)}(z)\right| = \left|\frac{1}{z^{7/2}}\right|$ is maximum when z is between 4 and 4.03. The maximum occurs when $z = 4$ because that value minimizes the denominator and hence, maximizes the value of the fraction. Since $\left|\frac{1}{z^{7/2}}\right| \leq \left|\frac{1}{4^{7/2}}\right| \approx 0.0078$ on (4, 4.03),

$|R_3(x)| \leq \left|\frac{5}{128}(0.0078)(0.03)^4\right| < 3 \times 10^{-10}$.

[35] Let $x = -2 - 0.2$ or $x + 2 = -0.2$. Then, $\frac{1}{x} = -\frac{1}{2.2} \approx$

$-\frac{1}{2} - \frac{1}{4}(-0.2) - \frac{1}{8}(-0.2)^2 - \frac{1}{16}(-0.2)^3 - \frac{1}{32}(-0.2)^4 - \frac{1}{64}(-0.2)^5 \approx -0.454545$.

We must determine where $\left|f^{(6)}(z)\right| = \left|\frac{1}{z^7}\right|$ is maximum when z is between -2.2 and -2. $\left|\frac{1}{z^7}\right|$ is maximized when $z = -2$. Thus, $|R_5(x)| \leq \left|-2^{-7}(-0.2)^6\right| = 5 \times 10^{-7}$.

[37] Let $x = 0.25$. Then, $\ln(1 + x) \approx x - \frac{1}{2}x^2 + \frac{1}{3}x^3 - \frac{1}{4}x^4 \Rightarrow$

$\ln 1.25 \approx 0.25 - \frac{1}{2}(0.25)^2 + \frac{1}{3}(0.25)^3 - \frac{1}{4}(0.25)^4 \approx 0.22298$.

We must determine where $\left|f^{(5)}(z)\right| = \left|\frac{24}{(z+1)^5}\right|$ is maximum when z is between 0 and 0.25. This occurs when $z = 0$. Thus, $|R_4(x)| \leq \left|\frac{(0.25)^5}{5(0+1)^5}\right| \approx 2 \times 10^{-4}$.

[41] Using Exercise 21 with $n = 3$, $\underline{\cos x} = 1 - \frac{x^2}{2!} + \frac{\cos z}{4!}x^4$.

Since $|\cos z| \leq 1$, $|x^4| \leq (0.1)^4$ when $-0.1 \leq x \leq 0.1$, and z is between 0 and x,

$|R_3(x)| \leq \left|\frac{(1)(0.1)^4}{24}\right| \approx 4.2 \times 10^{-6} < 0.5 \times 10^{-5} \Rightarrow$ five decimal places.

[45] Using Exercise 19 with $n = 3$, $\underline{\ln(x+1)} = x - \frac{1}{2}x^2 + \frac{1}{3}x^3 - \dfrac{x^4}{4(z+1)^4}$. Since

$\left|\dfrac{1}{4(z+1)^4}\right| \le \dfrac{1}{4(0.9)^4}$, $|x^4| \le (0.1)^4$ when $-0.1 \le x \le 0.1$, and z is between 0 and x,

$$|R_3(x)| \le \left|\frac{(-0.1)^4}{4(0.9)^4}\right| \approx 0.000038 < 0.5 \times 10^{-4} \Rightarrow \underline{\text{four decimal places}}.$$

Exercises 11.10

[1] (a) Let $k = \frac{1}{2}$ in (11.50). Then,

$$(1+x)^{1/2} = 1 + \tfrac{1}{2}x + \frac{(\frac{1}{2})(-\frac{1}{2})}{2!}x^2 + \frac{(\frac{1}{2})(-\frac{1}{2})(-\frac{3}{2})}{3!}x^3 + \cdots$$

$$= 1 + \tfrac{1}{2}x - \tfrac{1}{8}x^2 + \sum_{n=3}^{\infty}(-1)^{n-1}\frac{1\cdot3\cdot5\cdot\cdots\cdot(2n-3)}{2^n\,n!}x^n.$$

We started the summation notation with $n = 3$ because this is when the pattern started to become apparent. We could have started the summation notation with another value for n. The "$2n - 3$" was obtained by noticing that the pattern of the product in the numerator, $1\cdot3\cdot5\cdot\cdots$, will have a last term of the form $(2n + k)$ for each value of n. In particular, we see that $2n + k = 3$ when $n = 3$, and hence $6 + k = 3$, or $k = -3$, and our term is $2n - 3$. The series is valid when $|x| < 1$. Thus, $r = 1$.

(b) Substituting $-x^3$ for x in part (a) yields

$$(1-x^3)^{1/2} = 1 - \tfrac{1}{2}x^3 - \tfrac{1}{8}x^6 - \sum_{n=3}^{\infty}\frac{1\cdot3\cdot5\cdot\cdots\cdot(2n-3)}{2^n\,n!}x^{3n}.$$

The series is valid when $|-x^3| < 1 \Leftrightarrow |x| < 1$. Thus, $r = 1$.

[3] Let $k = -\frac{2}{3}$ in (11.50). Then,

$$(1+x)^{-2/3} = 1 - \tfrac{2}{3}x + \frac{(-\frac{2}{3})(-\frac{5}{3})}{2!}x^2 + \frac{(-\frac{2}{3})(-\frac{5}{3})(-\frac{8}{3})}{3!}x^3 + \cdots$$

$$= 1 - \tfrac{2}{3}x + \tfrac{5}{9}x^2 + \sum_{n=3}^{\infty}\frac{(-2)(-5)(-8)\cdots(1-3n)}{3^n\,n!}x^n;\ r = 1.$$

To obtain the "$1 - 3n$" term, we notice that each term of the product $(-2)(-5)\cdots$ is 3 *less* than its predecessor, so the general term must be of the form $-3n + k$. In particular, when $n = 3$, $-3n + k = -8 \Rightarrow -9 + k = -8 \Rightarrow k = 1$, and our term is $-3n + 1$, or, equivalently, $1 - 3n$.

[5] Let $k = \frac{3}{5}$ and substitute $(-x)$ for x in (11.50). Then,

$$(1-x)^{3/5} = 1 + \tfrac{3}{5}(-x) + \frac{(\frac{3}{5})(-\frac{2}{5})}{2!}(-x)^2 + \frac{(\frac{3}{5})(-\frac{2}{5})(-\frac{7}{5})}{3!}(-x)^3 + \cdots$$

$$= 1 - \tfrac{3}{5}x - \tfrac{3}{25}x^2 + \sum_{n=3}^{\infty}\frac{(3)(-2)(-7)\cdots(8-5n)}{5^n\,n!}(-x)^n;\ r = 1.$$

[11] In order to apply (11.50), we must write $\sqrt[3]{8 + x}$ in the form $(1 + x)^k$, where the "1" is obtained by factoring out 8. $(8 + x)^{1/3} = \left[8(1 + \frac{1}{8}x)\right]^{1/3} = 8^{1/3}(1 + \frac{1}{8}x)^{1/3} = 2(1 + \frac{1}{8}x)^{1/3}$. Let $k = \frac{1}{3}$ and substitute $\frac{1}{8}x$ for x in (11.50). Then,

$$2(1 + \tfrac{1}{8}x)^{1/3} = 2\left[1 + \tfrac{1}{3}(\tfrac{1}{8}x) + \frac{(\frac{1}{3})(-\frac{2}{3})}{2!}(\tfrac{1}{8}x)^2 + \frac{(\frac{1}{3})(-\frac{2}{3})(-\frac{5}{3})}{3!}(\tfrac{1}{8}x)^3 + \cdots\right]$$

$$= 2\left[1 + \tfrac{1}{24}x - \tfrac{1}{576}x^2 + \sum_{n=3}^{\infty} \frac{(-2)(-5)\cdots(4 - 3n)}{3^n\, n!\, 8^n} x^n\right]$$

$$= 2 + \tfrac{1}{12}x - \tfrac{1}{288}x^2 + 2\sum_{n=3}^{\infty} (-1)^{n-1} \frac{2 \cdot 5 \cdot \cdots \cdot (3n - 4)}{24^n\, n!} x^n;$$

$\left|\frac{1}{8}x\right| < 1 \Rightarrow r = 8.$

[13] (a) Let $k = -\frac{1}{2}$ and substitute $-t^2$ for x in (11.50). Then, $(1 - t^2)^{-1/2}$

$$= 1 + (-\tfrac{1}{2})(-t^2) + \frac{(-\frac{1}{2})(-\frac{3}{2})}{2!}(-t^2)^2 + \frac{(-\frac{1}{2})(-\frac{3}{2})(-\frac{5}{2})}{3!}(-t^2)^3 + \cdots$$

$$= 1 + \frac{1}{2^1\, 1!}t^2 + \frac{1 \cdot 3}{2^2\, 2!}t^4 + \frac{1 \cdot 3 \cdot 5}{2^3\, 3!}t^6 + \cdots$$

$$= 1 + \sum_{n=1}^{\infty} \frac{1 \cdot 3 \cdot 5 \cdot \cdots \cdot (2n - 1)}{2^n\, n!} t^{2n}.$$

To find a power series representation for $f(x)$, we will use (11.40)(ii).

$$\sin^{-1}x = \int_0^x (1 - t^2)^{-1/2}\, dt = x + \sum_{n=1}^{\infty} \frac{1 \cdot 3 \cdot 5 \cdot \cdots \cdot (2n - 1)}{2^n\, n!\, (2n + 1)} x^{2n+1}.$$

(b) $t^2 < 1 \Rightarrow |t| < 1 \Rightarrow |x| < 1 \Rightarrow r = 1.$

[15] Substituting x^3 for x in Exercise 1(a), $(1 + x^3)^{1/2} = 1 + \frac{1}{2}x^3 - \frac{1}{8}x^6 + \cdots \Rightarrow$

$$\int_0^{1/2} (1 + x^3)^{1/2}\, dx \approx \int_0^{1/2} (1 + \tfrac{1}{2}x^3)\, dx = \left[x + \tfrac{1}{8}x^4\right]_0^{1/2} \approx 0.508.$$

Since the series is alternating in sign and satisfies (11.30), (11.31) applies.

$$\text{Hence, } |\text{Error}| \le \int_0^{1/2} \left|-\tfrac{1}{8}x^6\right| dx = \left[\tfrac{1}{8(7)}x^7\right]_0^{1/2} = \tfrac{1}{8(7)}(\tfrac{1}{2})^7 < 0.5 \times 10^{-3}.$$

[17] Substituting x^2 for x in Exercise 5, $(1 - x^2)^{3/5} = 1 - \frac{3}{5}x^2 - \frac{3}{25}x^4 - \cdots \Rightarrow$

$$\int_0^{0.2} (1 - x^2)^{3/5}\, dx \approx \int_0^{0.2} (1 - \tfrac{3}{5}x^2)\, dx = \left[x - \tfrac{1}{5}x^3\right]_0^{0.2} \approx 0.198.$$

The actual value is approximately 0.198392.

[23] (a) Since $(1 - x)^{-1/2} \approx 1 - \frac{1}{2}(-x) = 1 + \frac{1}{2}x$,

let $x = k^2\sin^2 u$ and $\dfrac{1}{\sqrt{1 - k^2\sin^2 u}} = (1 - k^2\sin^2 u)^{-1/2} \approx 1 + \frac{1}{2}k^2\sin^2 u$. Then,

$$T \approx 4\sqrt{\frac{L}{g}}\int_0^{\pi/2}(1 + \tfrac{1}{2}k^2\sin^2 u)\,du$$

$$= 4\sqrt{\frac{L}{g}}\int_0^{\pi/2}\left[1 + \tfrac{1}{2}k^2\left(\frac{1 - \cos 2u}{2}\right)\right]du$$

$$= 4\sqrt{\frac{L}{g}}\int_0^{\pi/2}(1 + \tfrac{1}{4}k^2 - \tfrac{1}{4}k^2\cos 2u)\,du$$

$$= 4\sqrt{\frac{L}{g}}\left[u + \tfrac{1}{4}k^2u - \tfrac{1}{8}k^2\sin 2u\right]_0^{\pi/2}$$

$$= 4\sqrt{\frac{L}{g}}\left[\tfrac{\pi}{2} + \tfrac{1}{4}k^2(\tfrac{\pi}{2})\right] = 2\pi\sqrt{\frac{L}{g}}\left(1 + \frac{1}{4}k^2\right).$$

(b) $\theta_0 = \frac{\pi}{6} \Rightarrow k = \sin\frac{1}{2}\theta_0 = \sin(\frac{1}{2}\cdot\frac{\pi}{6}) = \sin\frac{\pi}{12}$.

$$T \approx 2\pi\sqrt{L/g}\,(1 + \tfrac{1}{4}\sin^2\tfrac{\pi}{12}) \approx 6.39\sqrt{L/g}.$$

11.11 Review Exercises

Note: Let AC denote Absolutely Convergent; CC, Conditionally Convergent; D, Divergent; C, Convergent; and AST, the Alternating Series Test.

[1] Using (11.5)(i), $\lim_{x\to\infty}\dfrac{\ln(x^2 + 1)}{x}\ \{\frac{\infty}{\infty}\} = \lim_{x\to\infty}\dfrac{1/(x^2 + 1)\cdot 2x}{1} = \lim_{x\to\infty}\dfrac{2x}{x^2 + 1} = 0$.

Thus, L = 0; C.

[5] $\lim_{n\to\infty}\left(\dfrac{n}{\sqrt{n} + 4} - \dfrac{n}{\sqrt{n} + 9}\right) = \lim_{n\to\infty}\dfrac{n(\sqrt{n} + 9) - n(\sqrt{n} + 4)}{(\sqrt{n} + 4)(\sqrt{n} + 9)} =$

$$\lim_{n\to\infty}\frac{5n}{(\sqrt{n} + 4)(\sqrt{n} + 9)} = \lim_{n\to\infty}\frac{5n}{n + 13\sqrt{n} + 36} = 5;\ \text{C.}$$

Note: In Exercises 7–32, let S denote the given series.

[7] Let $a_n = \dfrac{1}{\sqrt[3]{n(n + 1)(n + 2)}}$ and $b_n = \frac{1}{n}$. $\lim_{n\to\infty}\frac{a_n}{b_n} =$

$\lim_{n\to\infty}\dfrac{n}{\sqrt[3]{n(n + 1)(n + 2)}} = 1 > 0$. S diverges by (11.27) since $\sum b_n$ diverges.

[9] This is a geometric series with $r = -\frac{2}{3}$ and contains negative terms.

Since $|r| = \left|-\frac{2}{3}\right| < 1$, S is AC by (11.15).

[13] $\lim_{n\to\infty}\dfrac{a_{n+1}}{a_n} = \lim_{n\to\infty}\dfrac{(n + 1)!}{\ln(n + 2)}\cdot\dfrac{\ln(n + 1)}{n!} = \lim_{n\to\infty}\dfrac{(n + 1)\ln(n + 1)}{\ln(n + 2)} = \infty \Rightarrow$

S is D by (11.28).

[15] This series contains negative terms.

$$\lim_{n\to\infty}\left|\frac{a_{n+1}}{a_n}\right| = \lim_{n\to\infty}\left|\frac{\left[(n+1)^2+9\right](-2)^{1-(n+1)}}{(n^2+9)(-2)^{1-n}}\right|$$

$$= \lim_{n\to\infty}\left|\frac{(n+1)^2+9}{n^2+9}\cdot\frac{(-2)^{-n}}{(-2)^{1-n}}\right|$$

$$= \lim_{n\to\infty}\left|\frac{(n+1)^2+9}{n^2+9}\cdot\frac{(-2)^{n-1}}{(-2)^n}\right| = 1\cdot\tfrac{1}{2} = \tfrac{1}{2} < 1 \Rightarrow \text{S is AC by (11.35).}$$

[19] This series contains negative terms.

$$\lim_{n\to\infty}\left|(-1)^n\frac{1}{\sqrt[n]{n}}\right| = \lim_{n\to\infty}\frac{1}{n^{1/n}}\ \{\text{see Exer. 11.1.37}\} = 1 \neq 0 \Rightarrow \text{S is D by (11.17).}$$

[23] This series contains negative terms.

We will show that this series satisfies (11.30) and, therefore, converges.

$a_n = \dfrac{\sqrt{n}}{n+1} > 0.$ $f(x) = \dfrac{\sqrt{x}}{x+1} \Rightarrow f'(x) = \dfrac{1-x}{2\sqrt{x}(x+1)^2} < 0$ for $x > 1 \Rightarrow$

$a_n \geq a_{n+1}$, and $\lim\limits_{n\to\infty} a_n = 0$. S converges by AST. To show that this series does not converge absolutely, we will apply the LCT. If $b_n = \dfrac{1}{\sqrt{n}}$, $\lim\limits_{n\to\infty}\dfrac{a_n}{b_n} = 1 > 0$ and $\sum a_n$ diverges since $\sum b_n$ diverges by (11.25) $\{p = \frac{1}{2}\}$. Thus, S is CC.

[25] Note that $1 - \cos n$ is always nonnegative and less than or equal to 2. Using a basic comparison test, $\dfrac{1-\cos n}{n^2} \leq \dfrac{2}{n^2}$ and $2\sum\limits_{n=1}^{\infty}\dfrac{1}{n^2}$ converges by (11.25) $\Rightarrow$ S is C.

[27] $\lim\limits_{n\to\infty}\sqrt[n]{a_n} = \lim\limits_{n\to\infty}\sqrt[n]{\dfrac{(2n)^n}{n^{2n}}} = \lim\limits_{n\to\infty}\dfrac{2n}{n^2} = 0 < 1 \Rightarrow$ S is C by (11.29).

[31] $a_n = \dfrac{\sqrt{\ln n}}{n} > 0$ for $n > 1$. $f(x) = \dfrac{\sqrt{\ln x}}{x} \Rightarrow f'(x) = \dfrac{1 - 2\ln x}{2x^2\sqrt{\ln x}} < 0$ for

$x > e^{1/2}$ $\{\approx 1.65\}$ and $\lim\limits_{n\to\infty} a_n = 0 \Rightarrow$ S converges by AST.

However, $\dfrac{\sqrt{\ln n}}{n} > \dfrac{1}{n}$ for $n > e$ and $\sum\limits_{n=2}^{\infty}\dfrac{1}{n}$ diverges. Thus, S is CC.

Note: In Exer. 33–38, each $f(x)$ is positive, continuous, and decreasing on the interval of integration.

[33] Let $f(x) = \dfrac{1}{(3x+2)^3}$.

$$\int_1^{\infty} f(x)\,dx = \lim_{t\to\infty}\int_1^t (3x+2)^{-3}\,dx = \lim_{t\to\infty}\left[-\frac{1}{6(3x+2)^2}\right]_1^t = 0 - \left(-\tfrac{1}{150}\right);\ \text{C}$$

[37] Let $f(x) = \dfrac{10}{\sqrt[3]{x+8}}$.

$$\int_1^{\infty} f(x)\,dx = \lim_{t\to\infty}\int_1^t 10(x+8)^{-1/3}\,dx = \lim_{t\to\infty}\left[15(x+8)^{2/3}\right]_1^t = \infty;\ \text{D}$$

[39] This series satisfies (11.30) and therefore, we will apply (11.31). We need to find an n such that $a_n < 0.5 \times 10^{-3}$. $a_n = \dfrac{1}{(2n+1)!} < 0.0005 \Rightarrow (2n+1)! > 2000 \Rightarrow$ $2n + 1 \geq 7$ {since $6! = 720$ and $7! = 5040$} $\Rightarrow n \geq 3$.

Thus, $S \approx S_2 = \frac{1}{3!} - \frac{1}{5!} \approx 0.158$.

Note: In Exer. 41–46, let u_n denote the nth term of the power series.

[41] We will apply (11.35) to the given power series. $u_n = \dfrac{(n+1)x^n}{(-3)^n} \Rightarrow$

$$\lim_{n\to\infty}\left|\frac{u_{n+1}}{u_n}\right| = \lim_{n\to\infty}\left|\frac{(n+2)\,x^{n+1}}{(-3)^{n+1}} \cdot \frac{(-3)^n}{(n+1)\,x^n}\right|$$

$$= \lim_{n\to\infty}\left|\frac{n+2}{n+1} \cdot \frac{(-3)^n}{(-3)^{n+1}} \cdot \frac{x^{n+1}}{x^n}\right| = \tfrac{1}{3}|x|.\ \ \tfrac{1}{3}|x| < 1 \Leftrightarrow -3 < x < 3.$$

If $x = 3$, $\sum_{n=0}^{\infty} (-1)^n (n+1)$ is D. If $x = -3$, $\sum_{n=0}^{\infty} (n+1)$ is D. ★ $(-3, 3)$

[45] $$\lim_{n\to\infty}\left|\frac{u_{n+1}}{u_n}\right| = \lim_{n\to\infty}\left|\frac{(2n+2)!\,x^{n+1}}{(n+1)!\,(n+1)!} \cdot \frac{n!\,n!}{(2n)!\,x^n}\right|$$

$$= \lim_{n\to\infty}\left|\frac{(2n+2)(2n+1)(2n)!}{(2n)!} \cdot \frac{n!\,n!}{(n+1)\,n!\,(n+1)\,n!} \cdot \frac{x^{n+1}}{x^n}\right|$$

$$= \lim_{n\to\infty}\frac{(2n+2)(2n+1)}{(n+1)(n+1)}|x| = 4|x|.\ \ 4|x| < 1 \Leftrightarrow |x| < \tfrac{1}{4}.$$ ★ $r = \frac{1}{4}$.

[47] If $x \neq 0$, using (11.48)(b), we have

$$\frac{1 - \cos x}{x} = \frac{1 - \left(1 - \frac{x^2}{2!} + \frac{x^4}{4!} - \frac{x^6}{6!} + \cdots\right)}{x}$$

$$= \left(\frac{1}{x}\right)\left(\frac{x^2}{2!} - \frac{x^4}{4!} + \frac{x^6}{6!} - \cdots\right)$$

$$= \frac{x}{2!} - \frac{x^3}{4!} + \frac{x^5}{6!} - \cdots$$

$$= \sum_{n=1}^{\infty} (-1)^{n+1}\frac{x^{2n-1}}{(2n)!}.$$ At $x = 0$, the series equals 0.

Since (11.48)(b) converges for every x, it follows that $r = \infty$ for this series too.

[49] $\sin x \cos x = \frac{1}{2}(2\sin x \cos x) = \frac{1}{2}\sin 2x$.

By (11.48)(a), $\frac{1}{2}\sin 2x = \frac{1}{2}\sum_{n=0}^{\infty} (-1)^n \dfrac{(2x)^{2n+1}}{(2n+1)!} = \sum_{n=0}^{\infty} (-1)^n \dfrac{2^{2n}\,x^{2n+1}}{(2n+1)!}$. $r = \infty$.

[55] $\sqrt{x} = \sqrt{4 + (x - 4)} = \sqrt{4\left[1 + \frac{1}{4}(x - 1)\right]} = 2\left[1 + \frac{1}{4}(x - 4)\right]^{1/2}$.

Using the binomial series with $k = \frac{1}{2}$ and substituting $\frac{1}{4}(x - 4)$ for x gives us

$$\sqrt{x} = 2\left[1 + \tfrac{1}{8}(x - 4) + \frac{(\frac{1}{2})(-\frac{1}{2})}{2!}\left[\tfrac{1}{4}(x - 4)\right]^2 + \frac{(\frac{1}{2})(-\frac{1}{2})(-\frac{3}{2})}{3!}\left[\tfrac{1}{4}(x - 4)\right]^3 + \cdots\right]$$

$$= 2 + \tfrac{1}{4}(x - 4) + 2\sum_{n=2}^{\infty}(-1)^{n-1}\frac{1 \cdot 3 \cdot 5 \cdot \cdots \cdot (2n - 3)}{2^n\, n!\, 4^n}(x - 4)^n$$

$$= 2 + \tfrac{1}{4}(x - 4) + \sum_{n=2}^{\infty}(-1)^{n-1}\frac{1 \cdot 3 \cdot 5 \cdot \cdots \cdot (2n - 3)}{2^{3n-1}\, n!}(x - 4)^n.$$

$\{2^n 4^n = 2^n \cdot 2^{2n} = 2^{3n}\}$

[57] Substituting $(-x^2)$ for x in (11.41), we find that

$$\int_0^1 x^2 e^{-x^2}\,dx = \int_0^1 x^2\left[1 - x^2 + \frac{x^4}{2!} - \frac{x^6}{3!} + \frac{x^8}{4!} - \frac{x^{10}}{5!} + \frac{x^{12}}{6!} - \cdots\right]dx$$

$$= \int_0^1\left[x^2 - x^4 + \frac{x^6}{2!} - \frac{x^8}{3!} + \frac{x^{10}}{4!} - \frac{x^{12}}{5!} + \frac{x^{14}}{6!} - \cdots\right]dx$$

$$= \left[\tfrac{1}{3}x^3 - \tfrac{1}{5}x^5 + \frac{x^7}{7(2!)} - \frac{x^9}{9(3!)} + \frac{x^{11}}{11(4!)} - \frac{x^{13}}{13(5!)} + \frac{x^{15}}{15(6!)} - \cdots\right]_0^1.$$

Using the first 6 terms, $I \approx \frac{1}{3} - \frac{1}{5} + \frac{1}{14} - \frac{1}{54} + \frac{1}{264} - \frac{1}{1560} \approx 0.189$.

Since the series is alternating in sign and satisfies (11.30), (11.31) applies.

Hence, with $|\text{error}| \le \dfrac{1}{15(6!)} = \dfrac{1}{10{,}800} \approx 9.26 \times 10^{-5} < 0.5 \times 10^{-3}$.

[61] *DERIV*: $\ln\cos x$, $\ln\left(\frac{\sqrt{3}}{2}\right)$‡ $-\tan x$, $-\frac{\sqrt{3}}{3}$‡ $-\sec^2 x$, $-\frac{4}{3}$‡ $-2\sec^2 x\tan x$, $-\frac{8\sqrt{3}}{9}$‡

$-2\sec^4 z - 4\sec^2 z\tan^2 z$.

$$\underline{\ln\cos x} = f(\tfrac{\pi}{6}) + f'(\tfrac{\pi}{6})(x - \tfrac{\pi}{6}) + \frac{f''(\frac{\pi}{6})}{2!}(x - \tfrac{\pi}{6})^2 + \frac{f'''(\frac{\pi}{6})}{3!}(x - \tfrac{\pi}{6})^3 + \frac{f^{(4)}(z)}{4!}(x - \tfrac{\pi}{6})^4$$

$$= \ln\left(\tfrac{\sqrt{3}}{2}\right) - \tfrac{\sqrt{3}}{3}(x - \tfrac{\pi}{6}) - \tfrac{2}{3}(x - \tfrac{\pi}{6})^2 - \tfrac{4\sqrt{3}}{27}(x - \tfrac{\pi}{6})^3$$

$$- \tfrac{1}{12}(\sec^4 z + 2\sec^2 z\tan^2 z)(x - \tfrac{\pi}{6})^4,\ z \text{ is between } x \text{ and } \tfrac{\pi}{6}.$$

[63] *DERIV*: e^{-x^2}, 1‡ $-2xe^{-x^2}$, 0‡ $(4x^2 - 2)e^{-x^2}$, -2‡ $(-8x^3 + 12x)e^{-x^2}$, 0‡

$(16z^4 - 48z^2 + 12)e^{-z^2}$.

$$\underline{e^{-x^2}} = f(0) + f'(0)\,x + \frac{f''(0)}{2!}x^2 + \frac{f'''(0)}{3!}x^3 + \frac{f^{(4)}(z)}{4!}x^4$$

$$= 1 - x^2 + \tfrac{1}{6}(4z^4 - 12z^2 + 3)e^{-z^2}x^4,\ z \text{ is between } x \text{ and } 0.$$

65 Let $f(x) = \cos x$ and $c = \frac{\pi}{4}$. *DERIV*: $\cos x, \frac{1}{\sqrt{2}}$‡ $-\sin x, -\frac{1}{\sqrt{2}}$‡ $-\cos x, -\frac{1}{\sqrt{2}}$‡ $\sin z$.

$\underline{\cos x} = \frac{1}{\sqrt{2}} - \frac{1}{\sqrt{2}}(x - \frac{\pi}{4}) - \frac{1}{2\sqrt{2}}(x - \frac{\pi}{4})^2 + \frac{\sin z}{6}(x - \frac{\pi}{4})^3$, z is between x and $\frac{\pi}{4}$.

Since $43^\circ = 45^\circ - 2^\circ = \frac{\pi}{4} - \frac{\pi}{90}$, let $x = \frac{\pi}{4} - \frac{\pi}{90}$ and then $(x - \frac{\pi}{4}) = -\frac{\pi}{90}$.

Thus, $\cos x = \frac{1}{\sqrt{2}} - \frac{1}{\sqrt{2}}(-\frac{\pi}{90}) - \frac{1}{2\sqrt{2}}(-\frac{\pi}{90})^2 + \frac{\sin z}{6}(-\frac{\pi}{90})^3$.

Using the first three terms, $\cos 43^\circ \approx 0.7314$,

$$\text{with } |\text{error}| \leq |R_3(x)| = \left|\frac{\sin z}{6}(-\frac{\pi}{90})^3\right| < \frac{\pi^3}{6 \cdot 90^3} \approx 7.09 \times 10^{-6} < 0.5 \times 10^{-4}.$$

Chapter 12: Topics From Analytic Geometry

Exercises 12.1

Note: Let V, F, and l denote the vertex, focus, and directrix, respectively.

[1] The equation $y = -\frac{1}{12}x^2$ has the form $y = ax^2$ with $a = -\frac{1}{12}$.

$p = \dfrac{1}{4a} = \dfrac{1}{4(-\frac{1}{12})} = -3$. Referring to (12.2) with $p < 0$,

we see that the focus is 3 units below the origin and the directrix is 3 units above it.

Thus, the vertex is $V(0, 0)$, the focus is $F(0, -3)$, and the directrix is $y = 3$.

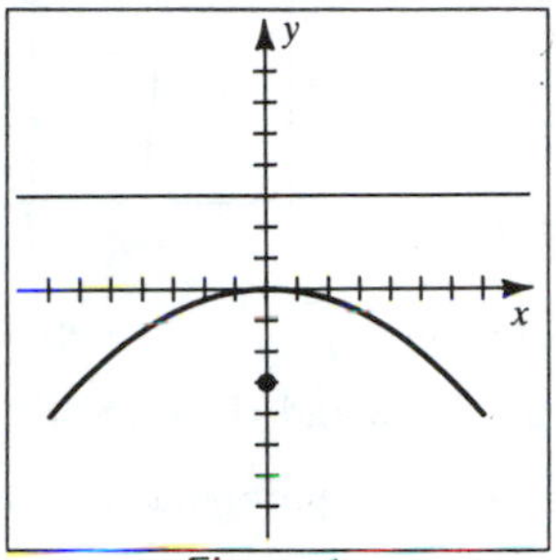

Figure 1

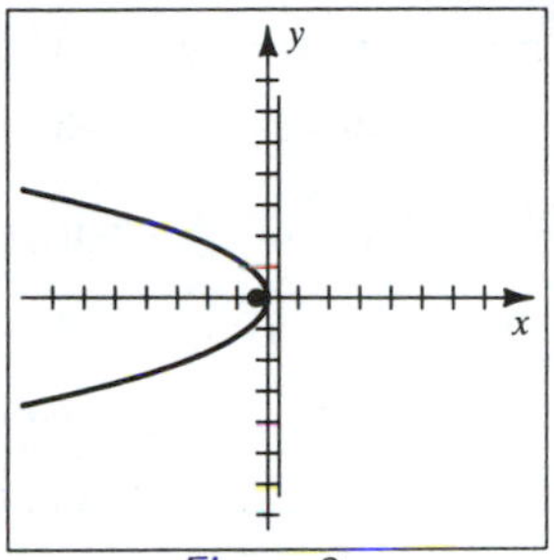

Figure 3

[3] We will write the equation in the form $x = ay^2$. $2y^2 = -3x \Rightarrow x = -\frac{2}{3}y^2 \Rightarrow$

$a = -\frac{2}{3}$. $p = \dfrac{1}{4(-\frac{2}{3})} = -\dfrac{3}{8}$. Referring to (12.2) with $p < 0$, we see that the focus is

$\frac{3}{8}$ units to the left of the origin and the directrix is $\frac{3}{8}$ units to the right.

$V(0, 0)$; $F(-\frac{3}{8}, 0)$; $x = \frac{3}{8}$

[7] As in the solution of Example 3, we will identify the vertex by finding the point at which the tangent line is horizontal. $y = x^2 - 4x + 2 \Rightarrow y' = 2x - 4$. $y' = 0 \Rightarrow$ $x = 2$ and hence $y = -2$. The vertex is $V(2, -2)$. $y = x^2 - 4x + 2 \Rightarrow a = 1$.

$p = \frac{1}{4(1)} = \frac{1}{4}$. Since p is positive, the focus is p units above the vertex, i.e.,

$F(2, -2 + \frac{1}{4}) = F(2, -\frac{7}{4})$.

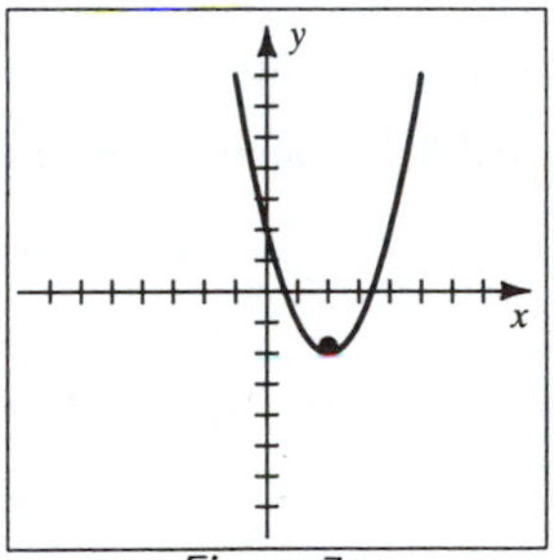

Figure 7

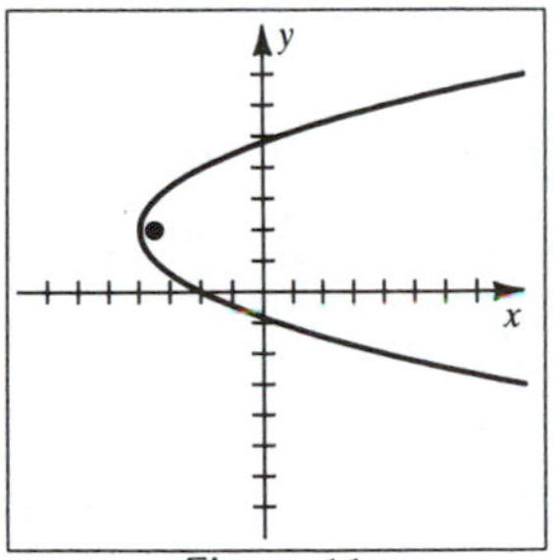

Figure 11

[11] Since the equation is linear in x, first solve for x and then locate the vertex by finding the point where the tangent line is parallel to the y-axis. By (12.4) the graph has a horizontal axis. $y^2 - 4y - 2x - 4 = 0 \Rightarrow x = \frac{1}{2}y^2 - 2y - 2 \Rightarrow x' = y - 2$.
$x' = 0 \Rightarrow y = 2 \Rightarrow V(-4, 2)$. $a = \frac{1}{2} \Rightarrow p = \dfrac{1}{4a} = \frac{1}{2}$. Since p is positive, the focus is p units to the right of the vertex, i.e., $F(-4 + \frac{1}{2}, 2) = F(-\frac{7}{2}, 2)$.

[13] Since the equation is linear in y, first solve for y and then locate the vertex by finding the point where the tangent line is horizontal. By (12.4) the graph has a vertical axis. $4x^2 + 40x + y + 106 = 0 \Rightarrow$
$y = -4x^2 - 40x - 106 \Rightarrow y' = -8x - 40$.
$y' = 0 \Rightarrow x = -5 \Rightarrow V(-5, -6)$. $a = -4 \Rightarrow$
$p = \dfrac{1}{4a} = -\frac{1}{16}$. Since p is negative, the focus is $|p|$ units below the vertex, i.e., $F(-5, -6 - \frac{1}{16}) = F(-5, -\frac{97}{16})$.

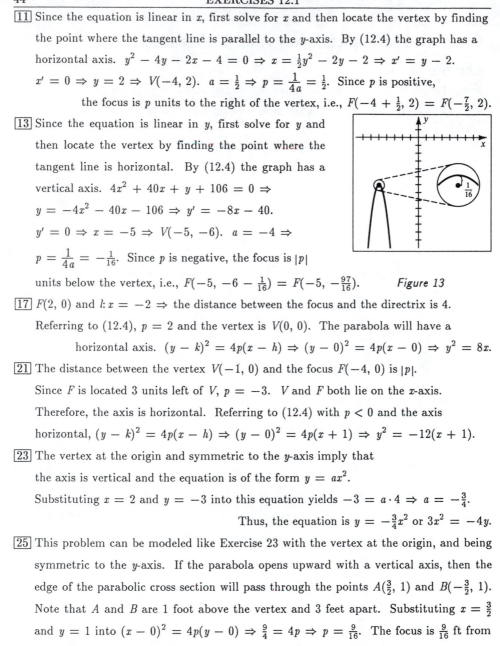

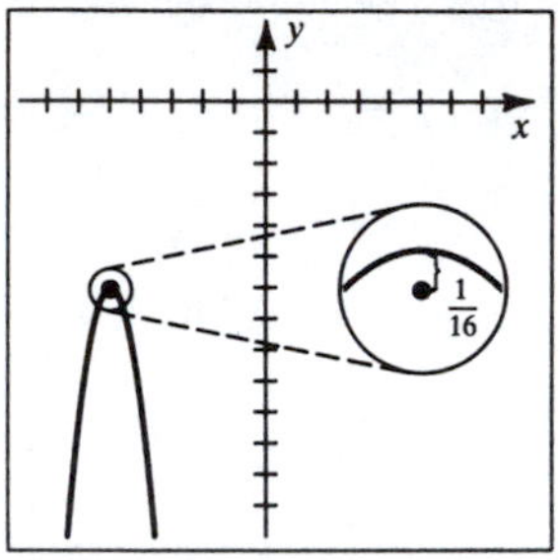

Figure 13

[17] $F(2, 0)$ and $l\colon x = -2 \Rightarrow$ the distance between the focus and the directrix is 4. Referring to (12.4), $p = 2$ and the vertex is $V(0, 0)$. The parabola will have a horizontal axis. $(y - k)^2 = 4p(x - h) \Rightarrow (y - 0)^2 = 4p(x - 0) \Rightarrow y^2 = 8x$.

[21] The distance between the vertex $V(-1, 0)$ and the focus $F(-4, 0)$ is $|p|$.
Since F is located 3 units left of V, $p = -3$. V and F both lie on the x-axis. Therefore, the axis is horizontal. Referring to (12.4) with $p < 0$ and the axis horizontal, $(y - k)^2 = 4p(x - h) \Rightarrow (y - 0)^2 = 4p(x + 1) \Rightarrow y^2 = -12(x + 1)$.

[23] The vertex at the origin and symmetric to the y-axis imply that the axis is vertical and the equation is of the form $y = ax^2$.
Substituting $x = 2$ and $y = -3$ into this equation yields $-3 = a \cdot 4 \Rightarrow a = -\frac{3}{4}$.
Thus, the equation is $y = -\frac{3}{4}x^2$ or $3x^2 = -4y$.

[25] This problem can be modeled like Exercise 23 with the vertex at the origin, and being symmetric to the y-axis. If the parabola opens upward with a vertical axis, then the edge of the parabolic cross section will pass through the points $A(\frac{3}{2}, 1)$ and $B(-\frac{3}{2}, 1)$. Note that A and B are 1 foot above the vertex and 3 feet apart. Substituting $x = \frac{3}{2}$ and $y = 1$ into $(x - 0)^2 = 4p(y - 0) \Rightarrow \frac{9}{4} = 4p \Rightarrow p = \frac{9}{16}$. The focus is $\frac{9}{16}$ ft from the vertex.

[27] The parabola has a vertical axis and l passes horizontally through the focus.
$x^2 = 4y \Rightarrow y = \frac{1}{4}x^2 \Rightarrow a = \frac{1}{4} \Rightarrow p = \dfrac{1}{4a} = 1$.
Thus, the focus is $F(0, 1)$ and the line l is $y = 1$.

(a) Using (6.1), $A = \displaystyle\int_{-2}^{2} (1 - \tfrac{1}{4}x^2)\,dx = 2\int_{0}^{2} (1 - \tfrac{1}{4}x^2)\,dx = 2\Big[x - \tfrac{1}{12}x^3\Big]_0^2 = \tfrac{8}{3}$.

(b) Using shells (6.11), $V = 2\pi\int_0^2 x(1 - \frac{1}{4}x^2)\,dx = 2\pi\left[\frac{1}{2}x^2 - \frac{1}{16}x^4\right]_0^2 = 2\pi.$

(c) Using washers (6.9), $V = \pi\int_{-2}^2 (1^2 - \frac{1}{16}x^4)\,dx = 2\pi\left[x - \frac{1}{80}x^5\right]_0^2 = \frac{16\pi}{5}.$

31 (a) The parabola has a vertical axis and is of the form $(x - h)^2 = a(y - k)$. Since the vertex is $V(0, 10)$, we have $h = 0$, $k = 10$, and the equation is $x^2 = a(y - 10)$. In the xy-plane, the tops of the two towers are located at the points $(\pm 200, 90)$. Substituting $x = 200$ and $y = 90$ into the equation of the parabola yields $(200)^2 = a(80)$ and hence $a = 500$. Thus, $x^2 = 500(y - 10)$.

(b) Use (6.14) with $y = f(x) = \frac{1}{500}x^2 + 10$.

Then, $f'(x) = \frac{1}{250}x$ and $L = \int_{-200}^{200} \sqrt{1 + (\frac{1}{250}x)^2}\,dx.$

(c) The spacing between each support is $\frac{400}{10} = 40$ ft. The length of cable required at location $(x, 0)$ is given by $f(x) = \frac{1}{500}x^2 + 10$. Using symmetry, the total cable length is $f(0) + 2\sum_{n=1}^{5} f(40n) = 10 + 2(13.2 + 22.8 + 38.8 + 61.2) = 282$ ft.

35 (a) Let the path be described by $y(x) = ax^2 + bx + c$. $y(0) = 0 \Rightarrow c = 0$. Also, $y'(x) = 2ax + b$ and $y'(0) = 1$ (since the ball is thrown up at a 45° angle) $\Rightarrow$ $b = 1$. Thus, $y(x) = ax^2 + x$. Now, let $P(x_0, y_0)$ be the point where the ball strikes the ground. Since the line representing ground level has a slope of $-\frac{3}{4}$ and passes through the origin, it follows that $y_0 = -\frac{3}{4}x_0$.

Then $x_0^2 + y_0^2 = 50^2 \Rightarrow x_0^2 + \frac{9}{16}x_0^2 = 2500 \Rightarrow x_0 = 40$ and $y_0 = -30$.

Thus, $y(40) = a(40)^2 + 40 = -30 \Rightarrow a = -\frac{70}{1600}$ and $y(x) = -\frac{7}{160}x^2 + x$.

(b) Let h represent the height of the ball *off the ground.*

Then, $h(x) = (-\frac{7}{160}x^2 + x) - (-\frac{3}{4}x) = -\frac{7}{160}x^2 + \frac{7}{4}x$.

$h'(x) = -\frac{7}{80}x + \frac{7}{4} = 0 \Rightarrow x = 20$. $h(20) = 17.5$ ft, which is a maximum.

37 (a) Note that the value of p completely determines the parabola. If (x_1, y_1) is a point on the parabola, then $y_1^2 = 4p(x_1 + p) \Rightarrow 4p^2 + 4x_1p - y_1^2 = 0$. This can be regarded as a quadratic equation in p with $a = 4$, $b = 4x_1$, and $c = -y_1^2$.

Using the quadratic formula, $p = \dfrac{-x_1 \pm \sqrt{x_1^2 + y_1^2}}{2}$. If $y_1 \neq 0$, $x_1^2 + y_1^2 \neq 0$.

Then there are exactly two values for p and hence, exactly two parabolas.

(b) $y^2 = 4p(x + p) \Rightarrow 2yy' = 4p \Rightarrow y' = \frac{2p}{y}$. Calculating the value of y' at (x_1, y_1) for each value of p, and then multiplying these together gives

$$\left[2\left(\frac{-x_1 + \sqrt{x_1^2 + y_1^2}}{2}\right)\cdot\frac{1}{y_1}\right]\cdot\left[2\left(\frac{-x_1 - \sqrt{x_1^2 + y_1^2}}{2}\right)\cdot\frac{1}{y_1}\right] =$$

$$\frac{-x_1 + \sqrt{x_1^2 + y_1^2}}{y_1}\cdot\frac{-x_1 - \sqrt{x_1^2 + y_1^2}}{y_1} = \frac{-y_1^2}{y_1^2} = -1.$$

Thus, the tangent lines are perpendicular.

[43] Without loss of generality, let the parabola have the equation $x^2 = 4py$ and P have coordinates (x_0, y_0). First, we find the equation of l and then the coordinates of Q. Let m equal the slope of l. Then, $y = \frac{1}{4p}x^2 \Rightarrow y' = \frac{x}{2p} \Rightarrow x_0 = 2y'p = 2mp$. Substituting (x_0, y_0) into $4py = x^2$ yields $4py_0 = x_0^2 = (2mp)^2 \Rightarrow y_0 = m^2p$. Thus, l has equation $y - y_0 = m(x - x_0) \Leftrightarrow y - m^2p = m(x - 2mp)$. Since the directrix has equation $y = -p$, setting y equal to $-p$ gives us the x-value where l intersects the directrix. $-p - m^2p = m(x - 2mp) \Rightarrow x = \frac{pm^2 - p}{m}$. Thus, $Q = \left(\frac{pm^2 - p}{m}, -p\right)$. Since F has coordinates $(0, p)$, the slope of segment QF is $\dfrac{p - (-p)}{0 - \frac{pm^2 - p}{m}} = -\frac{2pm}{pm^2 - p}$. P has coordinates $(x_0, y_0) = (2mp, pm^2)$ and the slope of segment PF is $\frac{p - pm^2}{0 - 2mp} = \frac{pm^2 - p}{2pm}$. Their product, $-\frac{2pm}{pm^2 - p}\cdot\frac{pm^2 - p}{2pm}$, equals -1 and hence, the two segments are perpendicular.

Exercises 12.2

Note: Let C, V, F, and M denote the center, the vertices, the foci, and the end points of the minor axis, respectively.

[1] Use (12.6) with $a^2 = 9$ and $b^2 = 4$. $c^2 = a^2 - b^2 = 9 - 4 = 5 \Rightarrow c = \pm\sqrt{5}$. The vertices are $(\pm a, 0) = V(\pm 3, 0)$. The end points of the minor axis are $(0, \pm b) = M(0, \pm 2)$. The foci are $(\pm c, 0) = F(\pm\sqrt{5}, 0)$. See Figure 12.15 in the text.

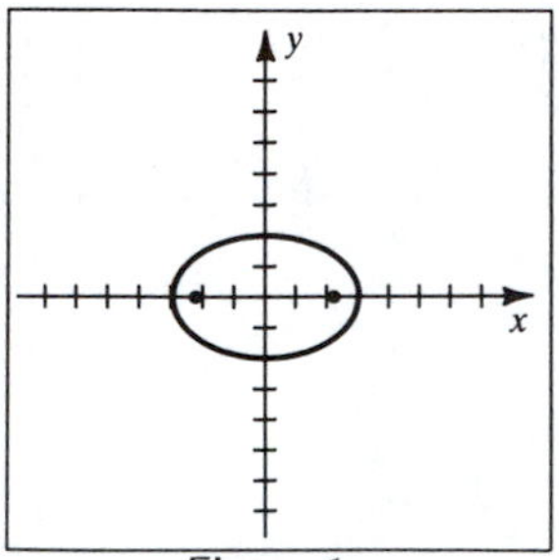

Figure 1

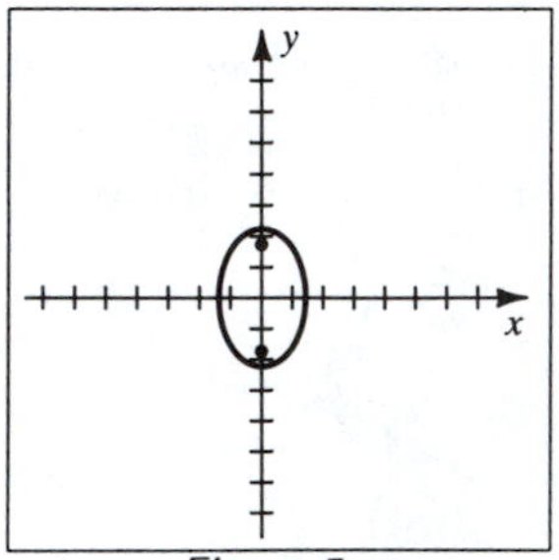

Figure 5

[5] $5x^2 + 2y^2 = 10 \Rightarrow \frac{x^2}{2} + \frac{y^2}{5} = 1$. Use (12.7) with $a^2 = 5$ and $b^2 = 2$.

$c^2 = a^2 - b^2 = 5 - 2 \Rightarrow c = \pm\sqrt{3}$. $V(0, \pm\sqrt{5})$; $F(0, \pm\sqrt{3})$; $M(\pm\sqrt{2}, 0)$.

See Figure 12.17 in the text.

[7] $4x^2 + 25y^2 = 1 \Rightarrow \frac{x^2}{\frac{1}{4}} + \frac{y^2}{\frac{1}{25}} = 1$. Use (12.6) with $a^2 = \frac{1}{4}$ and $b^2 = \frac{1}{25}$.

$c^2 = \frac{1}{4} - \frac{1}{25} = \frac{21}{100} \Rightarrow c = \pm\frac{1}{10}\sqrt{21}$. $V(\pm\frac{1}{2}, 0)$; $F(\pm\frac{1}{10}\sqrt{21}, 0)$; $M(0, \pm\frac{1}{5})$.

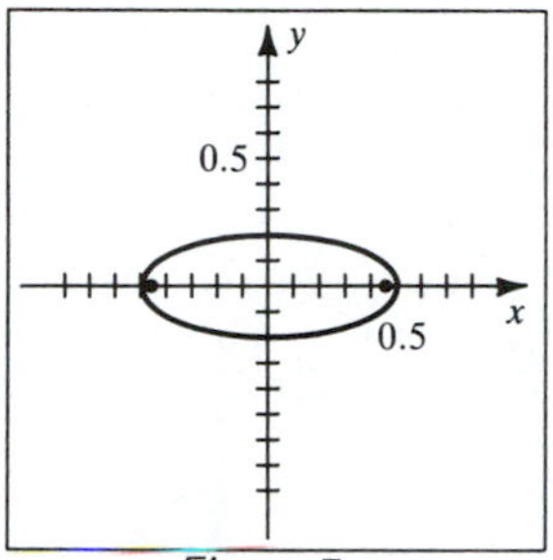

Figure 7

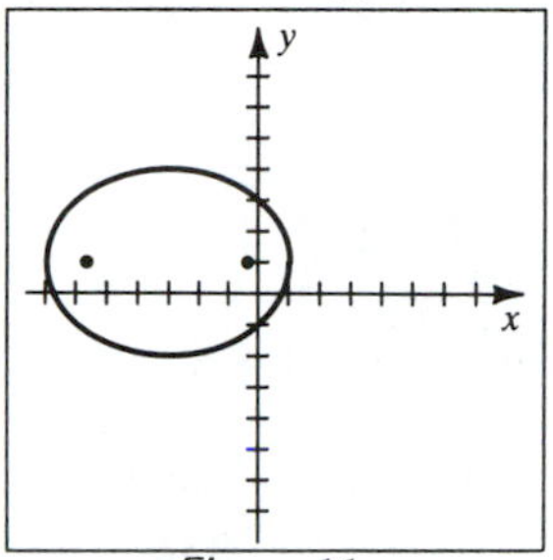

Figure 11

[11] As in the solution of Example 4, we first complete the squares on x and y to get the equation of the ellipse in the form $\frac{(x')^2}{a^2} + \frac{(y')^2}{b^2} = 1$. (See Figure 12.19 in the text.) $9x^2 + 16y^2 + 54x - 32y - 47 = 0 \Rightarrow$

$9(x^2 + 6x + \underline{9}) + 16(y^2 - 2y + \underline{1}) = 47 + \underline{81} + \underline{16} \Rightarrow$

$9(x + 3)^2 + 16(y - 1)^2 = 144 \Rightarrow \frac{(x + 3)^2}{16} + \frac{(y - 1)^2}{9} = 1$; $c^2 = 16 - 9 \Rightarrow$

$c = \pm\sqrt{7}$; $C(-3, 1)$; $V(-3 \pm 4, 1)$; $F(-3 \pm \sqrt{7}, 1)$; $M(-3, 1 \pm 3)$

[13] $25x^2 + 4y^2 - 250x - 16y + 541 = 0 \Rightarrow$

$25(x^2 - 10x + \underline{25}) + 4(y^2 - 4y + \underline{4}) =$

$-541 + \underline{625} + \underline{16} \Rightarrow$

$25(x - 5)^2 + 4(y - 2)^2 = 100 \Rightarrow$

$\frac{(x - 5)^2}{4} + \frac{(y - 2)^2}{25} = 1$;

$c^2 = 25 - 4 \Rightarrow c = \pm\sqrt{21}$; $C(5, 2)$;

$V(5, 2 \pm 5)$; $F(5, 2 \pm \sqrt{21})$; $M(5 \pm 2, 2)$

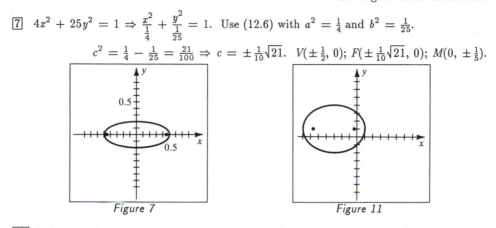

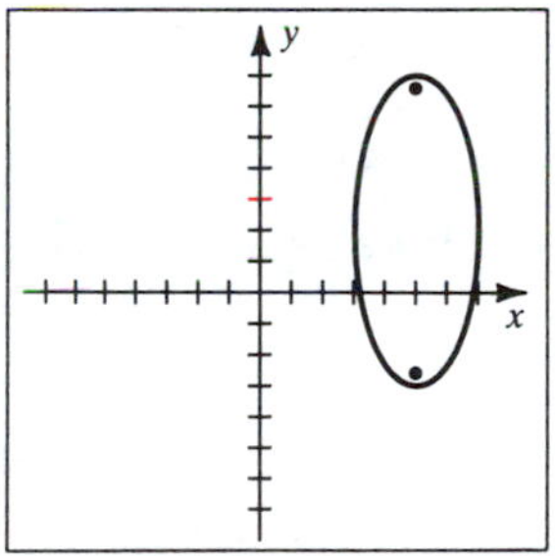

Figure 13

[15] Since the vertices are $V(\pm 8, 0)$, $a = 8$. Since the foci are $F(\pm 5, 0)$, $c = 5$.

$b^2 = a^2 - c^2 = 8^2 - 5^2 = 39$. An equation is $\frac{x^2}{a^2} + \frac{y^2}{b^2} = 1$, or $\frac{x^2}{64} + \frac{y^2}{39} = 1$.

[19] With the vertices at $(0, \pm 6)$, an equation of the ellipse is $\frac{x^2}{b^2} + \frac{y^2}{36} = 1$. Substituting $x = 3$ and $y = 2$ and solving for b^2 yields $\frac{9}{b^2} + \frac{4}{36} = 1 \Rightarrow \frac{9}{b^2} = \frac{8}{9} \Rightarrow b^2 = \frac{81}{8}$.

An equation is $\frac{8x^2}{81} + \frac{y^2}{36} = 1$.

[21] With vertices $V(0, \pm 4)$, an equation of the ellipse is $\frac{x^2}{b^2} + \frac{y^2}{16} = 1$.

By (12.8), $e = \frac{c}{a} = \frac{3}{4}$ and $a = 4 \Rightarrow c = 3$. Thus, $b^2 = a^2 - c^2 = 16 - 9 = 7$.

An equation is $\frac{x^2}{7} + \frac{y^2}{16} = 1$.

[25] Model this problem as an ellipse with center at the origin, $V(\pm 15, 0)$, and $M(0, \pm 10)$. Substituting $x = 6$ into $\frac{x^2}{15^2} + \frac{y^2}{10^2} = 1$ yields $\frac{y^2}{100} = \frac{189}{225} \Rightarrow y^2 = 84$.

The desired height is $\sqrt{84} = 2\sqrt{21} \approx 9.165$ ft.

[31] Let (x_1, y_1) be a point of tangency on $9x^2 + 4y^2 = 36$, or $\frac{x^2}{4} + \frac{y^2}{9} = 1$.

By Example 5(c), the tangent line has equation $\frac{x_1 x}{4} + \frac{y_1 y}{9} = 1$.

$(x, y) = (0, 6)$ on this line $\Rightarrow 0 + \frac{2}{3}y_1 = 1 \Rightarrow y_1 = \frac{3}{2}$. $9x_1^2 + 4y_1^2 = 36 \Rightarrow$
$x_1^2 = \frac{1}{9}(36 - 4y_1^2) = \frac{1}{9}(36 - 9) = 3$. There are two points: $(\pm\sqrt{3}, \frac{3}{2})$.

[33] Differentiating implicitly, $5x^2 + 4y^2 = 56 \Rightarrow 10x + 8yy' = 0 \Rightarrow y' = -\frac{5x}{4y}$.

At $P(-2, 3)$, $y' = \frac{5}{6}$ and an equation of the tangent line is $y - 3 = \frac{5}{6}(x + 2)$, or
$5x - 6y = -28$. The normal line is $y - 3 = -\frac{6}{5}(x + 2)$, or $6x + 5y = 3$.

[35] $b^2x^2 + a^2y^2 = a^2b^2 \Rightarrow a^2y^2 = a^2b^2 - b^2x^2 = b^2(a^2 - x^2) \Rightarrow y^2 = \frac{b^2}{a^2}(a^2 - x^2)$.

The upper half of the ellipse has equation $y = f(x) = \frac{b}{a}\sqrt{a^2 - x^2}$ for $-a \le x \le a$.

Using disks (6.5), $V = \int_{-a}^{a} \pi\left[f(x)\right]^2 = \pi \int_{-a}^{a} \frac{b^2}{a^2}(a^2 - x^2)\,dx = \frac{2b^2\pi}{a^2}\int_0^a (a^2 - x^2)\,dx$

$$= \frac{2b^2\pi}{a^2}\left[a^2x - \tfrac{1}{3}x^3\right]_0^a = \tfrac{4}{3}\pi ab^2.$$

[39] Let any cross section of the elliptical frustum be represented by $\frac{x^2}{a^2} + \frac{y^2}{b^2} = 1$, where a and b are functions of the distance h that the cross section is from the left end of the frustum in the figure. The area of this ellipse is πab by Example 6. a varies between a_1 when $h = 0$ and a_2 when $h = L$. Because of the shape of the sides of the frustum, a is a linear function of h passing through the points $(0, a_1)$ and (L, a_2). Using the slope-intercept form of a line ($y = mx + b$), $a = \frac{a_2 - a_1}{L}h + a_1$, where $h \in [0, L]$. Also, $\frac{a}{b} = k \Rightarrow b = \frac{a}{k}$. Thus, using volumes by cross sections (6.13),

$$V = \int_0^L (\pi ab)\, dh = \frac{\pi}{k}\int_0^L a^2\, dh = \frac{\pi}{k}\int_0^L \left(\frac{a_2 - a_1}{L} h + a_1\right)^2 dh$$

$$= \frac{\pi}{k}\int_0^L \left[\left(\frac{a_2 - a_1}{L}\right)^2 h^2 + 2a_1\left(\frac{a_2 - a_1}{L}\right)h + a_1^2\right] dh$$

$$= \frac{\pi}{k}\left[\left(\frac{a_2 - a_1}{L}\right)^2 \frac{h^3}{3} + 2a_1\left(\frac{a_2 - a_1}{L}\right)\frac{h^2}{2} + a_1^2 h\right]_0^L$$

$$= \frac{\pi}{k}\left[(a_2 - a_1)^2 \frac{L}{3} + a_1(a_2 - a_1)L + a_1^2 L\right]$$

$$= \frac{\pi L}{3k}\left[(a_2^2 - 2a_1a_2 + a_1^2) + (3a_1a_2 - 3a_1^2) + 3a_1^2\right] = \frac{\pi L}{3k}(a_1^2 + a_1a_2 + a_2^2).$$

[43] Since rotations and translations of the ellipse will not affect the result, let its equation be $\frac{x^2}{a^2} + \frac{y^2}{b^2} = 1$. Differentiating implicitly, $\frac{2x}{a^2} + \frac{2y}{b^2}y' = 0 \Rightarrow y' = -\frac{b^2x}{a^2y}$.

The slope of the normal line at (x_1, y_1) is the negative of the reciprocal of y' or $\frac{a^2y_1}{b^2x_1}$.

Since the normal line passes through $(0, 0)$ and (x_1, y_1), its slope must be $\frac{y_1}{x_1}$.

Now, $\frac{a^2y_1}{b^2x_1} = \frac{y_1}{x_1} \Rightarrow a^2 = b^2$. Thus, the ellipse is a circle.

Note: Depending on the type of software used, you may need to solve the given equation for y in order to graph the ellipses in Exercise 45.

[45] (a) $\frac{x^2}{2.9} + \frac{y^2}{2.1} = 1 \Rightarrow 2.1x^2 + 2.9y^2 = 6.09 \Rightarrow$

$y = \pm\sqrt{\frac{1}{2.9}(6.09 - 2.1x^2)}$.

$\frac{x^2}{4.3} + \frac{(y - 2.1)^2}{4.9} = 1 \Rightarrow$

$4.9x^2 + 4.3(y - 2.1)^2 = 21.07 \Rightarrow$

$y = 2.1 \pm \sqrt{\frac{1}{4.3}(21.07 - 4.9x^2)}$.

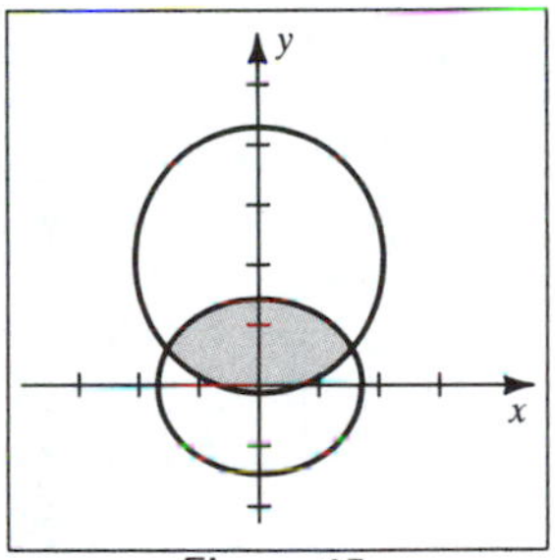

Figure 45

From the graph, the points of intersection are approximately $(\pm 1.540, 0.618)$.

(b) Area $\approx 2\int_0^{1.54}\left[\sqrt{\frac{1}{2.9}(6.09 - 2.1x^2)} - \left(2.1 - \sqrt{\frac{1}{4.3}(21.07 - 4.9x^2)}\right)\right] dx$

Exercises 12.3

Note: Let C, V, F, and W denote the center, the vertices, the foci, and the end points of the conjugate axis, respectively.

[1] Use (12.10) with $a^2 = 9$ and $b^2 = 4$. $c^2 = a^2 + b^2 = 9 + 4 = 13 \Rightarrow c = \pm\sqrt{13}$.

The vertices are $(\pm a, 0) = V(\pm 3, 0)$, the foci are $(\pm c, 0) = F(\pm\sqrt{13}, 0)$, and the end points of the conjugate axis are $(0, \pm b) = W(0, \pm 2)$.

The asymptotes are $y = \pm\frac{b}{a}x$, or $y = \pm\frac{2}{3}x$. See Figure 12.24 in the text.

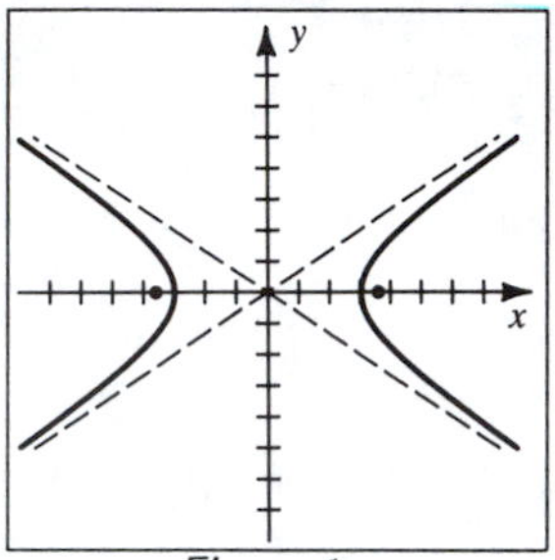

Figure 1

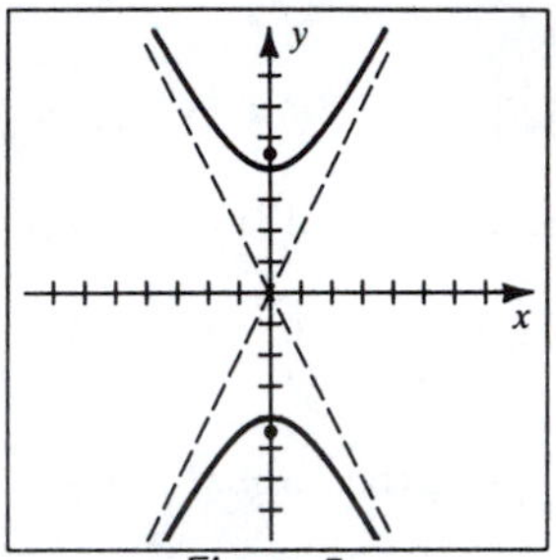

Figure 5

[5] $y^2 - 4x^2 = 16 \Rightarrow \dfrac{y^2}{16} - \dfrac{x^2}{4} = 1$. Use (12.11) with $a^2 = 16$ and $b^2 = 4$.

$c^2 = 16 + 4 \Rightarrow c = \pm 2\sqrt{5}$. See Figure 12.27 in the text.

$V(0, \pm 4)$; $F(0, \pm 2\sqrt{5})$; $W(\pm 2, 0)$; $y = \pm 2x$

[11] $3x^2 - y^2 = -3 \Rightarrow \dfrac{y^2}{3} - \dfrac{x^2}{1} = 1$. Use (12.11) with $a^2 = 3$ and $b^2 = 1$.

$c^2 = 3 + 1 \Rightarrow c = \pm 2$. See Figure 12.27.

$V(0, \pm\sqrt{3})$; $F(0, \pm 2)$; $W(\pm 1, 0)$; $y = \pm\sqrt{3}\,x$

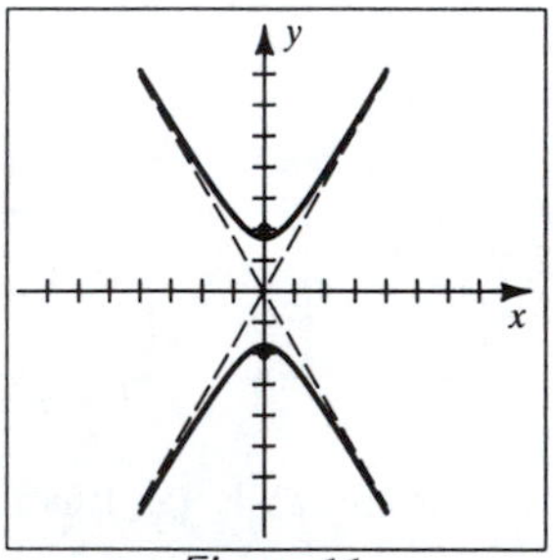

Figure 11

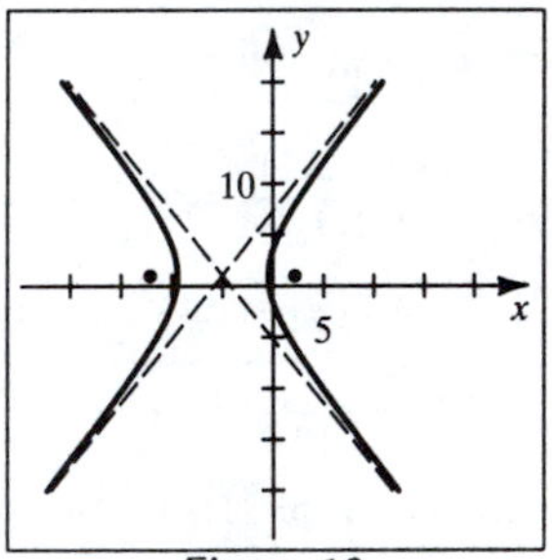

Figure 13

[13] As in the solution of Example 4, we first complete the squares on x and y to get the equation of the hyperbola in the form $\dfrac{(x')^2}{a^2} - \dfrac{(y')^2}{b^2} = 1$.

$25x^2 - 16y^2 + 250x + 32y + 109 = 0 \Rightarrow$

$25(x^2 + 10x + \underline{25}) - 16(y^2 - 2y + \underline{1}) = -109 + \underline{625} - \underline{16} \Rightarrow$

$25(x + 5)^2 - 16(y - 1)^2 = 500 \Rightarrow \dfrac{(x + 5)^2}{20} - \dfrac{(y - 1)^2}{\frac{125}{4}} = 1$;

$c^2 = 20 + \frac{125}{4} \Rightarrow c = \pm\frac{1}{2}\sqrt{205}$; $C(-5, 1)$;

$V(-5 \pm 2\sqrt{5}, 1)$; $F(-5 \pm \frac{1}{2}\sqrt{205}, 1)$; $W(-5, 1 \pm \frac{5}{2}\sqrt{5})$; $(y - 1) = \pm\frac{5}{4}(x + 5)$

[17] $9y^2 - x^2 - 36y + 12x - 36 = 0 \Rightarrow$

$9(y^2 - 4y + \underline{\ 4\ }) - (x^2 - 12x + \underline{\ 36\ }) =$

$36 + \underline{\ 36\ } - \underline{\ 36\ } \Rightarrow$

$9(y-2)^2 - (x-6)^2 = 36 \Rightarrow$

$\dfrac{(y-2)^2}{4} - \dfrac{(x-6)^2}{36} = 1;$

$c^2 = 4 + 36 \Rightarrow c = \pm 2\sqrt{10}$; $C(6, 2)$; $V(6, 2 \pm 2)$;

$F(6, 2 \pm 2\sqrt{10})$; $W(6 \pm 6, 2)$; $(y-2) = \pm\frac{1}{3}(x-6)$

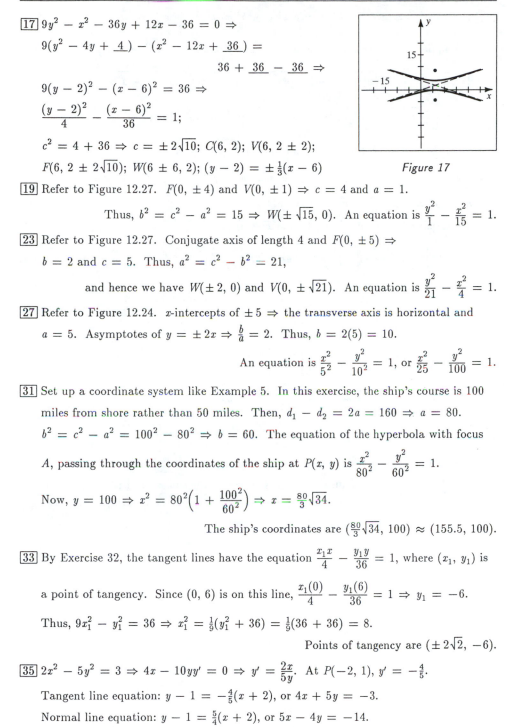

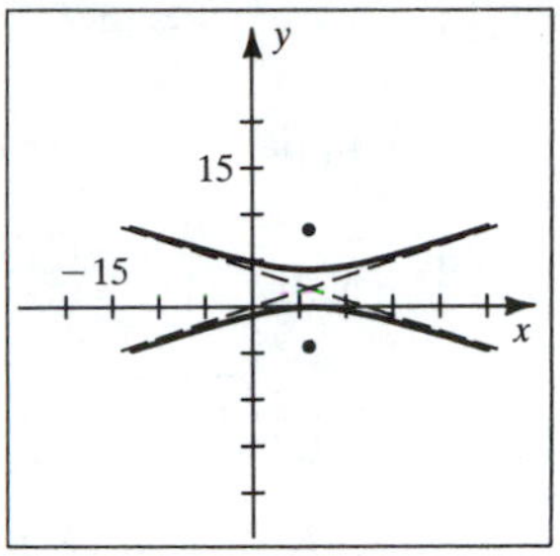

Figure 17

[19] Refer to Figure 12.27. $F(0, \pm 4)$ and $V(0, \pm 1) \Rightarrow c = 4$ and $a = 1$.

Thus, $b^2 = c^2 - a^2 = 15 \Rightarrow W(\pm\sqrt{15}, 0)$. An equation is $\dfrac{y^2}{1} - \dfrac{x^2}{15} = 1$.

[23] Refer to Figure 12.27. Conjugate axis of length 4 and $F(0, \pm 5) \Rightarrow$

$b = 2$ and $c = 5$. Thus, $a^2 = c^2 - b^2 = 21$,

and hence we have $W(\pm 2, 0)$ and $V(0, \pm\sqrt{21})$. An equation is $\dfrac{y^2}{21} - \dfrac{x^2}{4} = 1$.

[27] Refer to Figure 12.24. x-intercepts of $\pm 5 \Rightarrow$ the transverse axis is horizontal and

$a = 5$. Asymptotes of $y = \pm 2x \Rightarrow \frac{b}{a} = 2$. Thus, $b = 2(5) = 10$.

An equation is $\dfrac{x^2}{5^2} - \dfrac{y^2}{10^2} = 1$, or $\dfrac{x^2}{25} - \dfrac{y^2}{100} = 1$.

[31] Set up a coordinate system like Example 5. In this exercise, the ship's course is 100 miles from shore rather than 50 miles. Then, $d_1 - d_2 = 2a = 160 \Rightarrow a = 80$.

$b^2 = c^2 - a^2 = 100^2 - 80^2 \Rightarrow b = 60$. The equation of the hyperbola with focus A, passing through the coordinates of the ship at $P(x, y)$ is $\dfrac{x^2}{80^2} - \dfrac{y^2}{60^2} = 1$.

Now, $y = 100 \Rightarrow x^2 = 80^2\left(1 + \dfrac{100^2}{60^2}\right) \Rightarrow x = \frac{80}{3}\sqrt{34}$.

The ship's coordinates are $(\frac{80}{3}\sqrt{34}, 100) \approx (155.5, 100)$.

[33] By Exercise 32, the tangent lines have the equation $\dfrac{x_1 x}{4} - \dfrac{y_1 y}{36} = 1$, where (x_1, y_1) is a point of tangency. Since $(0, 6)$ is on this line, $\dfrac{x_1(0)}{4} - \dfrac{y_1(6)}{36} = 1 \Rightarrow y_1 = -6$.

Thus, $9x_1^2 - y_1^2 = 36 \Rightarrow x_1^2 = \frac{1}{9}(y_1^2 + 36) = \frac{1}{9}(36 + 36) = 8$.

Points of tangency are $(\pm 2\sqrt{2}, -6)$.

[35] $2x^2 - 5y^2 = 3 \Rightarrow 4x - 10yy' = 0 \Rightarrow y' = \dfrac{2x}{5y}$. At $P(-2, 1)$, $y' = -\frac{4}{5}$.

Tangent line equation: $y - 1 = -\frac{4}{5}(x + 2)$, or $4x + 5y = -3$.

Normal line equation: $y - 1 = \frac{5}{4}(x + 2)$, or $5x - 4y = -14$.

[37] $b^2x^2 - a^2y^2 = a^2b^2 \Rightarrow a^2y^2 = b^2x^2 - a^2b^2 = b^2(x^2 - a^2) \Rightarrow y^2 = \frac{b^2}{a^2}(x^2 - a^2)$.

The upper and lower boundaries of the region are $y = \pm\frac{b}{a}\sqrt{x^2 - a^2}$ for $a \le x \le c$.

Thus, $A = \int_a^c \left[\frac{b}{a}\sqrt{x^2 - a^2} - \left(-\frac{b}{a}\sqrt{x^2 - a^2}\right)\right] dx = \frac{2b}{a}\int_a^c \sqrt{x^2 - a^2}\, dx$

$= \frac{2b}{a}\left[\frac{x}{2}\sqrt{x^2 - a^2} - \frac{a^2}{2}\ln\left|x + \sqrt{x^2 - a^2}\right|\right]_a^c$ { Formula 39 or letting $x = a\sec\theta$ }

$= \frac{b}{a}\left[bc - a^2\ln(b + c) + a^2\ln a\right]$ since $\sqrt{c^2 - a^2} = b$.

[39] Refer to Figure 12.32(i). $x^2 - y^2 = 8 \Rightarrow y = \sqrt{x^2 - 8} \Rightarrow y' = \frac{x}{\sqrt{x^2 - 8}}$.

$a^2 = b^2 = 8 \Rightarrow c^2 = a^2 + b^2 = 16$.

Thus, using (6.19) with $V(\sqrt{8}, 0)$ and $F(4, 0) \Rightarrow$

$$S = 2\pi\int_{\sqrt{8}}^4 \sqrt{x^2 - 8}\sqrt{1 + \frac{x^2}{x^2 - 8}}\, dx = 2\pi\int_{\sqrt{8}}^4 \sqrt{2x^2 - 8}\, dx = 2\pi\sqrt{2}\int_{\sqrt{8}}^4 \sqrt{x^2 - 4}\, dx$$

$$= 2\pi\sqrt{2}\left[\frac{x}{2}\sqrt{x^2 - 4} - 2\ln\left|x + \sqrt{x^2 - 4}\right|\right]_{\sqrt{8}}^4 \quad \{\text{Formula 39}\}$$

$$= 2\pi\sqrt{2}\left[2\sqrt{12} - 2\ln\left|4 + \sqrt{12}\right| - 2\sqrt{2} + 2\ln\left|2 + \sqrt{8}\right|\right]$$

$$= 2\pi\sqrt{2}\left[4\sqrt{3} - 2\sqrt{2} + 2\ln\left(\frac{2 + 2\sqrt{2}}{4 + 2\sqrt{3}}\right)\right]$$

$$= 4\pi\sqrt{2}\left[2\sqrt{3} - \sqrt{2} + \ln\left(\frac{1 + \sqrt{2}}{2 + \sqrt{3}}\right)\right] \approx 28.69.$$

[41] $12x^2 + 24x - 4y^2 + 9 = 0 \Rightarrow 12(x + 1)^2 - 4y^2 = 3 \Rightarrow \frac{(x + 1)^2}{\frac{1}{4}} - \frac{y^2}{\frac{3}{4}} = 1 \Rightarrow$

$a = \frac{1}{2}$, $C(-1, 0) \Rightarrow V(-1 \pm \frac{1}{2}, 0)$. The right branch of the hyperbola has vertex $V(-\frac{1}{2}, 0)$. By Exercise 40, the comet is closest to the sun when it is at $V(-\frac{1}{2}, 0)$.

Distance $= \frac{1}{2}$AU.

Note: Depending on the type of software used, you may need to solve the given equation for y in order to graph the hyperbolas.

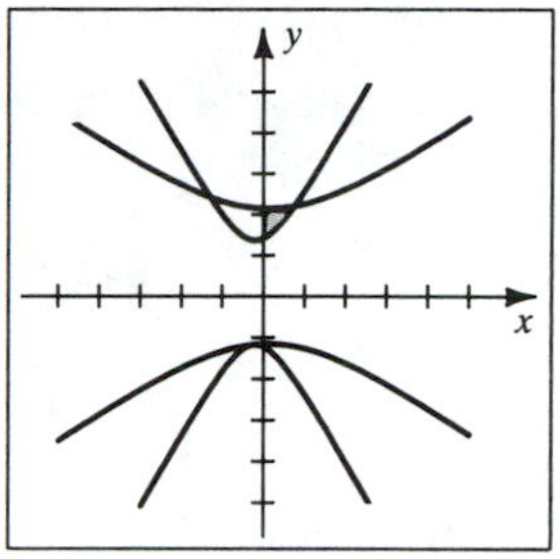

Figure 45

[45] (a) $\dfrac{(y-0.1)^2}{1.6} - \dfrac{(x+0.2)^2}{0.5} = 1 \Rightarrow$

$0.5(y-0.1)^2 - 1.6(x+0.2)^2 = 0.8 \Rightarrow$

$y = 0.1 \pm \sqrt{1.6 + 3.2(x+0.2)^2}$

$\dfrac{(y-0.5)^2}{2.7} - \dfrac{(x-0.1)^2}{5.3} = 1 \Rightarrow$

$5.3(y-0.5)^2 - 2.7(x-0.1)^2 = 14.31 \Rightarrow$

$y = 0.5 \pm \sqrt{\frac{1}{5.3}\left[14.31 + 2.7(x-0.1)^2\right]}$

From the graph,

the point of intersection in the first quadrant is approximately (0.741, 2.206).

(b) Area $\approx$

$$\int_0^{0.74} \left\{0.5 + \sqrt{\tfrac{1}{5.3}\left[14.31 + 2.7(x-0.1)^2\right]} - \left[0.1 + \sqrt{1.6 + 3.2(x+0.2)^2}\right]\right\} dx$$

Exercises 12.4

Note: Let D denote the value of the discriminant $B^2 - 4AC$. The following is a general outline of the solutions for Exercises 1-13 in this section and 39–40 in §12.5.

(a) Discriminant value and conic type are given.

(b) The 5 steps in part (b) are :

1) $\cot 2\phi = \dfrac{A-C}{B} \Rightarrow \phi = \frac{1}{2}\cot^{-1}\left(\dfrac{A-C}{B}\right)$, where the range of $\cot^{-1}$ is 0° to 180°. Note that the range of ϕ is $0^\circ < \phi < 90^\circ$.

2) $\cos 2\phi = \dfrac{\pm(A-C)}{\sqrt{(A-C)^2 + B^2}}$, $\sin\phi = \sqrt{\dfrac{1-\cos 2\phi}{2}}$, and $\cos\phi = \sqrt{\dfrac{1+\cos 2\phi}{2}}$;

Note that $\cos 2\phi$ will have the same sign as $\cot 2\phi$ since $\cot 2\phi = \dfrac{\cos 2\phi}{\sin 2\phi}$ and $\sin 2\phi$ is positive. Since ϕ is acute, $\sin\phi$ and $\cos\phi$ are positive.

3) The rotation of axes formulas are given. $\begin{cases} x = x'\cos\phi - y'\sin\phi \\ y = x'\sin\phi + y'\cos\phi \end{cases}$

4) The rotation of axes formulas are substituted into the original equation to obtain an equation in x' and y'. This equation is then simplified into a standard form.

5) The vertices (V') of the graph on the $x'y'$-plane are listed along with the corresponding vertices (V) of the graph of the original equation on the xy-plane.

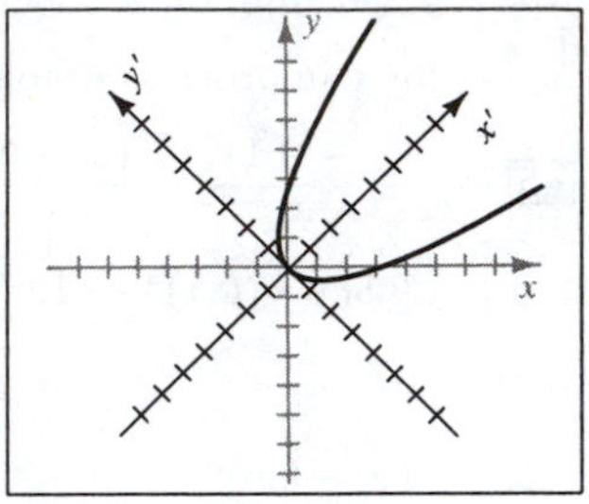

Figure 1

[1] (a) $D = (-2)^2 - 4(1)(1) = 0$, parabola

(b) 1) $\cot 2\phi = 0;\ \phi = 45°$

2) $\cos 2\phi = 0;\ \sin\phi = \frac{1}{2}\sqrt{2};\ \cos\phi = \frac{1}{2}\sqrt{2}$

3) $\begin{cases} x = \frac{1}{2}\sqrt{2}\,x' - \frac{1}{2}\sqrt{2}\,y' = \frac{1}{2}\sqrt{2}\,(x' - y') \\ y = \frac{1}{2}\sqrt{2}\,x' + \frac{1}{2}\sqrt{2}\,y' = \frac{1}{2}\sqrt{2}\,(x' + y') \end{cases}$

4) $2(y')^2 - 4x' = 0 \Rightarrow (y')^2 = 2(x')$

5) $V'(0, 0) \to V(0, 0)\ \{x\ \&\ y \text{ int. @ } 2\sqrt{2}\}$

It is helpful to arrange your work in a table format as follows. In the leftmost column we list each term of the original equation. We then make 5 columns — one for each of the terms $(x')^2$, $x'y'$, $(y')^2$, x', and y'. The column sums are the equation in the x'-y' system. The zero in the leftmost column total comes from the original equation.

x^2	$=$	$\frac{1}{2}(x')^2$	$-x'y'$	$+\frac{1}{2}(y')^2$		
$-2xy$	$=$	$-(x')^2$		$+(y')^2$		
y^2	$=$	$\frac{1}{2}(x')^2$	$+x'y'$	$+\frac{1}{2}(y')^2$		
$-2\sqrt{2}x$	$=$				$-2x'$	$+2y'$
$-2\sqrt{2}y$	$=$				$-2x'$	$-2y'$
0	$=$			$2(y')^2$	$-4(x')$	

This is equivalent to $(y')^2 = 2(x')$.

Note that it is easy to check that the $x'y'$ term is 0 in this format.

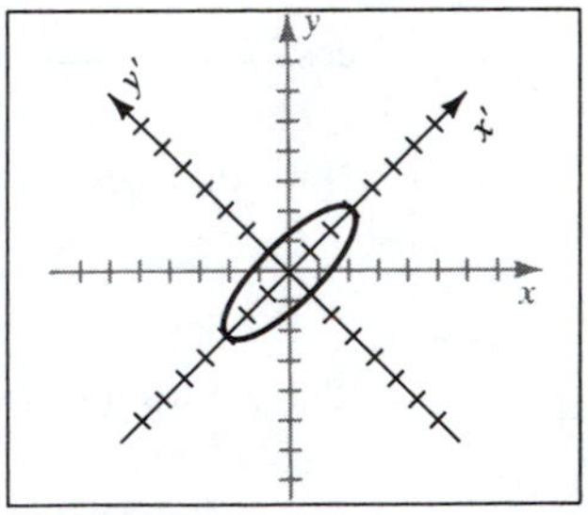

Figure 3

[3] (a) $D = (-8)^2 - 4(5)(5) = -36 < 0$, ellipse

(b) 1) $\cot 2\phi = 0;\ \phi = 45°$

2) $\cos 2\phi = 0;\ \sin\phi = \frac{1}{2}\sqrt{2};\ \cos\phi = \frac{1}{2}\sqrt{2}$

3) $\begin{cases} x = \frac{1}{2}\sqrt{2}\,x' - \frac{1}{2}\sqrt{2}\,y' = \frac{1}{2}\sqrt{2}\,(x' - y') \\ y = \frac{1}{2}\sqrt{2}\,x' + \frac{1}{2}\sqrt{2}\,y' = \frac{1}{2}\sqrt{2}\,(x' + y') \end{cases}$

4) $1(x')^2 + 9(y')^2 = 9 \Rightarrow \dfrac{(x')^2}{9} + \dfrac{(y')^2}{1} = 1$

5) $V'(\pm 3, 0) \to V(\pm\frac{3}{2}\sqrt{2}, \pm\frac{3}{2}\sqrt{2})$

[7] (a) $D = (-24)^2 - 4(16)(9) = 0$, parabola

(b) 1) $\cot 2\phi = -\frac{7}{24}$; $\phi \approx 53.13°$

2) $\cos 2\phi = -\frac{7}{25}$; $\sin\phi = \frac{4}{5}$; $\cos\phi = \frac{3}{5}$

3) $\begin{cases} x = \frac{3}{5}x' - \frac{4}{5}y' = \frac{1}{5}(3x' - 4y') \\ y = \frac{4}{5}x' + \frac{3}{5}y' = \frac{1}{5}(4x' + 3y') \end{cases}$

4) $\frac{625}{25}(y')^2 - 100\,x' + 100 = 0 \Rightarrow$

$(y')^2 = 4(x' - 1)$

5) $V'(1, 0) \rightarrow V(\frac{3}{5}, \frac{4}{5})$

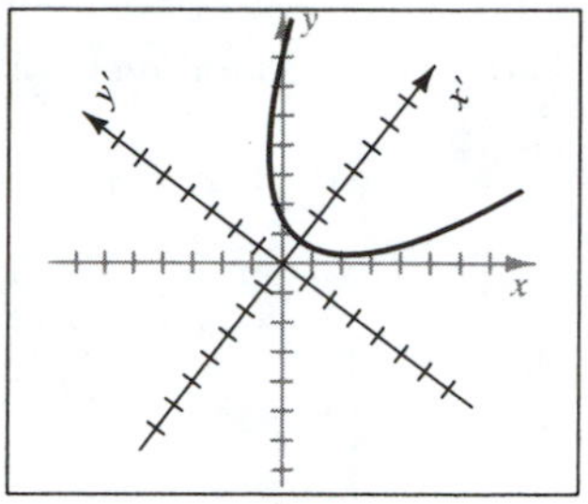

Figure 7

[11] (a) $D = (6\sqrt{3})^2 - 4(5)(-1) = 128 > 0$, hyperbola

(b) 1) $\cot 2\phi = \frac{1}{3}\sqrt{3}$; $\phi = 30°$

2) $\cos 2\phi = \frac{1}{2}$; $\sin\phi = \frac{1}{2}$; $\cos\phi = \frac{1}{2}\sqrt{3}$

3) $\begin{cases} x = \frac{1}{2}\sqrt{3}\,x' - \frac{1}{2}y' = \frac{1}{2}(\sqrt{3}\,x' - y') \\ y = \frac{1}{2}x' + \frac{1}{2}\sqrt{3}\,y' = \frac{1}{2}(x' + \sqrt{3}\,y') \end{cases}$

4) $\frac{32}{4}(x')^2 - \frac{16}{4}(y')^2 - 16\,y' - 12 = 0 \Rightarrow$

$\dfrac{(y' + 2)^2}{1} - \dfrac{(x')^2}{\frac{1}{2}} = 1$

5) $V'(0, -2 \pm 1) \rightarrow V(1 \mp \frac{1}{2}, -\sqrt{3} \pm \frac{1}{2}\sqrt{3})$

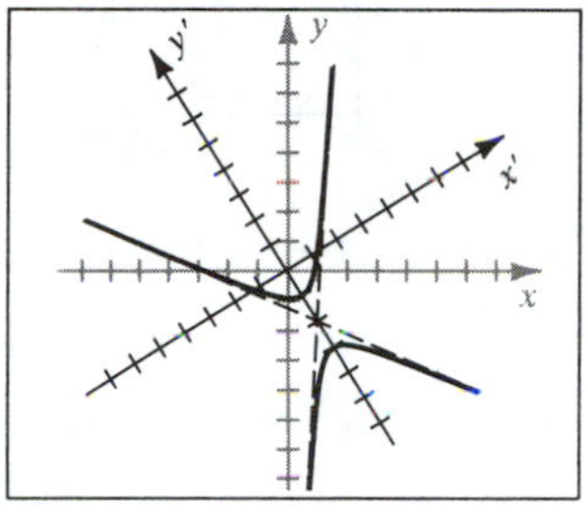

Figure 11

Note: Depending on the type of software used, you may need to solve the given equation for y in Exercise 15.

[15] $2.1x^2 - 4xy + 1.5y^2 - 4x + y - 1 = 0 \Rightarrow$

$1.5y^2 - (4x - 1)y + (2.1x^2 - 4x - 1) = 0 \Rightarrow$

$$y = \frac{(4x - 1) \pm \sqrt{(4x - 1)^2 - 6(2.1x^2 - 4x - 1)}}{3}$$

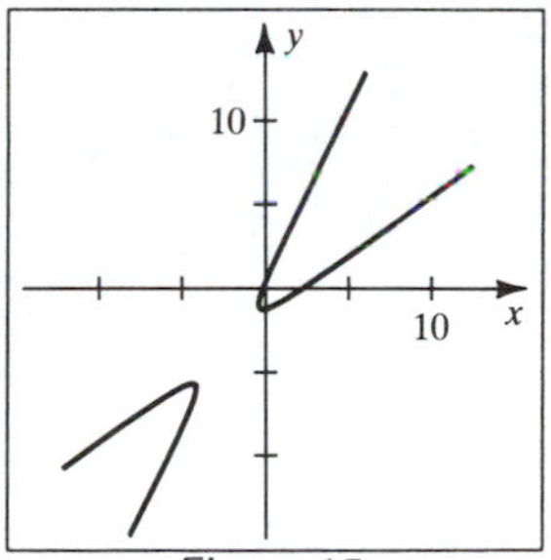

Figure 15

12.5 Review Exercises

Note: Let the notation be the same as in §12.1-12.4.

[1] $y^2 = 64x \Rightarrow x = \frac{1}{64}y^2 \Rightarrow a = \frac{1}{64}$. $p = \dfrac{1}{4(\frac{1}{64})} = 16$. The conic is a parabola opening to the right. See (12.2). $V(0, 0)$; $F(16, 0)$; $\ell\colon x = -16$

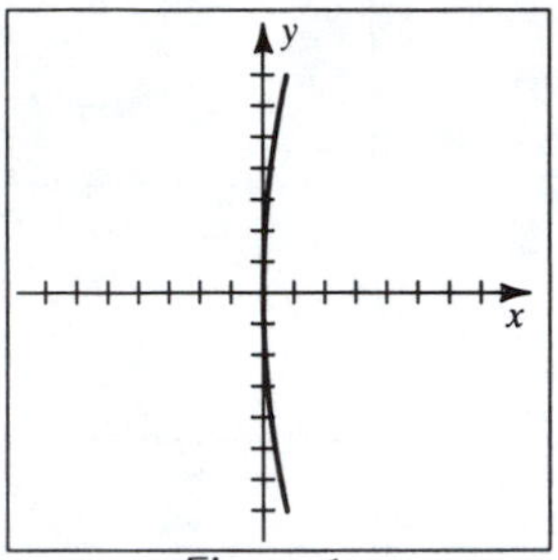

Figure 1

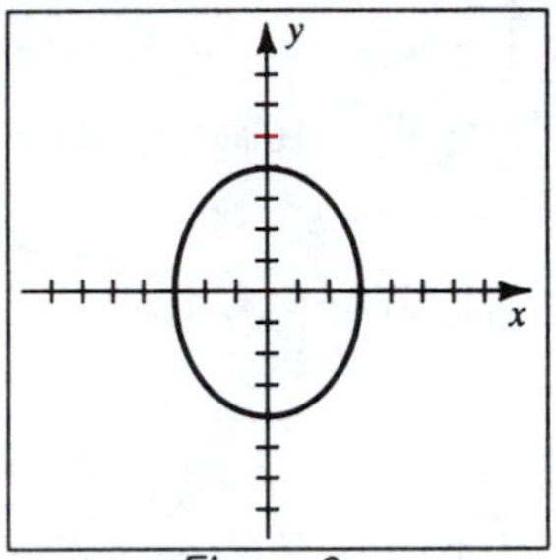

Figure 3

[3] $9y^2 = 144 - 16x^2 \Rightarrow \frac{x^2}{9} + \frac{y^2}{16} = 1$. The conic is an ellipse with vertices on the y-axis. $c^2 = 16 - 9 \Rightarrow c = \pm\sqrt{7}$. $V(0, \pm 4)$; $F(0, \pm\sqrt{7})$; $M(\pm 3, 0)$

[9] $x^2 - 9y^2 + 8x + 90y - 210 = 0 \Rightarrow$

$(x^2 + 8x + \underline{16}) - 9(y^2 - 10y + \underline{25}) = 210 + \underline{16} - \underline{225} \Rightarrow$

$(x + 4)^2 - 9(y - 5)^2 = 1 \Rightarrow \dfrac{(x+4)^2}{1} - \dfrac{(y-5)^2}{\frac{1}{9}} = 1$. The conic is a hyperbola with a horizontal transverse axis. $c^2 = 1 + \frac{1}{9} \Rightarrow c = \pm\frac{1}{3}\sqrt{10}$.

$C(-4, 5)$; $V(-4 \pm 1, 5)$; $F(-4 \pm \frac{1}{3}\sqrt{10}, 5)$; $W(-4, 5 \pm \frac{1}{3})$; $(y - 5) = \pm\frac{1}{3}(x + 4)$

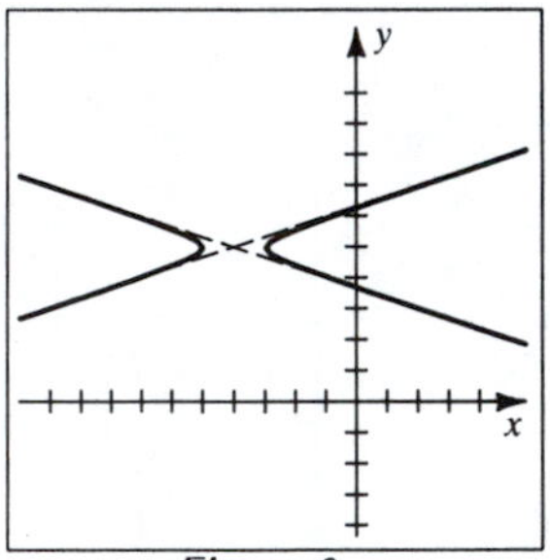

Figure 9

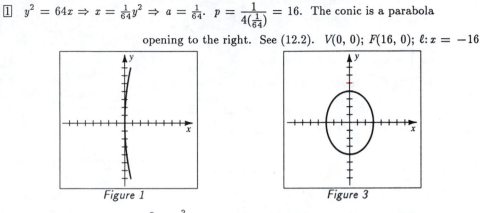

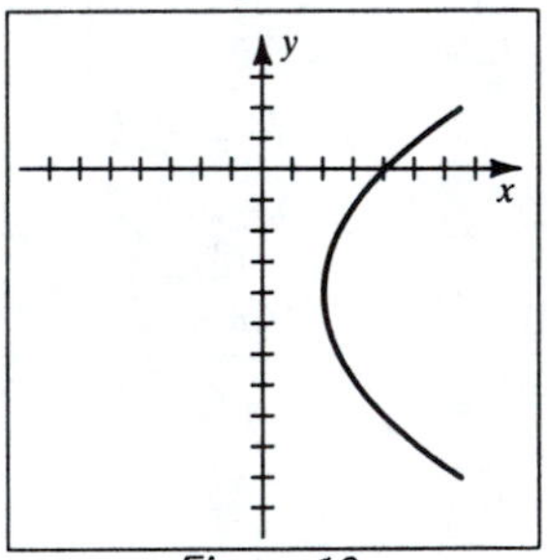

Figure 13

[13] $y^2 - 8x + 8y + 32 = 0 \Rightarrow x = \frac{1}{8}y^2 + y + 4$. The conic is a parabola opening to the right. $a = \frac{1}{8} \Rightarrow p = \dfrac{1}{4(\frac{1}{8})} = 2$. $x' = \frac{1}{4}y + 1 = 0 \Rightarrow y = -4$ and $x = 2$.

$V(2, -4)$; $F(4, -4)$; $\ell\colon x = 0$

[15] $x^2 - 9y^2 + 8x + 7 = 0 \Rightarrow$

$(x^2 + 8x + \underline{16}) - 9(y^2) = -7 + \underline{16} \Rightarrow$

$(x + 4)^2 - 9(y^2) = 9 \Rightarrow$

$\dfrac{(x+4)^2}{9} - \dfrac{y^2}{1} = 1$; $c^2 = 9 + 1 \Rightarrow c = \pm\sqrt{10}$;

$C(-4, 0)$; $V(-4 \pm 3, 0)$; $F(-4 \pm \sqrt{10}, 0)$;

$W(-4, 0 \pm 1)$; $y = \pm\frac{1}{3}(x + 4)$

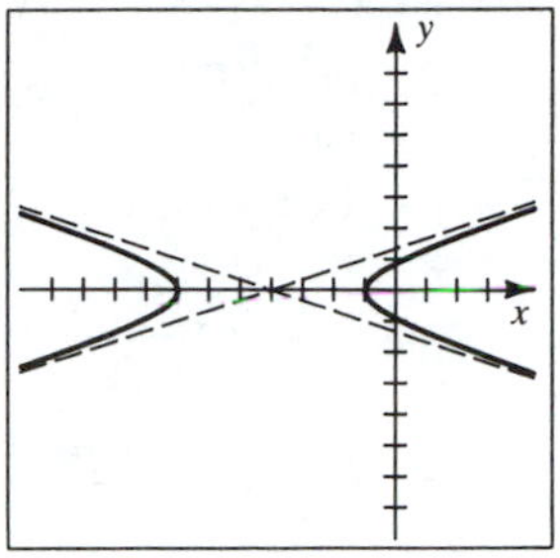

Figure 15

[17] The hyperbola has a vertical transverse axis and a horizontal conjugate axis.

Refer to Figure 12.27 with $a = 7$ and $b = 3$.

An equation is $\dfrac{y^2}{7^2} - \dfrac{x^2}{3^2} = 1$ or $\dfrac{y^2}{49} - \dfrac{x^2}{9} = 1$.

[19] $F(0, -10)$ and $\ell\colon y = 10 \Rightarrow p = -10$ and $V(0, 0)$.

By (12.2) with $p < 0$, $x^2 = 4py$, or $x^2 = -40y$.

[25] $M(\pm 5, 0) \Rightarrow b = 5$. $e = \frac{c}{a} = \dfrac{\sqrt{a^2 - b^2}}{a} = \dfrac{\sqrt{a^2 - 25}}{a} = \frac{2}{3} \Rightarrow \frac{2}{3}a = \sqrt{a^2 - 25} \Rightarrow$

$\frac{4}{9}a^2 = a^2 - 25 \Rightarrow \frac{5}{9}a^2 = 25 \Rightarrow a^2 = 45$. An equation is $\dfrac{x^2}{25} + \dfrac{y^2}{45} = 1$.

[27] (a) This problem can be modeled as an ellipse with $V(\pm 100, 0)$ and passing through the point $(25, 30)$. Substituting $x = 25$ and $y = 30$ into $\dfrac{x^2}{100^2} + \dfrac{y^2}{b^2} = 1$ yields $\dfrac{30^2}{b^2} = \dfrac{15}{16} \Rightarrow b^2 = 960$. An equation for the ellipse is $\dfrac{x^2}{10{,}000} + \dfrac{y^2}{960} = 1$.

An equation for the top half of the ellipse is $y = \sqrt{960\left(1 - \dfrac{x^2}{10{,}000}\right)}$.

(b) The height in the middle of the bridge occurs when $x = 0$ and is equal to

$\sqrt{960} = 8\sqrt{15} \approx 31$ ft.

[29] $4x^2 - 9y^2 - 8x + 6y - 36 = 0 \Rightarrow 8x - 18yy' - 8 + 6y' = 0 \Rightarrow$

$y' = \dfrac{8 - 8x}{6 - 18y} = -\dfrac{16}{15}$ at $P(-3, 2)$.

Tangent line: $y - 2 = -\frac{16}{15}(x + 3)$; Normal line: $y - 2 = \frac{15}{16}(x + 3)$.

[33] Without loss of generality, let $x^2 = 4py$.

We will solve for x and then integrate with respect to y. $x = \pm\sqrt{4py} \Rightarrow$

$$A = \int_0^p \left[\sqrt{4py} - (-\sqrt{4py})\right] dy = 2\int_0^p \sqrt{4py}\, dy = 4\sqrt{p}\left[\tfrac{2}{3}y^{3/2}\right]_0^p = \tfrac{8}{3}p^2.$$

[37] Let $\rho = 1$. $b^2y^2 - a^2x^2 = a^2b^2 \Rightarrow a^2x^2 = b^2y^2 - a^2b^2 = b^2(y^2 - a^2) \Rightarrow$

$x^2 = \frac{b^2}{a^2}\sqrt{y^2 - a^2}$. See Figure 12.27. The region described is an R_y region. Use (6.25) with the roles of x and y interchanged. Solving for x and integrating with respect to y to find m gives $x = \pm\frac{b}{a}\sqrt{y^2 - a^2} \Rightarrow m = \int_a^c \frac{2b}{a}\sqrt{y^2 - a^2}\,dy$

$= \frac{2b}{a}\Big[\frac{y}{2}\sqrt{y^2 - a^2} - \frac{a^2}{2}\ln\left|y + \sqrt{y^2 - a^2}\right|\Big]_a^c$ {Formula 39}

$= \frac{b}{a}\big[cb - a^2\ln(c + b) + a^2\ln a\big]$. Also, $M_x =$

$$\int_a^c y\left(\frac{2b}{a}\sqrt{y^2 - a^2}\right)dy = \frac{2b}{a}\Big[\tfrac{1}{3}(y^2 - a^2)^{3/2}\Big]_a^c = \frac{2b^4}{3a}\ \{b^2 = c^2 - a^2\}.$$

Thus, $\bar{y} = \frac{M_x}{m} = \frac{\frac{2}{3}b^3}{bc - a^2\big[\ln(c + b) - \ln a\big]}$ and $\bar{x} = 0$ by symmetry.

Note: See §12.4 for a discussion of the general solution outline for Exercises 39–40.

[39] $D = (-8)^2 - 4(1)(16) = 0$, parabola

1) $\cot 2\phi = \frac{15}{8}$; $\phi \approx 14.04^\circ$

2) $\cos 2\phi = \frac{15}{17}$; $\sin\phi = \frac{1}{17}\sqrt{17}$; $\cos\phi = \frac{4}{17}\sqrt{17}$

3) $\begin{cases} x = \frac{4}{17}\sqrt{17}\,x' - \frac{1}{17}\sqrt{17}\,y' = \frac{1}{17}\sqrt{17}\,(4x' - y') \\ y = \frac{1}{17}\sqrt{17}\,x' + \frac{4}{17}\sqrt{17}\,y' = \frac{1}{17}\sqrt{17}(x' + 4y') \end{cases}$

4) $\frac{289}{17}(y')^2 - 51(x') = 0 \Rightarrow (y')^2 = 3(x')$

5) $V'(0, 0) \rightarrow V(0, 0)$

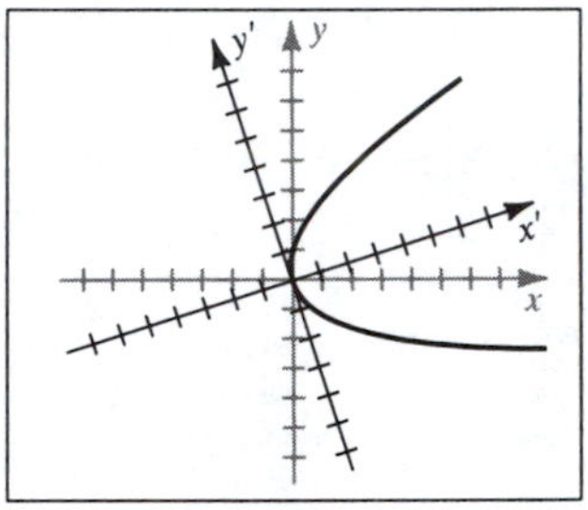

Figure 39

Chapter 13: Plane Curves and Polar Coordinates

Exercises 13.1

1 For this exercise (and others), we solve for t in terms of x, and then substitute that expression for t in the equation that relates y and t. $x = t - 2 \Rightarrow t = x + 2$. $y = 2t + 3 = 2(x + 2) + 3 = 2x + 7$.

As t varies from 0 to 5, (x, y) varies from $(-2, 3)$ to $(3, 13)$.

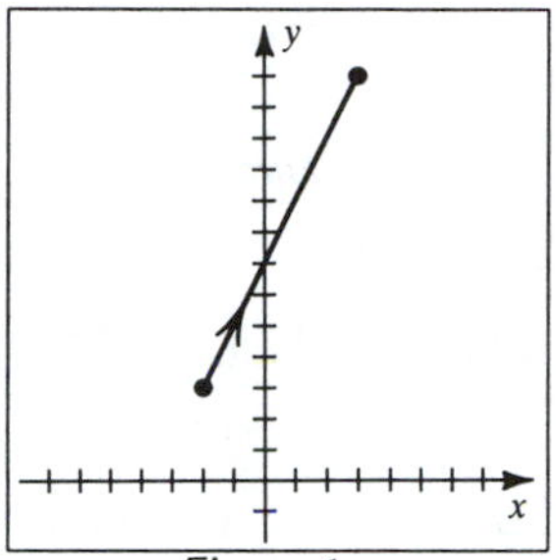

Figure 1

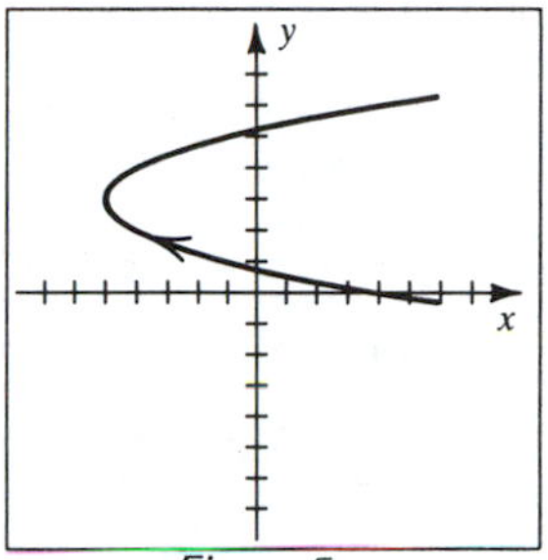

Figure 5

5 Since y is linear in t, it is easier to solve the second equation for t than it is to solve the first equation for t. Hence, we solve for t in terms of y, and then substitute that expression for t in the equation that relates x and t. $y = 2t + 3 \Rightarrow t = \frac{1}{2}(y - 3)$. $x = 4\left[\frac{1}{2}(y - 3)\right]^2 - 5 \Rightarrow (y - 3)^2 = x + 5$. This is a parabola with vertex at $(-5, 3)$. Since t takes on all real values, so does y, and the curve C is the entire parabola.

9 $x = 2\sin t$ and $y = 3\cos t \Rightarrow \frac{x}{2} = \sin t$ and $\frac{y}{3} = \cos t \Rightarrow$ $\frac{x^2}{4} + \frac{y^2}{9} = \sin^2 t + \cos^2 t = 1$. As t varies from 0 to 2π,

(x, y) traces the ellipse from $(0, 3)$ in a clockwise direction back to $(0, 3)$.

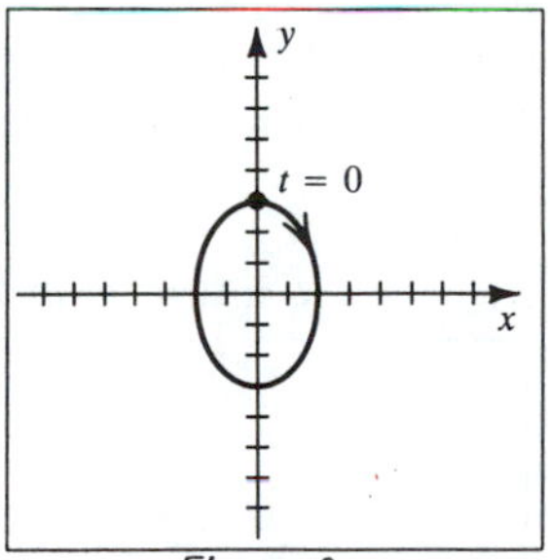

Figure 9

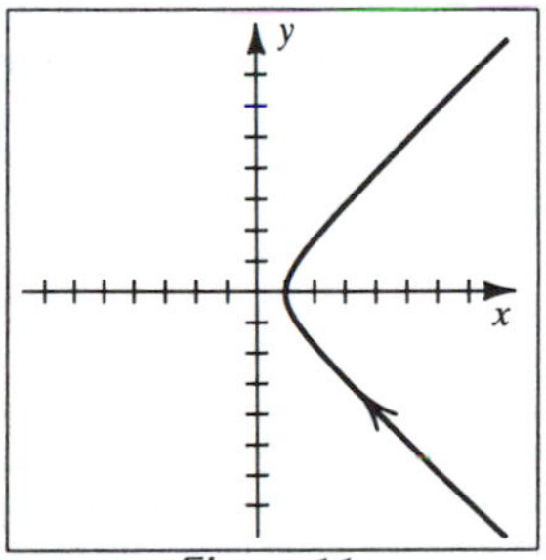

Figure 11

11 $x = \sec t$ and $y = \tan t \Rightarrow x^2 - y^2 = \sec^2 t - \tan^2 t = 1$.

As t varies from $-\frac{\pi}{2}$ to $\frac{\pi}{2}$, (x, y) traces the right branch of the hyperbola along the asymptote $y = -x$ to $(1, 0)$ and then along the asymptote $y = x$.

17 $x = \cosh t$ and $y = \sinh t \Rightarrow x^2 - y^2 = \cosh^2 t - \sinh^2 t = 1$ $(x \geq 1)$.

As t varies from $-\infty$ to ∞, (x, y) traces the right branch of the hyperbola along the asymptote $y = -x$ to $(1, 0)$ and then along the asymptote $y = x$.

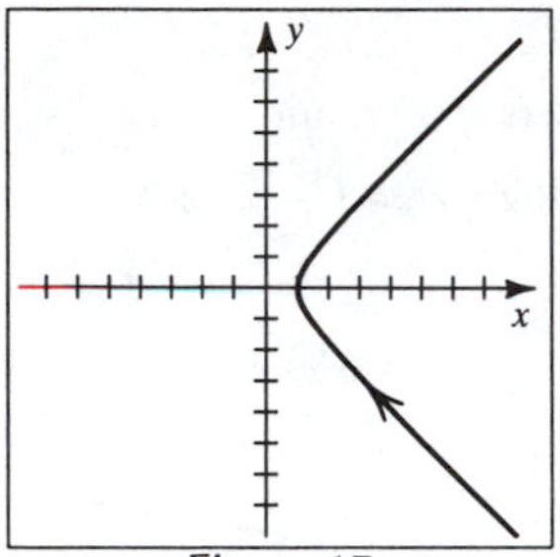

Figure 17

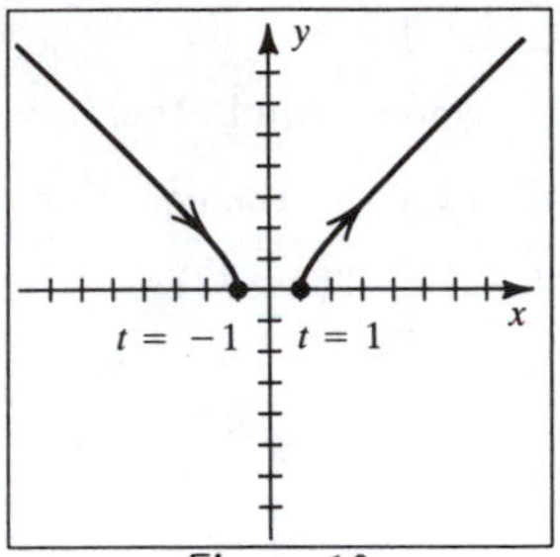

Figure 19

19 $x = t$ and $y = \sqrt{t^2 - 1} \Rightarrow y = \sqrt{x^2 - 1} \Rightarrow x^2 - y^2 = 1$.

Since y is nonnegative, the graph is the top half of both branches of the hyperbola.

21 $x = t$ and $y = \sqrt{t^2 - 2t + 1} \Rightarrow y = \sqrt{x^2 - 2x + 1} = \sqrt{(x - 1)^2} = |x - 1|$.

As t varies from 0 to 4, (x, y) traces $y = |x - 1|$ from $(0, 1)$ to $(4, 3)$.

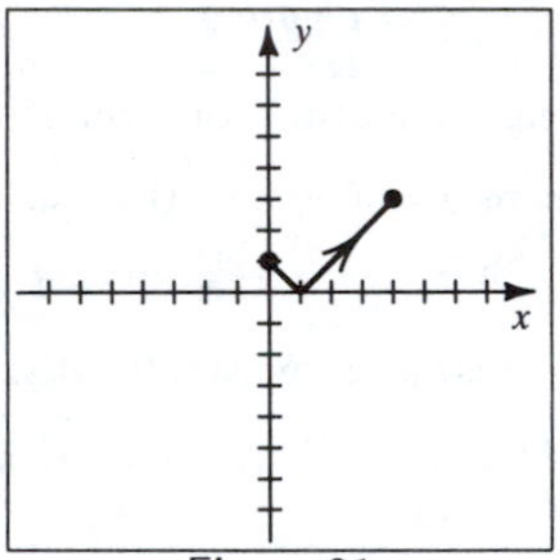

Figure 21

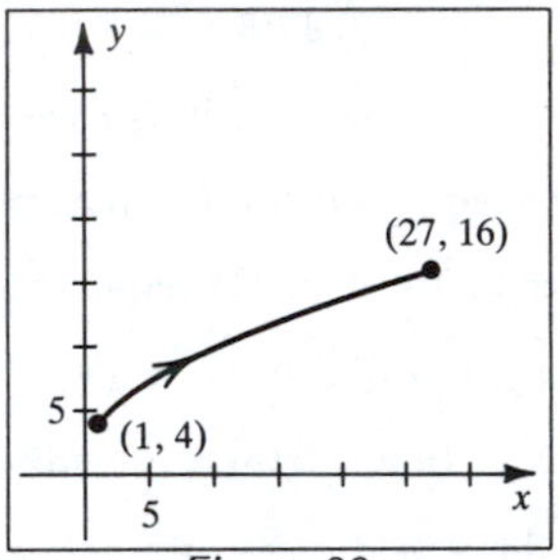

Figure 23

23 $x = (t + 1)^3 \Rightarrow t = x^{1/3} - 1$. $y = (t + 2)^2 = (x^{1/3} + 1)^2$. This is an unfamiliar graph. The graph of $y = x^{1/3}$ is similar to $y = x^{1/2}$ $\{y = \sqrt{x}\}$ but is symmetric with respect to the origin. The " + 1" shifts $y = x^{1/3}$ up 1 unit and the "squaring" makes all y values positive. Since we have restrictions on the variable t, we only have a portion of this graph. As t varies from 0 to 2, (x, y) varies from $(1, 4)$ to $(27, 16)$.

25 All of the curves are a portion of the parabola $x = y^2$.

C_1: $x = t^2 = y^2$. y takes on all real values and we have the entire parabola.

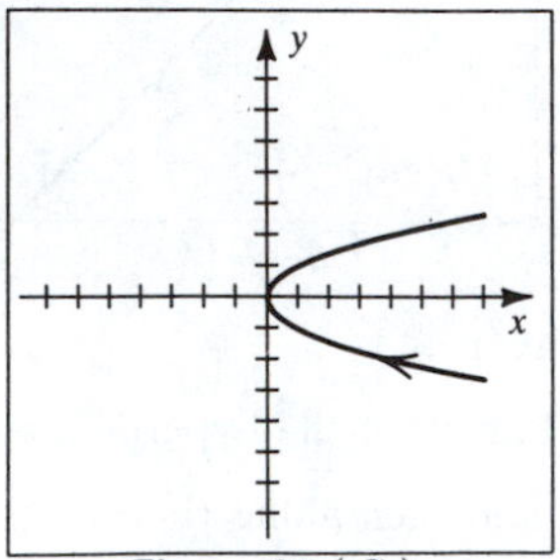

Figure 25 (C_1)

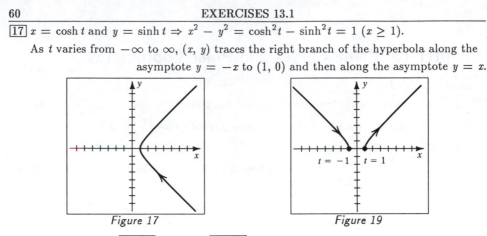

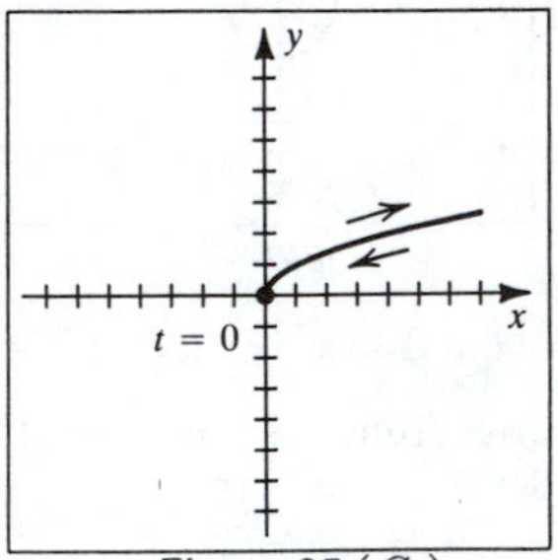

Figure 25 (C_2)

C_2: $x = t^4 = (t^2)^2 = y^2$. C_2 is only the top half since $y = t^2$ is nonnegative.

As t varies from $-\infty$ to ∞, the top portion is traced twice.

C_3: $x = \sin^2 t = (\sin t)^2 = y^2$. C_3 is the portion of the curve from $(1, -1)$ to $(1, 1)$.

The point $(1, 1)$ is reached at $t = \frac{\pi}{2} + 2\pi n$ and the point $(1, -1)$ when

$t = \frac{3\pi}{2} + 2\pi n$.

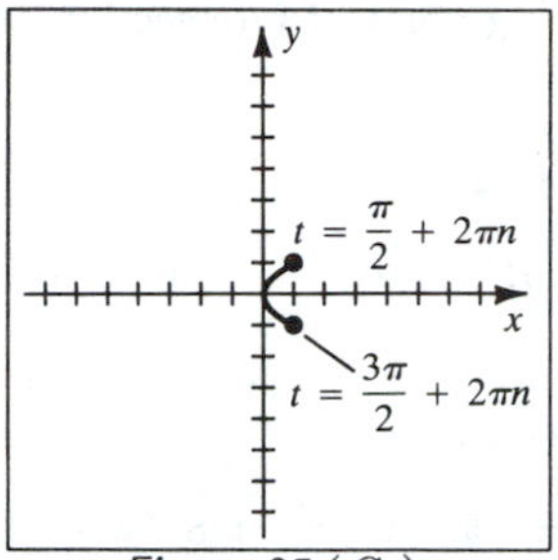

Figure 25 (C_3)

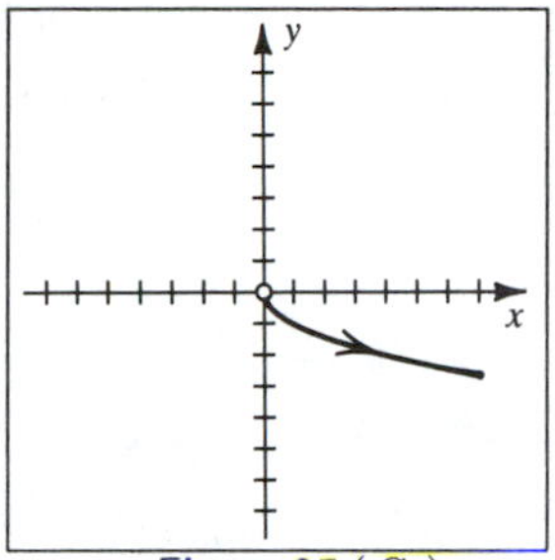

Figure 25 (C_4)

C_4: $x = e^{2t} = (e^t)^2 = (-e^t)^2 = y^2$. C_4 is the bottom half of the parabola since y

is negative. As t approaches $-\infty$, the parabola approaches the origin.

[27] In each part, the motion is on the unit circle since $x^2 + y^2 = 1$.

(a) For $0 \leq t \leq \pi$, x varies from 1 to -1 and y is nonnegative.

$P(x, y)$ moves from $(1, 0)$ counterclockwise to $(-1, 0)$.

(b) For $0 \leq t \leq \pi$, y varies from 1 to -1 and x is nonnegative.

$P(x, y)$ moves from $(0, 1)$ clockwise to $(0, -1)$.

(c) For $-1 \leq t \leq 1$, x varies from -1 to 1 and y is nonnegative.

$P(x, y)$ moves from $(-1, 0)$ clockwise to $(1, 0)$.

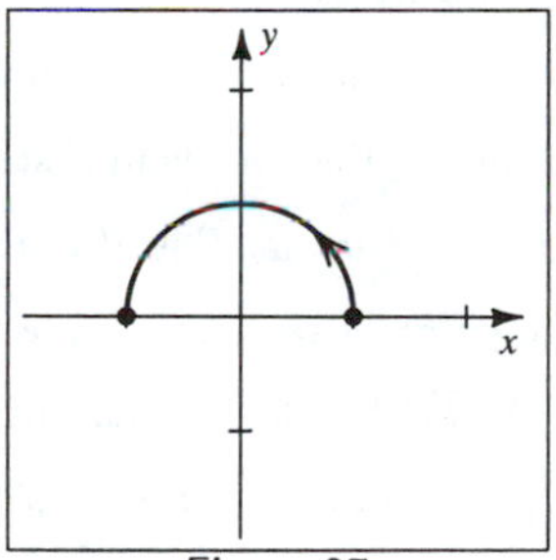

Figure 27a

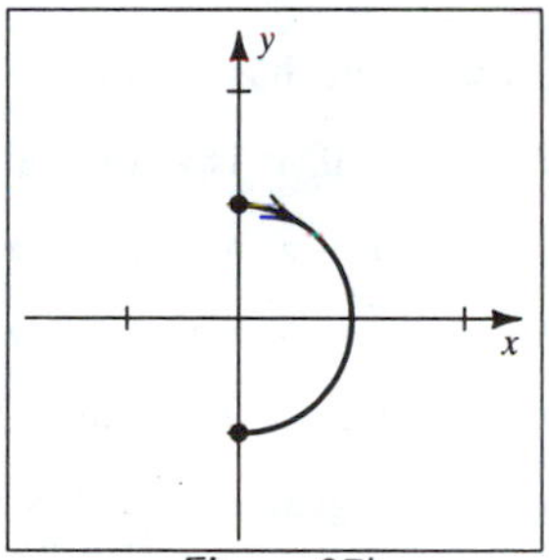

Figure 27b

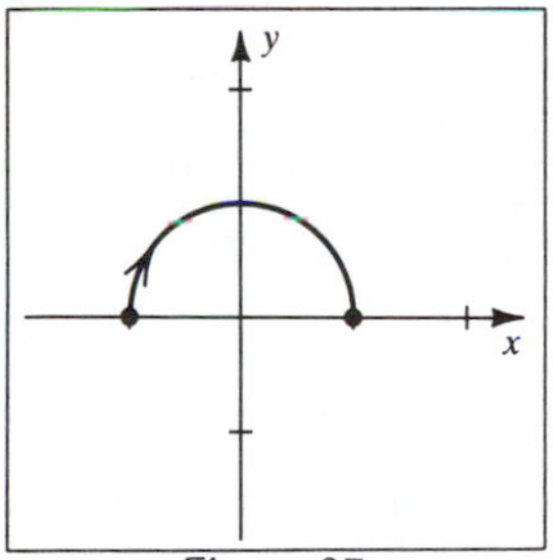

Figure 27c

[31] Solving the first equation for t we have $t = \dfrac{x - x_1}{x_2 - x_1}$.

Substituting this expression into the second equation yields

$y = (y_2 - y_1)\left(\dfrac{x - x_1}{x_2 - x_1}\right) + y_1$, or $y - y_1 = \left(\dfrac{y_2 - y_1}{x_2 - x_1}\right)(x - x_1)$, which is the

point-slope formula (1.8)(ii) for the equation of a line through P_1 and P_2.

[35] $b = \frac{1}{3}a \Rightarrow a = 3b$. Substituting into the equations from Exercise 33 yields:

$$x = (3b + b)\cos t - b\cos\left(\frac{3b + b}{b}t\right) = 4b\cos t - b\cos 4t$$

$$y = (3b + b)\sin t - b\sin\left(\frac{3b + b}{b}t\right) = 4b\sin t - b\sin 4t$$

As an aid in graphing, to determine where the path of the smaller circle will intersect the path of the larger circle (for the original starting point of intersection at $A(a, 0)$), we can solve $x^2 + y^2 = a^2$ for t.

$$x^2 + y^2 = 16b^2\cos^2 t - 8b^2\cos t\cos 4t + b^2\cos^2 4t + 16b^2\sin^2 t - 8b^2\sin t\sin 4t + b^2\sin^2 4t$$

$$= 17b^2 - 8b^2(\cos t\cos 4t + \sin t\sin 4t)$$

$$= 17b^2 - 8b^2[\cos(t - 4t)] = 17b^2 - 8b^2\cos 3t.$$

Thus, $x^2 + y^2 = a^2 \Rightarrow 17b^2 - 8b^2\cos 3t = a^2 = 9b^2 \Rightarrow 8b^2 = 8b^2\cos 3t \Rightarrow 1 = \cos 3t \Rightarrow 3t = 2\pi n \Rightarrow t = \frac{2\pi}{3}n$.

It follows that the intersection points are at $t = \frac{2\pi}{3}, \frac{4\pi}{3}$, and 2π.

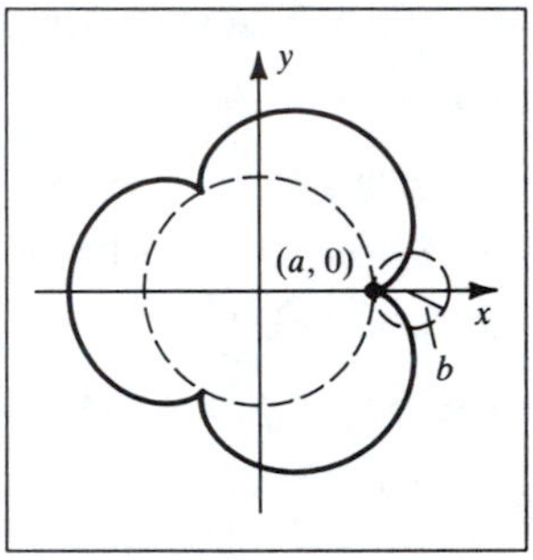

Figure 35

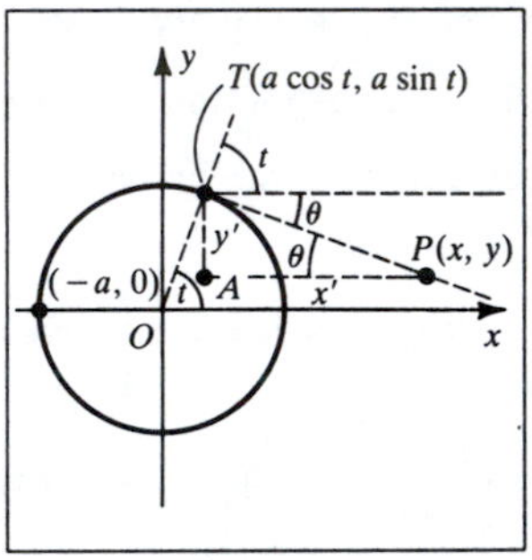

Figure 37

[37] Consider *Figure 37*. Since the circle has radius a, the coordinates of the point of tangency are $(a\cos t, a\sin t)$. Now if P is a typical point on the unraveling string, then the position of P is $x = a\cos t + x'$ and $y = a\sin t - y'$; that is, P is x' units to the right of the point of tangency and y' units below the point of tangency. We now seek to determine x' and y' in terms of t. Using (1.14), $\overline{TP}$ has length ta, the arc length on the circle. Since the tangent line TP is perpendicular to the radius OT, $\theta = \frac{\pi}{2} - t$. Hence, in the right triangle TAP,

$$\cos\theta = \frac{x'}{ta} \Rightarrow x' = at\cos\left(\tfrac{\pi}{2} - t\right) = at\sin t$$

and

$$\sin\theta = \frac{y'}{ta} \Rightarrow y' = at\sin\left(\tfrac{\pi}{2} - t\right) = at\cos t.$$

Thus, $x = a\cos t + at\sin t$ and $y = a\sin t - at\cos t$.

Factoring, $x = a(\cos t + t\sin t)$ and $y = a(\sin t - t\cos t)$.

[39] (a) $x = a\sin\omega t$ and $y = b\cos\omega t \Rightarrow \frac{x}{a} = \sin\omega t$ and $\frac{y}{b} = \cos\omega t \Rightarrow \frac{x^2}{a^2} + \frac{y^2}{b^2} = 1$.

The figure is an ellipse with center (0, 0) and axes of lengths $2a$ and $2b$.

(b) $f(t + p) = a\sin\left[\omega_1(t + p)\right] = a\sin\left[\omega_1 t + \omega_1 p\right] = a\sin\left[\omega_1 t + 2\pi n\right] =$

$a\sin\omega_1 t = f(t)$.

$g(t + p) = b\cos\left[\omega_2(t + p)\right] = b\cos\left[\omega_2 t + \frac{\omega_2}{\omega_1}2\pi n\right] = b\cos\left[\omega_2 t + \frac{m}{n}2\pi n\right] =$

$b\cos\left[\omega_2 t + 2\pi m\right] = b\cos\omega_2 t = g(t)$.

Since f and g are periodic with period p,

the curve retraces itself every p units of time.

[41] $x = 3\sin^5 t$ and $y = 3\cos^5 t \Rightarrow \frac{x}{3} = \sin^5 t$ and $\frac{y}{3} = \cos^5 t \Rightarrow$

$(\frac{x}{3})^{1/5} = \sin t$ and $(\frac{y}{3})^{1/5} = \cos t \Rightarrow (\frac{x}{3})^{2/5} = \sin^2 t$ and $(\frac{y}{3})^{2/5} = \cos^2 t \Rightarrow$

$(\frac{x}{3})^{2/5} + (\frac{y}{3})^{2/5} = \sin^2 t + \cos^2 t = 1$.

The graph traces an astroid (hypocycloid of four cusps).

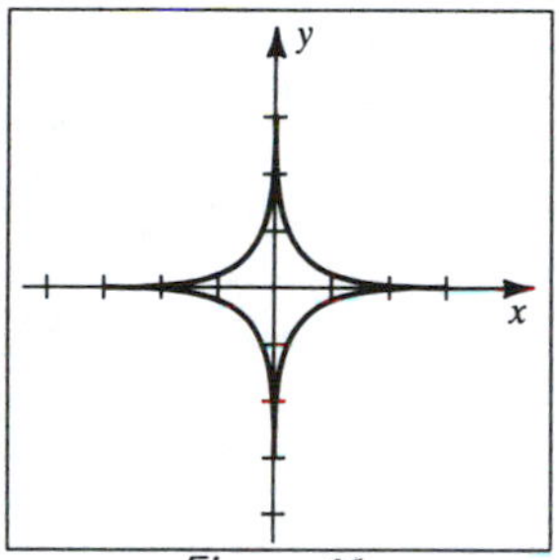

Figure 41

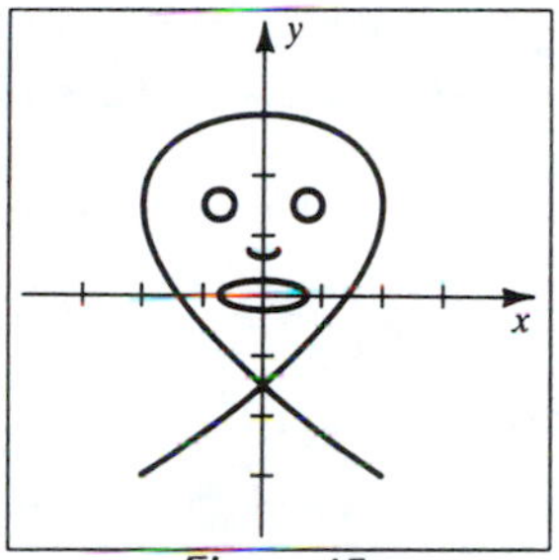

Figure 45

[45] C_1: This curve is a Lissajous figure that resembles a mask.

C_2: $x = \frac{1}{4}\cos t + \frac{3}{4}$, $y = \frac{1}{4}\sin t + \frac{3}{2} \Rightarrow (x - \frac{3}{4})^2 = \frac{1}{16}\cos^2 t$, $(y - \frac{3}{2})^2 = \frac{1}{16}\sin^2 t \Rightarrow$ $(x - \frac{3}{4})^2 + (y - \frac{3}{2})^2 = \frac{1}{16}$. This is a circle centered at $(\frac{3}{4}, \frac{3}{2})$ with radius $\frac{1}{4}$ that traces the right eye.

C_3: Similar to C_2,

this is a circle centered at $(-\frac{3}{4}, \frac{3}{2})$ with radius $\frac{1}{4}$ that traces the left eye.

C_4: $x = \frac{3}{4}\cos t$, $y = \frac{1}{4}\sin t \Rightarrow \frac{x^2}{9/16} + \frac{y^2}{1/16} = \cos^2 t + \sin^2 t = 1$.

This is an ellipse centered at the origin that traces the mouth.

C_5: $x = \frac{1}{4}\cos t$, $y = \frac{1}{8}\sin t + \frac{3}{4} \Rightarrow \frac{x^2}{1/16} + \frac{(y - \frac{3}{4})^2}{1/64} = \cos^2 t + \sin^2 t = 1$.

This is the lower half an ellipse centered at $(0, \frac{3}{4})$ and traces the nose.

The figure is a mask with a mouth, nose, and eyes.

Exercises 13.2

[1] Using (13.3) to find the slope of the tangent line,

we find that $\frac{dx}{dt} = 2t$ and $\frac{dy}{dt} = 2t \Rightarrow \frac{dy}{dx} = \frac{dy/dt}{dx/dt} = \frac{2t}{2t} = 1$ for all $t \neq 0$.

Remembering that the slope of the normal line is the negative of the reciprocal of the slope of the tangent line, we see that it has slope -1.

[5] $\frac{dx}{dt} = e^t$ and $\frac{dy}{dt} = -2e^{-2t} \Rightarrow \frac{dy}{dx} = \frac{-2e^{-2t}}{e^t} = -\frac{2}{e^{3t}} = -\frac{2}{e^3}$ at $t = 1$. Normal: $\frac{e^3}{2}$

[9] We must find all values of t such that $\frac{dy}{dx} = 2$. Using (13.3), $\frac{dx}{dt} = -3t^2$ and

$\frac{dy}{dt} = -12t - 18 \Rightarrow \frac{dy}{dx} = \frac{-12t - 18}{-3t^2}$. $\frac{dy}{dx} = 2 \Rightarrow \frac{-12t - 18}{-3t^2} = 2 \Rightarrow$

$-12t - 18 = -6t^2 \Rightarrow t^2 - 2t - 3 = 0 \Rightarrow t = 3, -1$. $t = 3 \Rightarrow x = -27$ and

$y = -108$. $t = -1 \Rightarrow x = 1$ and $y = 12$. ★ $(-27, -108)$, $(1, 12)$

Note: In Exer. 11–20, let *HT* denote Horizontal Tangent and *VT* denote Vertical Tangent.

[11] (a) Horizontal tangents occur when $\frac{dy}{dx} = 0$. Vertical tangents occur when $\frac{dy}{dx}$ becomes unbounded (this usually happens if the denominator of $\frac{dy}{dx}$ approaches zero, but the numerator does not). $\frac{dx}{dt} = 8t$ and $\frac{dy}{dt} = 3t^2 - 12 \Rightarrow$ $\frac{dy}{dx} = \frac{3t^2 - 12}{8t} = 0$ at $t = \pm 2$. *HT* at $(16, \pm 16)$. $\frac{dy}{dx}$ becomes unbounded when $8t = 0$ or $t = 0$. *VT* at $(0, 0)$.

(b) Using (13.4) and part (a) for y',

$$\frac{d^2y}{dx^2} = \frac{dy'/dt}{dx/dt} = \frac{D_t\left[\frac{3}{8}t - \frac{3}{2t}\right]}{8t} = \frac{\frac{3}{8} + \frac{3}{2t^2}}{8t} = \frac{3t^2 + 12}{64t^3}.$$

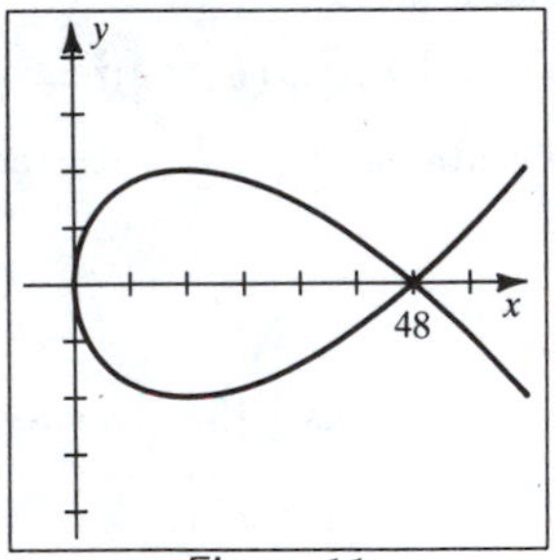

Figure 11

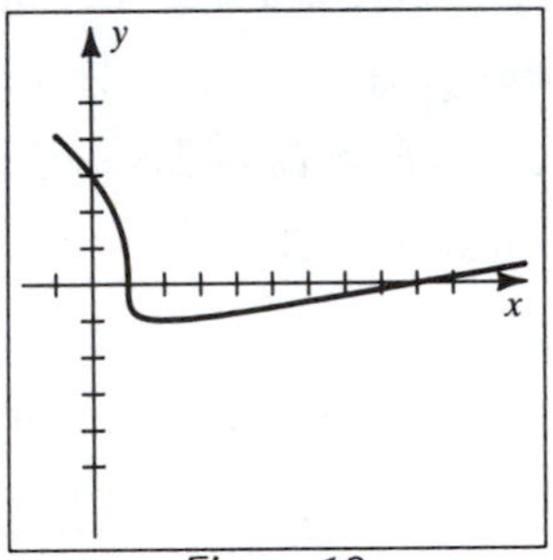

Figure 13

[13] (a) $\frac{dx}{dt} = 3t^2$ and $\frac{dy}{dt} = 2t - 2 \Rightarrow \frac{dy}{dx} = \frac{2t - 2}{3t^2} = 0$ at $t = 1$.

HT at $(2, -1)$; *VT* at $t = 0$ or $(1, 0)$.

(b) $$\frac{d^2y}{dx^2} = \frac{D_t\left[\frac{2}{3t} - \frac{2}{3t^2}\right]}{3t^2} = \frac{-\frac{2}{3t^2} + \frac{4}{3t^3}}{3t^2} = \frac{-2t + 4}{9t^5}.$$

[17] (a) $\frac{dx}{dt} = -3\cos^2 t \sin t$ and $\frac{dy}{dt} = 3\sin^2 t \cos t \Rightarrow$

$\frac{dy}{dx} = -\frac{\sin t}{\cos t} = -\tan t = 0$ at $t = 0, \pi, 2\pi$.

HT at $(\pm 1, 0)$. Since $\cos t = 0$ and $\sin t \neq 0$ at $t = \frac{\pi}{2}$ and $\frac{3\pi}{2}$, there are VT at $(0, \pm 1)$.

(b) $\frac{d^2y}{dx^2} = \frac{D_t[-\tan t]}{-3\cos^2 t \sin t} = \frac{-\sec^2 t}{-3\cos^2 t \sin t}$

$= \frac{1}{3}\sec^4 t \csc t.$

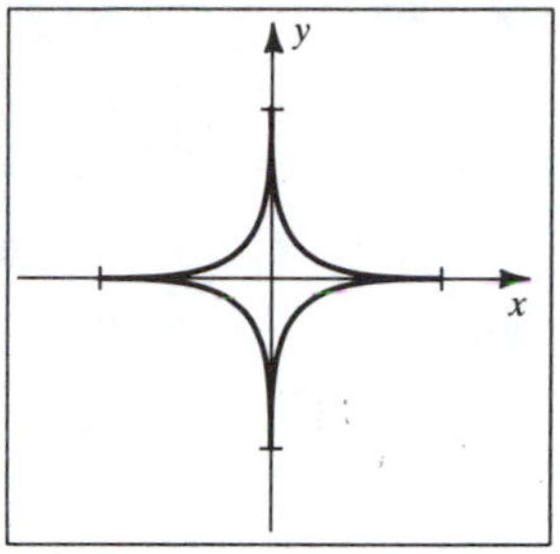

Figure 17

[19] From the figure, we see that there are six points where the tangent is horizontal and four points where the tangent is vertical. Notice the symmetry in their locations.

$\frac{dx}{dt} = 8\cos 2t$ and $\frac{dy}{dt} = -6\sin 3t \Rightarrow \frac{dy}{dx} = -\frac{3\sin 3t}{4\cos 2t} = 0 \Rightarrow \sin 3t = 0 \Rightarrow 3t = \pi n$, or $t = \frac{\pi}{3}n$. Since $\sin 2t$ has period π and $\cos 3t$ has period $\frac{2\pi}{3}$, the curve will retrace itself every 2π units; that is, the period will be the $\mathrm{lcm}(\pi, \frac{2\pi}{3}) = 2\pi$ (also see Exercise 39(b), §13.1). On $[0, 2\pi)$, $t = 0, \frac{\pi}{3}, \frac{2\pi}{3}, \pi, \frac{4\pi}{3}, \frac{5\pi}{3}$ gives HT at $(0, \pm 2)$, $(2\sqrt{3}, \pm 2)$, $(-2\sqrt{3}, \pm 2)$. VT occur when $\cos 2t = 0$ (and $\sin 3t \neq 0$) $\Rightarrow 2t = \frac{\pi}{2} + \pi n \Rightarrow$

$t = \frac{\pi}{4} + \frac{\pi}{2}n$. On $[0, 2\pi)$, $t = \frac{\pi}{4}, \frac{3\pi}{4}, \frac{5\pi}{4}, \frac{7\pi}{4}$ gives VT at $(4, \pm\sqrt{2})$, $(-4, \pm\sqrt{2})$.

[21] Using (13.5), $\frac{dx}{dt} = 10t$ and $\frac{dy}{dt} = 6t^2 \Rightarrow$

$$L = \int_0^1 \sqrt{(10t)^2 + (6t^2)^2}\,dt = \int_0^1 \sqrt{4t^2(25 + 9t^2)}\,dt = \int_0^1 2t\sqrt{25 + 9t^2}\,dt =$$

$$\tfrac{1}{9}\Big[\tfrac{2}{3}(25 + 9t^2)^{3/2}\Big]_0^1 = \tfrac{2}{27}(34^{3/2} - 125) \approx 5.43$$

[23] $\frac{dx}{dt} = e^t(\cos t - \sin t)$ and $\frac{dy}{dt} = e^t(\cos t + \sin t) \Rightarrow$

$$L = \int_0^{\pi/2} \sqrt{e^{2t}(\cos t - \sin t)^2 + e^{2t}(\cos t + \sin t)^2}\,dt$$

$$= \int_0^{\pi/2} \sqrt{e^{2t}\big[(\cos^2 t - 2\cos t \sin t + \sin^2 t) + (\cos^2 t + 2\cos t \sin t + \sin^2 t)\big]}\,dt$$

$$= \int_0^{\pi/2} \sqrt{e^{2t}(2\cos^2 t + 2\sin^2 t)}\,dt$$

$$= \int_0^{\pi/2} \sqrt{(e^t)^2(2)}\,dt = \int_0^{\pi/2} e^t\sqrt{2}\,dt = \sqrt{2}\Big[e^t\Big]_0^{\pi/2} = \sqrt{2}(e^{\pi/2} - 1) \approx 5.39$$

[27] $\frac{dx}{dt} = -2\sin t$ and $\frac{dy}{dt} = 3\cos t$. $L = \int_0^{2\pi} \sqrt{4\sin^2 t + 9\cos^2 t}\,dt$.

Let $f(t) = \sqrt{4\sin^2 t + 9\cos^2 t}$. $L \approx S =$

$$\tfrac{2\pi - 0}{3(6)}\Big[f(0) + 4f(\tfrac{\pi}{3}) + 2f(\tfrac{2\pi}{3}) + 4f(\pi) + 2f(\tfrac{4\pi}{3}) + 4f(\tfrac{5\pi}{3}) + f(2\pi)\Big] \approx 15.881.$$

[29] Using (13.7) with $x = f(t) = t^2$ and $y = g(t) = 2t$, $\frac{dx}{dt} = 2t$ and $\frac{dy}{dt} = 2 \Rightarrow$

$$S = 2\pi \int_0^4 2t\sqrt{4t^2 + 4}\, dt = 4\pi \int_0^4 2t\sqrt{t^2 + 1}\, dt =$$

$$4\pi\left[\tfrac{2}{3}(t^2 + 1)^{3/2}\right]_0^4 = \tfrac{8\pi}{3}(17^{3/2} - 1) \approx 578.83$$

[33] Using (13.7) with $x = f(t) = t - \sin t$ and $y = g(t) = 1 - \cos t$,

$\frac{dx}{dt} = 1 - \cos t$ and $\frac{dy}{dt} = \sin t \Rightarrow$

$$S = 2\pi \int_0^{2\pi} (1 - \cos t)\sqrt{(1 - \cos t)^2 + \sin^2 t}\, dt$$

$$= 2\pi \int_0^{2\pi} (1 - \cos t)\sqrt{1 - 2\cos t + \cos^2 t + \sin^2 t}\, dt$$

$$= 2\pi \int_0^{2\pi} (1 - \cos t)\sqrt{2 - 2\cos t}\, dt$$

$$= 2\sqrt{2}\,\pi \int_0^{2\pi} (1 - \cos t)\sqrt{1 - \cos t}\, dt$$

$$= 2\sqrt{2}\pi \int_0^{2\pi} (1 - \cos t)^{3/2}\, dt$$

$$= 2\sqrt{2}\pi \int_0^{2\pi} \left[1 - \{1 - 2\sin^2(\tfrac{1}{2}t)\}\right]^{3/2} dt \quad \{\cos t = 1 - 2\sin^2(\tfrac{1}{2}t)\}$$

$$= 2\sqrt{2}\pi \int_0^{2\pi} 2^{3/2}\sin^3(\tfrac{1}{2}t)\, dt$$

$$= 2^{3/2}\pi \int_0^{2\pi} 2^{3/2}\sin^2(\tfrac{1}{2}t)\sin(\tfrac{1}{2}t)\, dt$$

$$= 8\pi \int_0^{2\pi} \left[1 - \cos^2(\tfrac{1}{2}t)\right]\sin(\tfrac{1}{2}t)\, dt$$

$$= 8\pi(-2)\int_1^{-1} (1 - u^2)\, du \quad \{u = \cos(\tfrac{1}{2}t),\ -2\, du = \sin(\tfrac{1}{2}t)\, dt\}$$

$$= 16\pi \cdot 2\int_0^1 (1 - u^2)\, du = 32\pi\left[u - \tfrac{1}{3}u^3\right]_0^1 = \tfrac{64\pi}{3} \approx 67.02$$

[35] Using the formula in blue that is below (13.7), $\frac{dx}{dt} = 2t^{-1/2}$ and $\frac{dy}{dt} = t - t^{-2} \Rightarrow$

$$S = 2\pi \int_1^4 4t^{1/2}\sqrt{4t^{-1} + (t - t^{-2})^2}\, dt$$

$$= 8\pi \int_1^4 t^{1/2}\sqrt{4t^{-1} + (t^2 - 2t^{-1} + t^{-4})}\, dt$$

$$= 8\pi \int_1^4 t^{1/2}\sqrt{(t^2 + 2t^{-1} + t^{-4})}\, dt$$

$$= 8\pi \int_1^4 t^{1/2}\sqrt{(t + t^{-2})^2}\, dt$$

$$= 8\pi \int_1^4 t^{1/2}(t + t^{-2})\, dt$$

$$= 8\pi \int_1^4 (t^{3/2} + t^{-3/2})\, dt = 8\pi\left[\tfrac{2}{5}t^{5/2} - 2t^{-1/2}\right]_1^4 = 8\pi(\tfrac{67}{5}) = \tfrac{536\pi}{5} \approx 336.78$$

Exercises 13.3

Note: For the following exercises, the substitutions $y = r\sin\theta$, $x = r\cos\theta$, $r^2 = x^2 + y^2$, and $\tan\theta = \frac{y}{x}$ are used without mention. We have found it helpful to find the "pole" values {when the graph intersects the pole} to determine which values of θ should be used in the construction of an r-θ chart. The numbers listed on each line of the r-θ chart correspond to the numbers labeled on the figures.

[3] $\theta = -\frac{\pi}{6}$ and $r \in \mathbb{R}$. Since $\tan\theta = \frac{y}{x}$, the line is $y = (\tan\theta)\, x$ or $y = -\frac{1}{3}\sqrt{3}\, x$.

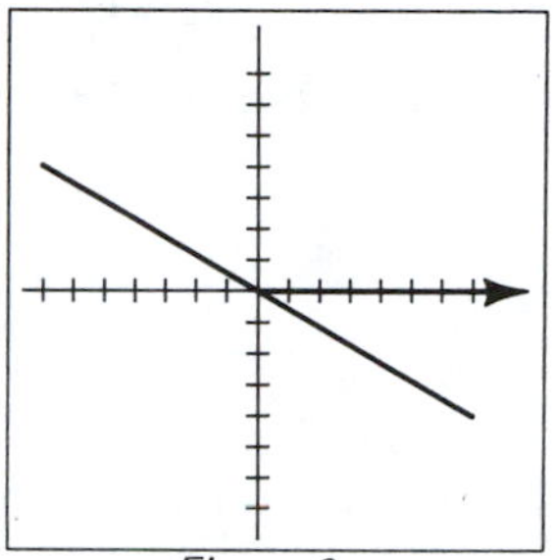

Figure 3

[5] $r = 3\cos\theta \Rightarrow$ {multiply by r to obtain r^2}

$r^2 = 3r\cos\theta \Rightarrow x^2 + y^2 = 3x \Rightarrow$

{recognize this as an equation of a circle and complete the square}

$(x^2 - 3x + \underline{\tfrac{9}{4}}) + y^2 = \underline{\tfrac{9}{4}} \Rightarrow (x - \frac{3}{2})^2 + y^2 = \frac{9}{4}$.

This is a circle with center $(\frac{3}{2}, 0)$ and radius $\sqrt{\frac{9}{4}} = \frac{3}{2}$. From the table, we see that as θ varies from 0 to $\frac{\pi}{2}$, r will vary from 3 to 0. This corresponds to the portion of the circle in the first quadrant. As θ varies from $\frac{\pi}{2}$ to π, r varies from 0 to -3. Remember that -3 in the π direction is the same as 3 in the 0 direction.

This corresponds to the portion of the circle in the fourth quadrant.

	Range of θ	Range of r
1)	$0 \rightarrow \frac{\pi}{2}$	$3 \rightarrow 0$
2)	$\frac{\pi}{2} \rightarrow \pi$	$0 \rightarrow -3$

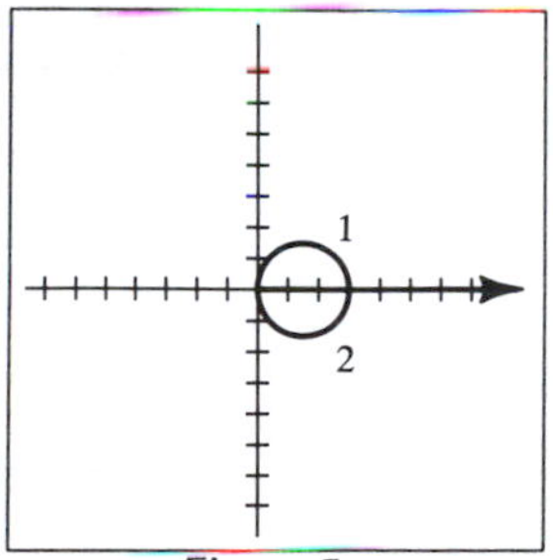

Figure 5

[7] $r = 4 - 4\sin\theta$ is a cardioid since the coefficient of $\sin\theta$ has the same magnitude as the constant term. $0 = 4 - 4\sin\theta \Rightarrow \sin\theta = 1 \Rightarrow \theta = \frac{\pi}{2} + 2\pi n$.

The "v" in the heart-shaped curve corresponds to the pole value $\frac{\pi}{2}$.

	Range of θ	Range of r
1)	$0 \rightarrow \frac{\pi}{2}$	$4 \rightarrow 0$
2)	$\frac{\pi}{2} \rightarrow \pi$	$0 \rightarrow 4$
3)	$\pi \rightarrow \frac{3\pi}{2}$	$4 \rightarrow 8$
4)	$\frac{3\pi}{2} \rightarrow 2\pi$	$8 \rightarrow 4$

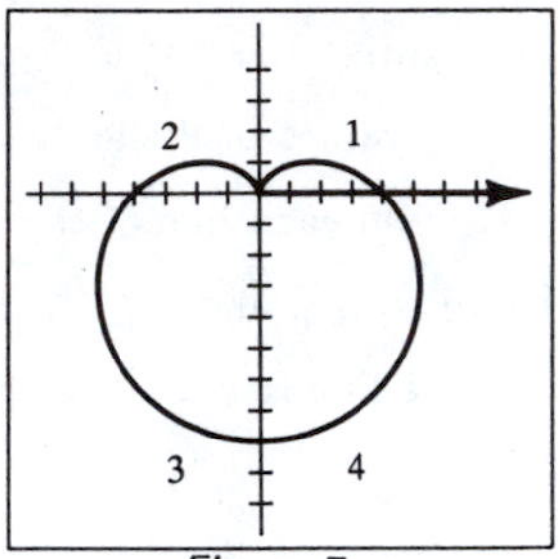

Figure 7

[9] $r = 2 + 4\sin\theta$ is a limaçon with a loop since the constant term has a smaller magnitude than the coefficient of $\sin\theta$. $0 = 2 + 4\sin\theta \Rightarrow \sin\theta = -\frac{1}{2} \Rightarrow$ $\theta = \frac{7\pi}{6} + 2\pi n, \frac{11\pi}{6} + 2\pi n$. Trace through the table and the figure to make sure you understand what values of θ form the loop ($\frac{7\pi}{6}$ to $\frac{11\pi}{6}$).

	Range of θ	Range of r
1)	$0 \rightarrow \frac{\pi}{2}$	$2 \rightarrow 6$
2)	$\frac{\pi}{2} \rightarrow \pi$	$6 \rightarrow 2$
3)	$\pi \rightarrow \frac{7\pi}{6}$	$2 \rightarrow 0$
4)	$\frac{7\pi}{6} \rightarrow \frac{3\pi}{2}$	$0 \rightarrow -2$
5)	$\frac{3\pi}{2} \rightarrow \frac{11\pi}{6}$	$-2 \rightarrow 0$
6)	$\frac{11\pi}{6} \rightarrow 2\pi$	$0 \rightarrow 2$

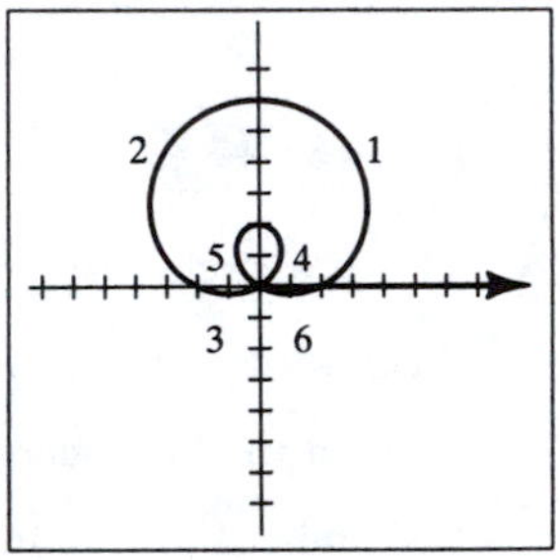

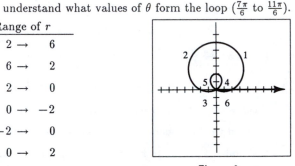

Figure 9

[11] $0 = 2 - \cos\theta \Rightarrow \cos\theta = 2 \Rightarrow$ no pole values.

	Range of θ	Range of r
1)	$0 \rightarrow \frac{\pi}{2}$	$1 \rightarrow 2$
2)	$\frac{\pi}{2} \rightarrow \pi$	$2 \rightarrow 3$
3)	$\pi \rightarrow \frac{3\pi}{2}$	$3 \rightarrow 2$
4)	$\frac{3\pi}{2} \rightarrow 2\pi$	$2 \rightarrow 1$

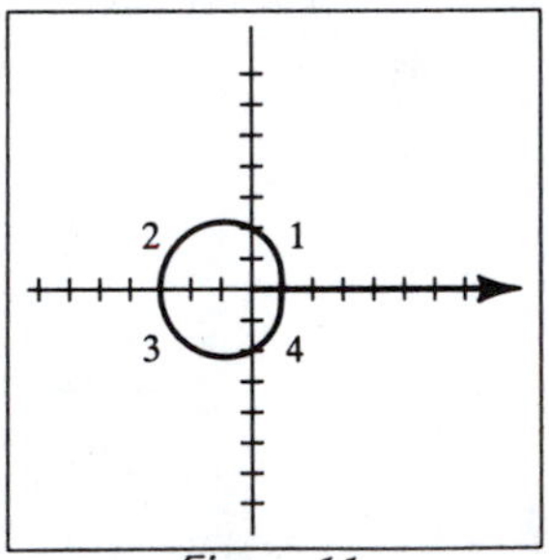

Figure 11

13 $r = 4\csc\theta \Rightarrow r\sin\theta = 4 \Rightarrow y = 4.$

r is undefined at $\theta = \pi n$.

This is a horizontal line with y-intercept (0, 4).

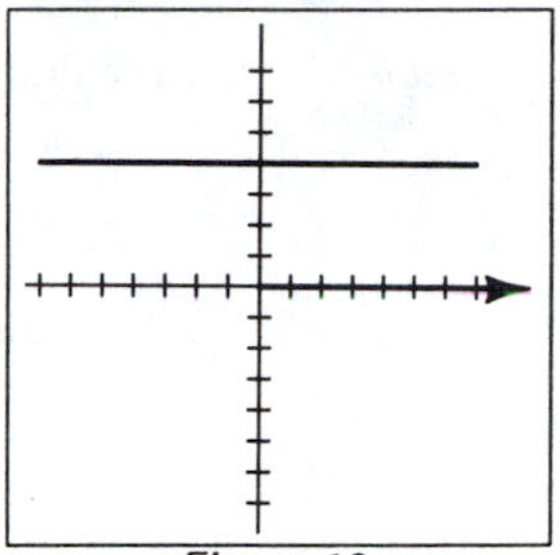

Figure 13

17 $r = 3\sin 2\theta$ is a 4-leafed rose. $0 = 3\sin 2\theta \Rightarrow \sin 2\theta = 0 \Rightarrow 2\theta = \pi n \Rightarrow \theta = \frac{\pi}{2}n.$

	Range of θ	Range of r
1)	$0 \to \frac{\pi}{4}$	$0 \to 3$
2)	$\frac{\pi}{4} \to \frac{\pi}{2}$	$3 \to 0$
3)	$\frac{\pi}{2} \to \frac{3\pi}{4}$	$0 \to -3$
4)	$\frac{3\pi}{4} \to \pi$	$-3 \to 0$
5)	$\pi \to \frac{5\pi}{4}$	$0 \to 3$
6)	$\frac{5\pi}{4} \to \frac{3\pi}{2}$	$3 \to 0$
7)	$\frac{3\pi}{2} \to \frac{7\pi}{4}$	$0 \to -3$
8)	$\frac{7\pi}{4} \to 2\pi$	$-3 \to 0$

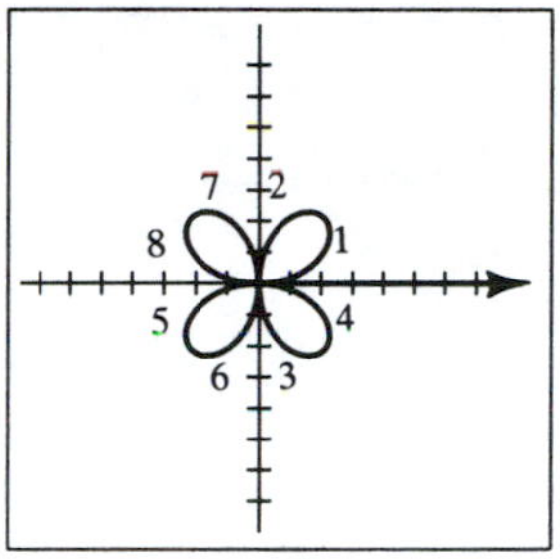

Figure 17

19 Note that since $r^2 \geq 0$, we must have $\cos 2\theta \geq 0$. Hence, 2θ must be in QI or QIV, or else r is undefined. $0 = 4\cos 2\theta \Rightarrow \cos 2\theta = 0 \Rightarrow 2\theta = \frac{\pi}{2} + \pi n \Rightarrow \theta = \frac{\pi}{4} + \frac{\pi}{2}n.$

	Range of θ	Range of r
1)	$0 \to \frac{\pi}{4}$	$\pm 2 \to 0$
2)	$\frac{\pi}{4} \to \frac{\pi}{2}$	undefined
3)	$\frac{\pi}{2} \to \frac{3\pi}{4}$	undefined
4)	$\frac{3\pi}{4} \to \pi$	$0 \to \pm 2$

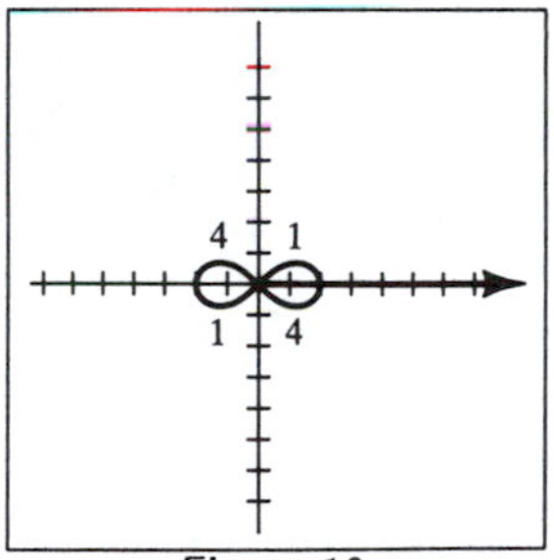

Figure 19

[25] Note that $r = 2\sec\theta \Leftrightarrow r\cos\theta = 2 \Leftrightarrow x = 2$. If $\theta \in (0, \frac{\pi}{2})$ or $\theta \in (\frac{3\pi}{2}, 2\pi)$, then $\sec\theta > 0$ and the graph of $r = 2 + 2\sec\theta$ is to the right of $x = 2$. If $\theta \in (\frac{\pi}{2}, \frac{3\pi}{2})$, $\sec\theta < 0$ and $r = 2 + 2\sec\theta$ is to the left of $x = 2$. r is undefined at $\theta = \frac{\pi}{2} + \pi n$.

$$0 = 2 + 2\sec\theta \Rightarrow \sec\theta = -1 \Rightarrow \theta = \pi + 2\pi n.$$

	Range of θ	Range of r
1)	$0 \rightarrow \frac{\pi}{2}$	$4 \rightarrow \infty$
2)	$\frac{\pi}{2} \rightarrow \pi$	$-\infty \rightarrow 0$
3)	$\pi \rightarrow \frac{3\pi}{2}$	$0 \rightarrow -\infty$
4)	$\frac{3\pi}{2} \rightarrow 2\pi$	$\infty \rightarrow 4$

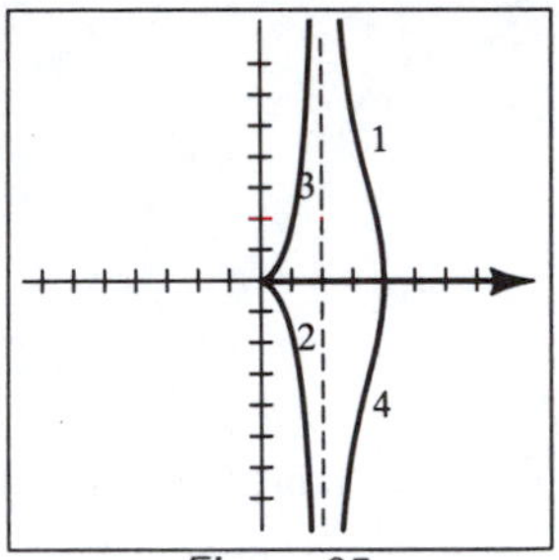

Figure 25

[27] $x = -3 \Rightarrow r\cos\theta = -3 \Rightarrow r = \dfrac{-3}{\cos\theta} \Rightarrow r = -3\sec\theta$

[31] $2y = -x \Rightarrow \frac{y}{x} = -\frac{1}{2} \Rightarrow \tan\theta = -\frac{1}{2} \Rightarrow \theta = \tan^{-1}(-\frac{1}{2})$

[33] $y^2 - x^2 = 4 \Rightarrow r^2\sin^2\theta - r^2\cos^2\theta = 4 \Rightarrow -r^2(\cos^2\theta - \sin^2\theta) = 4 \Rightarrow$

$$-r^2\cos 2\theta = 4 \Rightarrow r^2 = \frac{-4}{\cos 2\theta} \Rightarrow r^2 = -4\sec 2\theta$$

[35] $(x^2 + y^2)\tan^{-1}(\frac{y}{x}) = ay \Rightarrow r^2\theta = ar\sin\theta \Rightarrow r\theta = a\sin\theta$ or $r = 0 \Rightarrow r\theta = a\sin\theta$

since $\theta = 0$ gives $r = 0$ as one solution. Note that $\theta = \tan^{-1}(\frac{y}{x}) \Rightarrow \theta \in (-\frac{\pi}{2}, \frac{\pi}{2})$.

[39] $r = -3\csc\theta \Rightarrow r\sin\theta = -3 \Rightarrow y = -3$. θ is undefined at πn.

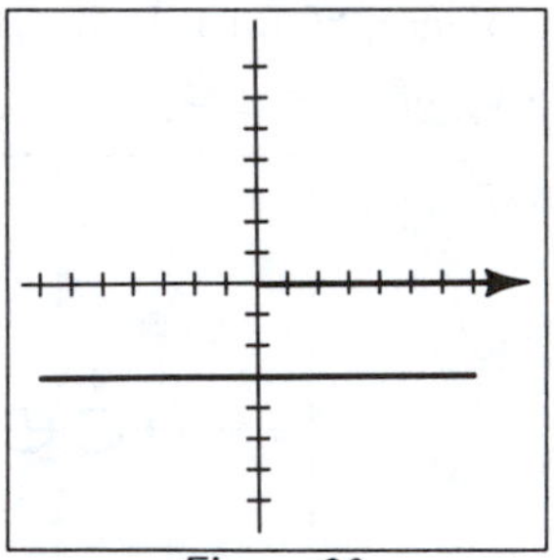
Figure 39

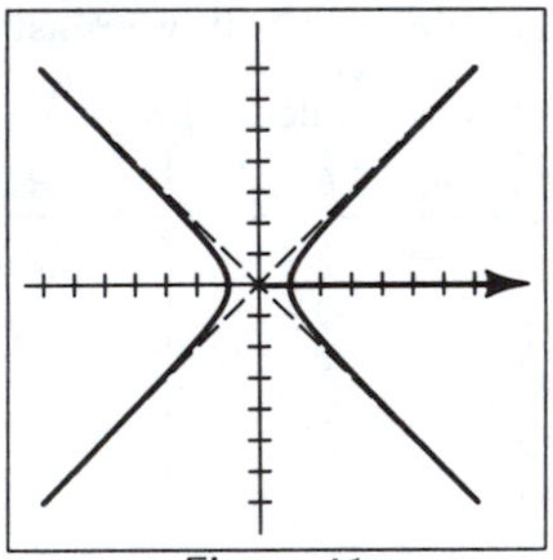
Figure 41

[41] $r^2\cos 2\theta = 1 \Rightarrow r^2(\cos^2\theta - \sin^2\theta) = 1 \Rightarrow r^2\cos^2\theta - r^2\sin^2\theta = 1 \Rightarrow x^2 - y^2 = 1$.

[47] $r = 8\sin\theta - 2\cos\theta \Rightarrow r^2 = 8r\sin\theta - 2r\cos\theta$ {multiply by r} $\Rightarrow$

$x^2 + y^2 = 8y - 2x \Rightarrow x^2 + 2x + \underline{1} + y^2 - 8y + \underline{16} = \underline{1} + \underline{16} \Rightarrow$

$(x + 1)^2 + (y - 4)^2 = 17.$

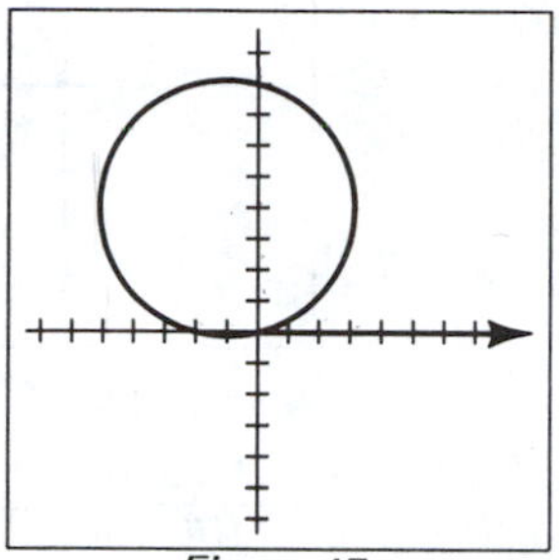

Figure 47

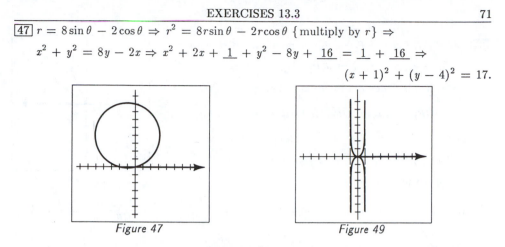

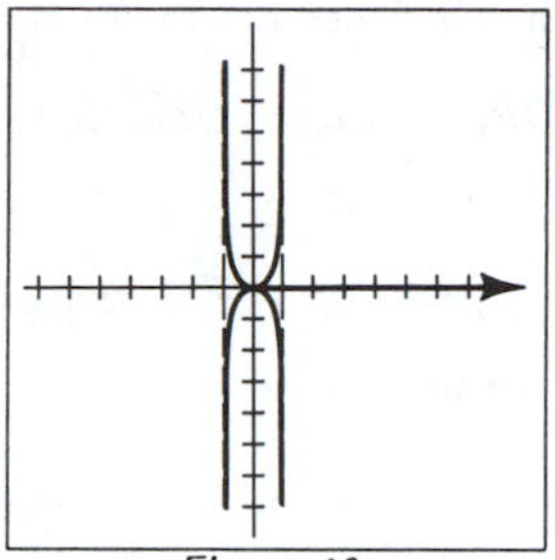

Figure 49

[49] $r = \tan\theta \Rightarrow r^2 = \tan^2\theta \Rightarrow x^2 + y^2 = \dfrac{y^2}{x^2} \Rightarrow x^4 + x^2y^2 = y^2 \Rightarrow$

$y^2 - x^2y^2 = x^4 \Rightarrow y^2(1 - x^2) = x^4 \Rightarrow y^2 = \dfrac{x^4}{1 - x^2}.$

This is a tough one even after getting the equation in x and y.

Since we have solved for y^2, the right side must be positive.

Since x^4 is always nonnegative, we must have $1 - x^2 > 0$, or equivalently, $|x| < 1$.

We have vertical asymptotes at $x = \pm 1$, that is, when the denominator is 0.

Note: Exer. 51–60: With $r = f(\theta)$ and θ given, we will list r', r' evaluated at θ, and r evaluated at θ. Substitute these into the formula $m = \dfrac{r'\sin\theta + r\cos\theta}{r'\cos\theta - r\sin\theta}$ and simplify to obtain the value of m given.

[51] $r' = -2\sin\theta$, $r'(\frac{\pi}{3}) = -\sqrt{3}$, $r(\frac{\pi}{3}) = 1$.

Using (13.10), $m = \dfrac{(-\sqrt{3})\sin\frac{\pi}{3} + (1)\cos\frac{\pi}{3}}{(-\sqrt{3})\cos\frac{\pi}{3} - (1)\sin\frac{\pi}{3}} = \dfrac{(-\sqrt{3})(\sqrt{3}/2) + (1)(1/2)}{(-\sqrt{3})(1/2) - (1)(\sqrt{3}/2)} = \sqrt{3}/3.$

[57] Using implicit differentiation, $r^2 = 4\cos 2\theta \Rightarrow 2rr' = -8\sin 2\theta \Rightarrow r' = -\dfrac{4\sin 2\theta}{r}.$

Since r' is in terms of r, we compute $r(\theta)$ first.

$r^2(\frac{\pi}{6}) = 4\cos(2\cdot\frac{\pi}{6}) \Rightarrow r(\frac{\pi}{6}) = \pm 2\sqrt{\cos\frac{\pi}{3}} = \pm 2\sqrt{\frac{1}{2}} = \pm 2\cdot\frac{\sqrt{2}}{2} = \pm\sqrt{2}.$

$r'(\frac{\pi}{6}) = -\left[\dfrac{4\sin\frac{\pi}{3}}{\pm\sqrt{2}}\right] = \mp 2\sqrt{2}\cdot\dfrac{\sqrt{3}}{2} = \mp\sqrt{6}.$

$m = \dfrac{(\mp\sqrt{6})\sin\frac{\pi}{6} + (\pm\sqrt{2})\cos\frac{\pi}{6}}{(\mp\sqrt{6})\cos\frac{\pi}{6} - (\pm\sqrt{2})\sin\frac{\pi}{6}} = 0$ in either case.

[59] $r' = 2^\theta(\ln 2)$, $r'(\pi) = 2^\pi(\ln 2)$, $r(\pi) = 2^\pi$,

$m = \dfrac{(2^\pi\ln 2)\sin\pi + (2^\pi)\cos\pi}{(2^\pi\ln 2)\cos\pi - (2^\pi)\sin\pi} = \dfrac{-2^\pi}{-2^\pi\ln 2} = \dfrac{1}{\ln 2}$

[61] Let $P_1(r_1, \theta_1)$ and $P_2(r_2, \theta_2)$ be points in an $r\theta$-plane. Let $a = r_1$, $b = r_2$, $c = d(P_1, P_2)$, and $\gamma = \theta_2 - \theta_1$. Note that if O denotes the origin, then $\gamma = \angle P_1OP_2$, $r_1 = \overline{OP_1}$, and $r_2 = \overline{OP_2}$. Substituting into the law of cosines, $c^2 = a^2 + b^2 - 2ab\cos\gamma$, gives us $[d(P_1, P_2)]^2 = r_1^2 + r_2^2 - 2r_1r_2\cos(\theta_2 - \theta_1)$, which is the required formula.

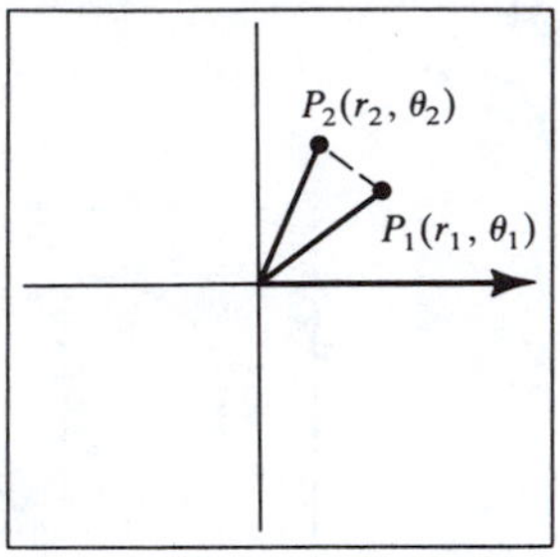

Figure 61

[63] By (13.10), the slope of one tangent line is $m_1 = \dfrac{f'(\theta)\sin\theta + f(\theta)\cos\theta}{f'(\theta)\cos\theta - f(\theta)\sin\theta} = \dfrac{\text{A}}{\text{B}}$ and the slope of the other is $m_2 = \dfrac{g'(\theta)\sin\theta + g(\theta)\cos\theta}{g'(\theta)\cos\theta - g(\theta)\sin\theta} = \dfrac{\text{C}}{\text{D}}$.

Now, $m_1 \cdot m_2 = -1 \Leftrightarrow \dfrac{\text{A}\cdot\text{C}}{\text{B}\cdot\text{D}} = -1 \Leftrightarrow \dfrac{\text{A}\cdot\text{C} + \text{B}\cdot\text{D}}{\text{B}\cdot\text{D}} = 0 \Leftrightarrow \text{A}\cdot\text{C} + \text{B}\cdot\text{D} = 0.$

This is equivalent to $\left[f'(\theta)\sin\theta + f(\theta)\cos\theta\right]\left[g'(\theta)\sin\theta + g(\theta)\cos\theta\right]$

$+ \left[f'(\theta)\cos\theta - f(\theta)\sin\theta\right]\left[g'(\theta)\cos\theta - g(\theta)\sin\theta\right] = 0 \Leftrightarrow$

$f'(\theta)\,g'(\theta)\left[\sin^2\theta + \cos^2\theta\right] + f(\theta)\,g(\theta)\left[\sin^2\theta + \cos^2\theta\right] = 0 \Leftrightarrow$

$f'(\theta)\,g'(\theta) + f(\theta)\,g(\theta) = 0.$

[65] $m = \dfrac{\dfrac{dr}{d\theta}\sin\theta + r\cos\theta}{\dfrac{dr}{d\theta}\cos\theta - r\sin\theta}$ {from 13.10}

$= \dfrac{\dfrac{dr}{d\theta}\cdot\dfrac{\sin\theta}{\cos\theta} + r\dfrac{\cos\theta}{\cos\theta}}{\dfrac{dr}{d\theta}\cdot\dfrac{\cos\theta}{\cos\theta} - r\dfrac{\sin\theta}{\cos\theta}}$ {divide each term by $\cos\theta$}

$= \dfrac{\dfrac{dr}{d\theta}\tan\theta + r}{\dfrac{dr}{d\theta} - r\tan\theta}$ {simplify}

Exercises 13.4

[1] See Figure 13.25. By symmetry and (13.11),

$$A = 2\int_0^{\pi/2} \tfrac{1}{2}(2\cos\theta)^2\,d\theta = 4\int_0^{\pi/2}\cos^2\theta\,d\theta = 4\int_0^{\pi/2}\left(\frac{1+\cos 2\theta}{2}\right)d\theta =$$

$$2\int_0^{\pi/2}(1+\cos 2\theta)\,d\theta = 2\left[\theta + \tfrac{1}{2}\sin 2\theta\right]_0^{\pi/2} = \pi.$$

[5] The graph is a four-leafed rose. See Figure 13.21. Using symmetry,

$$A = 4\int_0^{\pi/2}\tfrac{1}{2}\sin^2 2\theta\,d\theta = \int_0^{\pi/2}(1-\cos 4\theta)\,d\theta = \left[\theta - \tfrac{1}{4}\sin 4\theta\right]_0^{\pi/2} = \tfrac{\pi}{2}.$$

[7] $A = \displaystyle\int_0^{\pi/2}\tfrac{1}{2}e^{2\theta}\,d\theta = \left[\tfrac{1}{4}e^{2\theta}\right]_0^{\pi/2} = \tfrac{1}{4}(e^{\pi} - 1) \approx 5.54$

9 One loop is traced out for $-\frac{\pi}{4} \leq \theta \leq \frac{\pi}{4}$ {see Exer. 19, §13.3}.

Thus, by symmetry, $A = 2\int_0^{\pi/4} \frac{1}{2}(4\cos 2\theta)\,d\theta = 4\left[\frac{1}{2}\sin 2\theta\right]_0^{\pi/4} = 2.$

13 In the figure, r is first bounded by the line $x = 4$ and then by the circle $x^2 + y^2 = 25$. θ varies between 0 and $\frac{\pi}{2}$. The boundary of r changes when $\tan\theta = \frac{3}{4}$, or $\theta = \arctan\frac{3}{4}$. $x = 4 \Leftrightarrow r\cos\theta = 4 \Leftrightarrow r = 4\sec\theta$.

$x^2 + y^2 = 25 \Leftrightarrow r = 5$. If $0 \leq \theta \leq \arctan\frac{3}{4}$, then $r = 4\sec\theta$. If $\arctan\frac{3}{4} \leq \theta \leq \frac{\pi}{2}$, then $r = 5$. $A = \int_0^{\arctan(3/4)} \frac{1}{2}(4\sec\theta)^2\,d\theta + \int_{\arctan(3/4)}^{\pi/2} \frac{1}{2}(5)^2\,d\theta.$

15 This region is an example of the type of region shown in Figure 13.34. The outer region is bounded by the line $y = 4$ and the inner region is bounded by the circle $x^2 + y^2 = 4$. $\tan\theta$ varies between 1 and 3, which are the slopes of the lines $y = 1x$ and $y = 3x$, respectively. $x^2 + y^2 = 4 \Leftrightarrow r = 2$. $y = x \Rightarrow \frac{y}{x} = 1 \Rightarrow \tan^{-1}\frac{y}{x} = \tan^{-1}1 \Rightarrow \theta = \frac{\pi}{4}$. $y = 4 \Leftrightarrow r\sin\theta = 4 \Leftrightarrow r = 4\csc\theta$. $y = 3x \Rightarrow \frac{y}{x} = 3 \Rightarrow \tan^{-1}\frac{y}{x} = \tan^{-1}3 \Rightarrow \theta = \arctan 3$. If $\frac{\pi}{4} \leq \theta \leq \arctan 3$, then the outer radius is $4\csc\theta$ and the inner radius is 2. Using the formula associated with Figure 13.34, $A = \int_{\pi/4}^{\arctan 3} \frac{1}{2}(4\csc\theta)^2\,d\theta - \int_{\pi/4}^{\arctan 3} \frac{1}{2}(2)^2\,d\theta =$

$$\int_{\pi/4}^{\arctan 3} \frac{1}{2}\left[(4\csc\theta)^2 - (2)^2\right]d\theta.$$

17 (a) Consider only the blue region in the first quadrant bounded below by the polar axis. The boundaries of the blue and green regions intersect when $4\cos 2\theta = 2 \Rightarrow \cos 2\theta = \frac{1}{2} \Rightarrow 2\theta = \frac{\pi}{3} \Rightarrow \theta = \frac{\pi}{6}$.

If $0 \leq \theta \leq \frac{\pi}{6}$, then the outer radius is $4\cos 2\theta$ and the inner radius is 2.

Using symmetry, $A = 8\int_0^{\pi/6} \frac{1}{2}\left[(4\cos 2\theta)^2 - (2)^2\right]d\theta.$

(b) Consider only the green region in the first quadrant bounded below by the polar axis. r is bounded by two different functions in this region. Note that $\frac{\pi}{4}$ is a pole value for $r = 4\cos 2\theta$ since $4\cos 2\theta = 0 \Rightarrow \cos 2\theta = 0 \Rightarrow 2\theta = \frac{\pi}{2} + \pi n \Rightarrow \theta = \frac{\pi}{4} + \frac{\pi}{2}n$. If $0 \leq \theta \leq \frac{\pi}{6}$, then $r = 2$. If $\frac{\pi}{6} \leq \theta \leq \frac{\pi}{4}$, then $r = 4\cos 2\theta$.

Using symmetry, $A = 8\left[\int_0^{\pi/6} \frac{1}{2}(2)^2\,d\theta + \int_{\pi/6}^{\pi/4} \frac{1}{2}(4\cos 2\theta)^2\,d\theta\right].$

[19] See Figure 13.35. Using symmetry,

$$A = 2\int_{\pi/3}^{\pi} \tfrac{1}{2}\left[3^2 - (2 + 2\cos\theta)^2\right] d\theta$$

$$= \int_{\pi/3}^{\pi} (5 - 8\cos\theta - 4\cos^2\theta)\, d\theta$$

$$= \int_{\pi/3}^{\pi} \left[5 - 8\cos\theta - 4\left(\frac{1 + \cos 2\theta}{2}\right)\right] d\theta$$

$$= \int_{\pi/3}^{\pi} (3 - 8\cos\theta - 2\cos 2\theta)\, d\theta$$

$$= \left[3\theta - 8\sin\theta - \sin 2\theta\right]_{\pi/3}^{\pi} = 2\pi + \tfrac{9}{2}\sqrt{3} \approx 14.08.$$

[23] $\sin\theta = \sqrt{3}\cos\theta \Rightarrow \tan\theta = \sqrt{3}$ at $\theta = \frac{\pi}{3}$. For $0 \le \theta \le \frac{\pi}{3}$, the boundary is $r = \sin\theta$ and for $\frac{\pi}{3} \le \theta \le \frac{\pi}{2}$, the boundary is $r = \sqrt{3}\cos\theta$. Thus,

$$A = \tfrac{1}{2}\int_0^{\pi/3} \sin^2\theta\, d\theta + \tfrac{1}{2}\int_{\pi/3}^{\pi/2} 3\cos^2\theta\, d\theta$$

$$= \tfrac{1}{4}\int_0^{\pi/3} (1 - \cos 2\theta)\, d\theta + \tfrac{3}{4}\int_{\pi/3}^{\pi/2} (1 + \cos 2\theta)\, d\theta$$

$$= \tfrac{1}{4}\left[\theta - \tfrac{1}{2}\sin 2\theta\right]_0^{\pi/3} + \tfrac{3}{4}\left[\theta + \tfrac{1}{2}\sin 2\theta\right]_{\pi/3}^{\pi/2}$$

$$= \tfrac{1}{4}\left[\tfrac{\pi}{3} - \tfrac{\sqrt{3}}{4}\right] + \tfrac{3}{4}\left[\tfrac{\pi}{2} - \left(\tfrac{\pi}{3} + \tfrac{\sqrt{3}}{4}\right)\right] = \tfrac{5\pi}{24} - \tfrac{\sqrt{3}}{4} \approx 0.22.$$

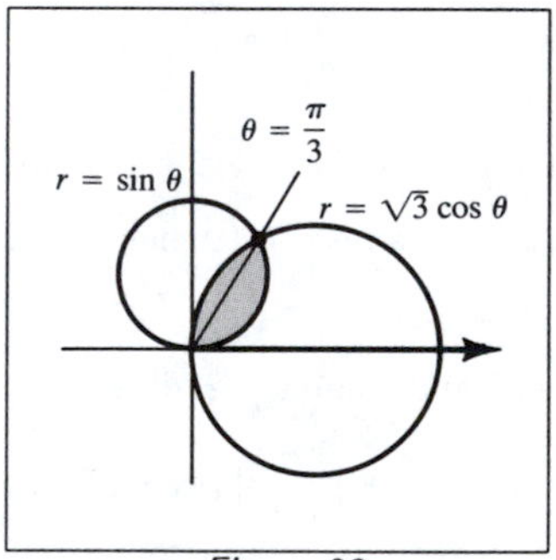

Figure 23

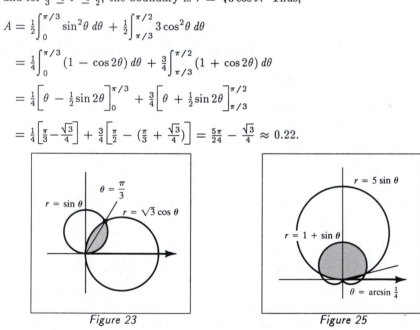

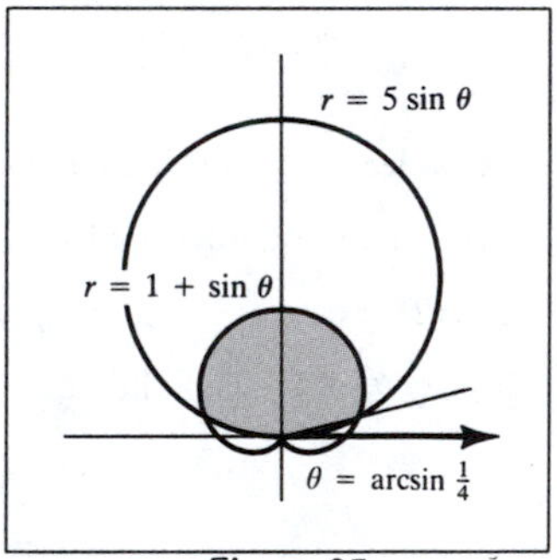

Figure 25

[25] $1 + \sin\theta = 5\sin\theta \Rightarrow \sin\theta = \frac{1}{4}$ at $\theta = \arcsin\frac{1}{4}$ {call this α} and $(\pi - \alpha)$. For $0 \le \theta \le \alpha$, the boundary is $r = 5\sin\theta$ and for $\alpha \le \theta \le \frac{\pi}{2}$, the boundary is $r = 1 + \sin\theta$. By symmetry,

$$A = 2\int_0^{\alpha} \tfrac{1}{2}(5\sin\theta)^2\, d\theta + 2\int_{\alpha}^{\pi/2} \tfrac{1}{2}(1 + \sin\theta)^2\, d\theta$$

$$= 25\int_0^{\alpha} \sin^2\theta\, d\theta + \int_{\alpha}^{\pi/2} \left(1 + 2\sin\theta + \frac{1 - \cos 2\theta}{2}\right) d\theta$$

$$= \tfrac{25}{2}\int_0^{\alpha} (1 - \cos 2\theta)\, d\theta + \int_{\alpha}^{\pi/2} \left(\tfrac{3}{2} + 2\sin\theta - \tfrac{1}{2}\cos 2\theta\right) d\theta$$

$$= \tfrac{25}{2}\Big[\theta - \tfrac{1}{2}\sin 2\theta\Big]_0^{\alpha} + \Big[\tfrac{3}{2}\theta - 2\cos\theta - \tfrac{1}{4}\sin 2\theta\Big]_{\alpha}^{\pi/2}$$

$$\{\sin 2\alpha = 2\sin\alpha\cos\alpha = 2\cdot\tfrac{1}{4}\cdot\tfrac{\sqrt{15}}{4} = \tfrac{\sqrt{15}}{8}\}$$

$$= \tfrac{25}{2}\Big[(\alpha - \tfrac{1}{2}\cdot\tfrac{1}{8}\sqrt{15}) - (0)\Big] + \Big[(\tfrac{3\pi}{4}) - (\tfrac{3}{2}\alpha - 2\cdot\tfrac{1}{4}\sqrt{15} - \tfrac{1}{4}\cdot\tfrac{1}{8}\sqrt{15})\Big] =$$

$$11\arcsin\tfrac{1}{4} - \tfrac{1}{4}\sqrt{15} + \tfrac{3\pi}{4} \approx 4.17$$

27 $f(\theta) = e^{-\theta} \Rightarrow f'(\theta) = -e^{-\theta}$. Using (13.13),

$$L = \int_0^{2\pi} ds = \int_0^{2\pi}\sqrt{[f(\theta)]^2 + [f'(\theta)]^2}\,d\theta = \int_0^{2\pi}\sqrt{e^{-2\theta} + e^{-2\theta}}\,d\theta$$

$$= \int_0^{2\pi}\sqrt{2(e^{-\theta})^2}\,d\theta = \sqrt{2}\int_0^{2\pi} e^{-\theta}\,d\theta = \sqrt{2}\Big[-e^{-\theta}\Big]_0^{2\pi} = \sqrt{2}(1 - e^{-2\pi}) \approx 1.41.$$

31 $r = \sin^3(\tfrac{1}{3}\theta)$ has a period of 6π. However, since $\sin^3\big[\tfrac{1}{3}(\theta + 3\pi)\big] = \sin^3(\tfrac{1}{3}\theta + \pi) = -\sin^3(\tfrac{1}{3}\theta) = -r$, the graph will be traced out twice on $[0, 6\pi]$.

Note that (r, θ) and $(-r, \theta + 3\pi)$ denote the same point.

Thus, $f(\theta) = \sin^3(\tfrac{1}{3}\theta) \Rightarrow f'(\theta) = \sin^2(\tfrac{1}{3}\theta)\cos(\tfrac{1}{3}\theta)$ and

$$L = \int_0^{3\pi}\sqrt{[f(\theta)]^2 + [f'(\theta)]^2}\,d\theta$$

$$= \int_0^{3\pi}\sqrt{\sin^6(\tfrac{1}{3}\theta) + \sin^4(\tfrac{1}{3}\theta)\cos^2(\tfrac{1}{3}\theta)}\,d\theta$$

$$= \int_0^{3\pi}\sqrt{\sin^4(\tfrac{1}{3}\theta)\big[\sin^2(\tfrac{1}{3}\theta) + \cos^2(\tfrac{1}{3}\theta)\big]}\,d\theta$$

$$= \int_0^{3\pi}\sin^2(\tfrac{1}{3}\theta)\,d\theta = \int_0^{3\pi}\frac{1 - \cos(2\cdot\tfrac{1}{3}\theta)}{2}\,d\theta \ \{\text{half-angle formula}\}$$

$$= \tfrac{1}{2}\int_0^{3\pi}\big[1 - \cos\tfrac{2}{3}\theta\big]\,d\theta = \tfrac{1}{2}\Big[\theta - \tfrac{3}{2}\sin\tfrac{2}{3}\theta\Big]_0^{3\pi} = \tfrac{3\pi}{2}.$$

35 $f(\theta) = 2 + 2\cos\theta \Rightarrow f'(\theta) = -2\sin\theta$. Using (13.14),

$$S = \int_0^{\pi} 2\pi r\sin\theta\,ds = \int_0^{\pi} 2\pi(2 + 2\cos\theta)\sin\theta\sqrt{(2 + 2\cos\theta)^2 + (-2\sin\theta)^2}\,d\theta$$

$$= 4\pi\int_0^{\pi}(1 + \cos\theta)\sqrt{4 + 8\cos\theta + 4\cos^2\theta + 4\sin^2\theta}\sin\theta\,d\theta$$

$$= 4\pi\int_0^{\pi}(1 + \cos\theta)\sqrt{8(1 + \cos\theta)}\sin\theta\,d\theta = 8\pi\sqrt{2}\int_0^{\pi}(1 + \cos\theta)^{3/2}\sin\theta\,d\theta$$

$$= 8\pi\sqrt{2}\Big[-\tfrac{2}{5}(1 + \cos\theta)^{5/2}\Big]_0^{\pi} = -\tfrac{16}{5}\pi\sqrt{2}(-2^{5/2}) = \tfrac{128\pi}{5} \approx 80.42.$$

41 The radius of every point on the circle is a, so $f(\theta) = a \Rightarrow f'(\theta) = 0$.

Let (x, y) be any point on the circle. Since $x = a\cos\theta$, the average radius of the frustum generated by revolving a small section of the circle about the line $x = b$ is $b - a\cos\theta$. Thus, applying (13.14) with the line $x = b$,

$$S = \int_0^{2\pi}(2\pi)(b - a\cos\theta)\sqrt{a^2 + 0^2}\,d\theta = 2\pi a\Big[b\theta - a\sin\theta\Big]_0^{2\pi} = 4\pi^2 ab.$$

43 $f(\theta) = e^{-\theta} \Rightarrow f'(\theta) = -e^{-\theta}$. Using (13.14),

$$S = \int_0^{\pi/2} 2\pi r\cos\theta \, ds = 2\pi \int_0^{\pi/2} e^{-\theta}\cos\theta \sqrt{(e^{-\theta})^2 + (-e^{-\theta})^2}\, d\theta$$

$$= 2\pi \int_0^{\pi/2} e^{-\theta}\cos\theta \sqrt{2e^{-2\theta}}\, d\theta = 2\pi\sqrt{2}\int_0^{\pi/2} e^{-\theta}\cos\theta\,(e^{-\theta})\, d\theta$$

$$= 2\pi\sqrt{2}\int_0^{\pi/2} e^{-2\theta}\cos\theta\, d\theta$$

$$= 2\pi\sqrt{2}\left[\frac{e^{-2\theta}}{5}(\sin\theta - 2\cos\theta)\right]_0^{\pi/2} \{\text{Formula } 99\} = \tfrac{2}{5}\pi\sqrt{2}(e^{-\pi} + 2) \approx 3.63.$$

Exercises 13.5

Note: For the ellipse, the major axis is vertical if the denominator contains $\sin\theta$, horizontal if the denominator contains $\cos\theta$. For the hyperbola, the transverse axis is vertical if the denominator contains $\sin\theta$, horizontal if the denominator contains $\cos\theta$. The focus at the pole is called F and V is the vertex associated with (or closest to) F. $d(V, F)$ denotes the distance from the vertex to the focus. The foci are not asked for in the directions, but are listed. For the parabola, the directrix is on the right, on the left, above, or below the focus depending on the term "+cos", "−cos", "+sin", or "−sin", respectively, appearing in the denominator.

1 Divide the numerator and denominator by the constant term in the denominator, i.e., 6. $r = \dfrac{12}{6 + 2\sin\theta} = \dfrac{2}{1 + \frac{1}{3}\sin\theta} \Rightarrow e = \frac{1}{3} < 1$, ellipse. From the last note, we see that the denominator has "$+\sin\theta$" and we have vertices when $\theta = \frac{\pi}{2}$ and $\frac{3\pi}{2}$. They are $V(\frac{3}{2}, \frac{\pi}{2})$ and $V'(3, \frac{3\pi}{2})$. The distance from the focus at the pole to the vertex V is $\frac{3}{2}$. The distance from V' to F' must also be $\frac{3}{2}$ and we see that $F' = (\frac{3}{2}, \frac{3\pi}{2})$. We will use the following notation to summarize this in future problems :

$d(V, F) = \frac{3}{2} \Rightarrow F' = (\frac{3}{2}, \frac{3\pi}{2})$.

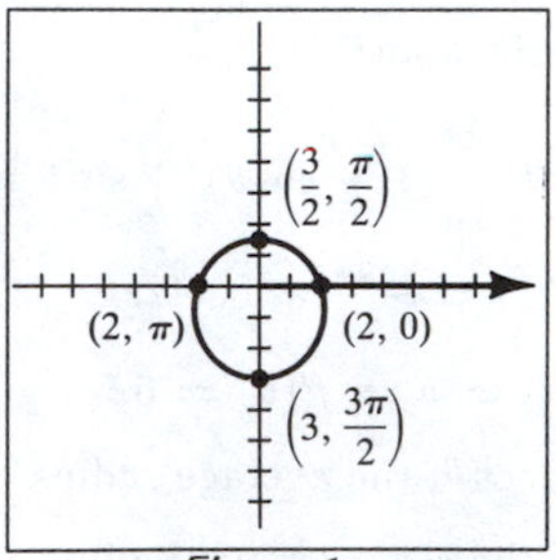

Figure 1

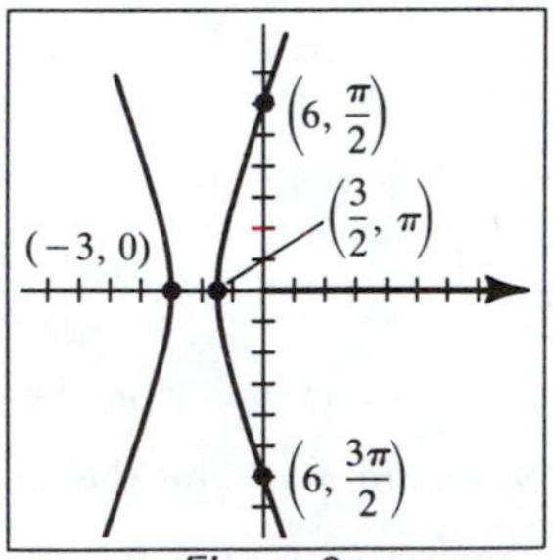

Figure 3

3 $r = \dfrac{12}{2 - 6\cos\theta} = \dfrac{6}{1 - 3\cos\theta} \Rightarrow e = 3 > 1$, hyperbola.

$V(\frac{3}{2}, \pi)$ and $V'(-3, 0)$. $d(V, F) = \frac{3}{2} \Rightarrow F' = (-\frac{9}{2}, 0)$.

[5] $r = \dfrac{3}{2 + 2\cos\theta} = \dfrac{\frac{3}{2}}{1 + 1\cos\theta} \Rightarrow e = 1$, parabola. Note that the expression is undefined in the $\theta = \pi$ direction. The vertex is in the $\theta = 0$ direction, $V(\frac{3}{4}, 0)$.

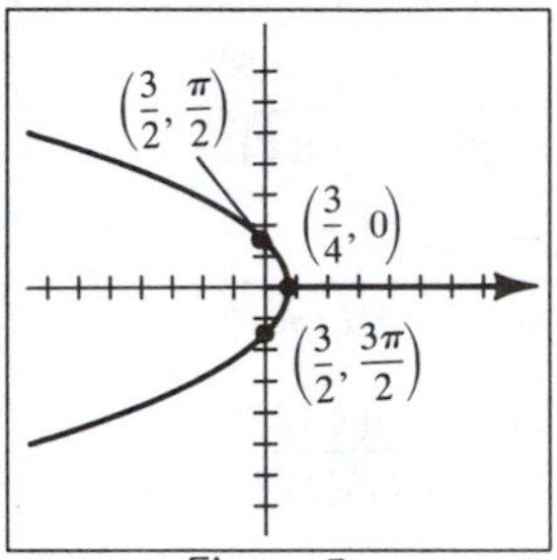

Figure 5

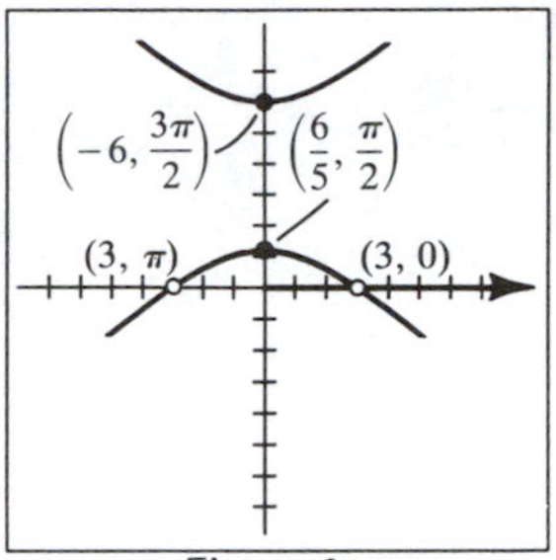

Figure 9

[9] $r = \dfrac{6\csc\theta}{2\csc\theta + 3} \cdot \dfrac{\sin\theta}{\sin\theta} = \dfrac{6}{2 + 3\sin\theta} = \dfrac{3}{1 + \frac{3}{2}\sin\theta} \Rightarrow e = \frac{3}{2} > 1$, hyperbola.

$V(\frac{6}{5}, \frac{\pi}{2})$ and $V'(-6, \frac{3\pi}{2})$. $d(V, F) = \frac{6}{5} \Rightarrow F' = (-\frac{36}{5}, \frac{3\pi}{2})$.

Since the original equation is undefined when $\csc\theta$ is undefined, the points $(3, 0)$ and $(3, \pi)$ are excluded from the graph.

Note: For the following exercises, the substitutions $x = r\cos\theta$, $y = r\sin\theta$, and $r^2 = x^2 + y^2$ are made without mention.

[11] $r = \dfrac{12}{6 + 2\sin\theta} \Rightarrow 6r + 2r\sin\theta = 12 \Rightarrow 6r + 2y = 12 \Rightarrow 3r = 6 - y \Rightarrow$
$(3r)^2 = (6 - y)^2 \Rightarrow 9r^2 = 36 - 12y + y^2 \Rightarrow 9(x^2 + y^2) = y^2 - 12y + 36 \Rightarrow$
$9x^2 + 8y^2 + 12y - 36 = 0$

[15] $r = \dfrac{3}{2 + 2\cos\theta} \Rightarrow 2r + 2r\cos\theta = 3 \Rightarrow 2r + 2x = 3 \Rightarrow 2r = 3 - 2x \Rightarrow$
$4r^2 = 4x^2 - 12x + 9 \Rightarrow 4y^2 + 12x - 9 = 0$

[19] $r = \dfrac{6\csc\theta}{2\csc\theta + 3} \cdot \dfrac{\sin\theta}{\sin\theta} = \dfrac{6}{2 + 3\sin\theta} \Rightarrow 2r + 3y = 6 \Rightarrow 2r = 6 - 3y \Rightarrow$
$4r^2 = 36 - 36y + 9y^2 \Rightarrow 4x^2 - 5y^2 + 36y - 36 = 0$.
r is undefined when $\theta = 0$ or π. For the rectangular equation, these points correspond to $y = 0$ (or $r\sin\theta = 0$). Substituting $y = 0$ into the above rectangular equation yields $4x^2 = 36$ or $x = \pm 3$. $\therefore$ exclude $(\pm 3, 0)$

[21] $r = 2\sec\theta \Rightarrow r\cos\theta = 2 \Rightarrow x = 2$. Thus, $d = 2$ and since the directrix is on the right of the focus at the pole, we use "$+\cos\theta$". $r = \dfrac{2(\frac{1}{3})}{1 + \frac{1}{3}\cos\theta} \cdot \dfrac{3}{3} = \dfrac{2}{3 + \cos\theta}$.

[25] $r\cos\theta = 5 \Rightarrow x = 5 \Rightarrow d = 5$ and use "$+\cos\theta$". $r = \dfrac{5(1)}{1 + 1\cos\theta} = \dfrac{5}{1 + \cos\theta}$.

[27] For a parabola, $e = 1$. The vertex is 4 units on top of the focus at the pole so $d = 2(4)$ and we should use "$+\sin\theta$" in the denominator. $r = \dfrac{8}{1 + \sin\theta}$

Note: Exer. 29–32: With $r = f(\theta)$ and θ given, we will list r', r' evaluated at θ, and r evaluated at θ. Substitute these into the formula $m = \dfrac{r'\sin\theta + r\cos\theta}{r'\cos\theta - r\sin\theta}$ and simplify to obtain the value of m given.

[29] $r = \dfrac{12}{6 + 2\sin\theta} \Rightarrow r' = -12 \cdot \dfrac{2\cos\theta}{(6 + 2\sin\theta)^2} = -\dfrac{24\cos\theta}{(6 + 2\sin\theta)^2}$.

$r'(\frac{\pi}{6}) = -\frac{12}{49}\sqrt{3}$ and $r(\frac{\pi}{6}) = \frac{12}{7}$. $m = \dfrac{(-\frac{12}{49}\sqrt{3})\sin\frac{\pi}{6} + (\frac{12}{7})\cos\frac{\pi}{6}}{(-\frac{12}{49}\sqrt{3})\cos\frac{\pi}{6} - (\frac{12}{7})\sin\frac{\pi}{6}} =$

$$\frac{(-\sqrt{3})(1/2) + (7)(\sqrt{3}/2)}{(-\sqrt{3})(\sqrt{3}/2) - (7)(1/2)} = \frac{(6\sqrt{3})/2}{(-10)/2} = -\tfrac{3}{5}\sqrt{3}.$$

[33] $r = 2\sec\theta \Leftrightarrow r\cos\theta = 2 \Leftrightarrow x = 2$, a vertical line.

Using (13.11), $A = \displaystyle\int_{\pi/6}^{\pi/3} \tfrac{1}{2}(2\sec\theta)^2\, d\theta$.

[37] (a) Let V and C denote the vertex closest to the sun and the center of the ellipse, respectively. Let s denote the distance from V to the directrix to the left of V.

$d(O, V) = d(C, V) - d(C, O) = a - c = a - ea = a(1 - e)$.

Also, by (13.15), $\dfrac{d(O, V)}{s} = e \Rightarrow s = \dfrac{d(O, V)}{e} = \dfrac{a(1 - e)}{e}$. Now,

$d = s + d(O, V) = \dfrac{a(1 - e)}{e} + a(1 - e) = \dfrac{a(1 - e^2)}{e}$ and $de = a(1 - e^2)$.

Thus, the equation of the orbit is $r = \dfrac{(1 - e^2)a}{1 - e\cos\theta}$.

(b) The minimum distance occurs when $\theta = \pi$. $r_{\text{per}} = \dfrac{(1 - e^2)a}{1 - e(-1)} = a(1 - e)$.

The maximum distance occurs when $\theta = 0$. $r_{\text{aph}} = \dfrac{(1 - e^2)a}{1 - e(1)} = a(1 + e)$.

13.6 Review Exercises

[1] $x = \frac{1}{t} + 1 \Rightarrow t = \dfrac{1}{x - 1}$ and

$y = 2(x - 1) - \left(\dfrac{1}{x - 1}\right) = \dfrac{2(x^2 - 2x + 1) - 1}{x - 1} = \dfrac{2x^2 - 4x + 1}{x - 1}$. This is a rational function with a vertical asymptote at $x = 1$ and an oblique asymptote of $y = 2x - 2$. The graph has a minimum point at $(\frac{5}{4}, -\frac{7}{2})$ when $t = 4$ and then approaches the oblique asymptote as t approaches 0.

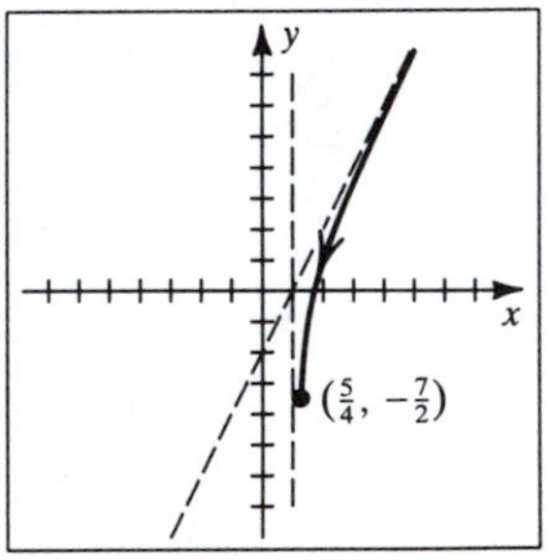

Figure 1

5 All of the curves are a portion of the circle $x^2 + y^2 = 16$.

C_1: $y = \sqrt{16 - t^2} = \sqrt{16 - x^2}$. Since $y = \sqrt{16 - t^2}$,

y must be nonnegative and we have the top half of the circle.

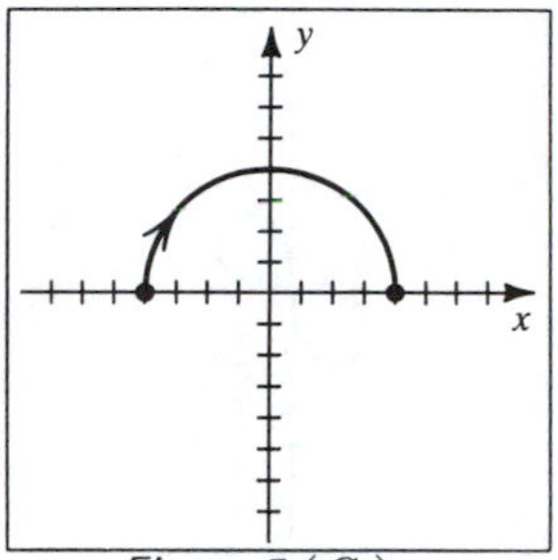

Figure 5 (C_1)

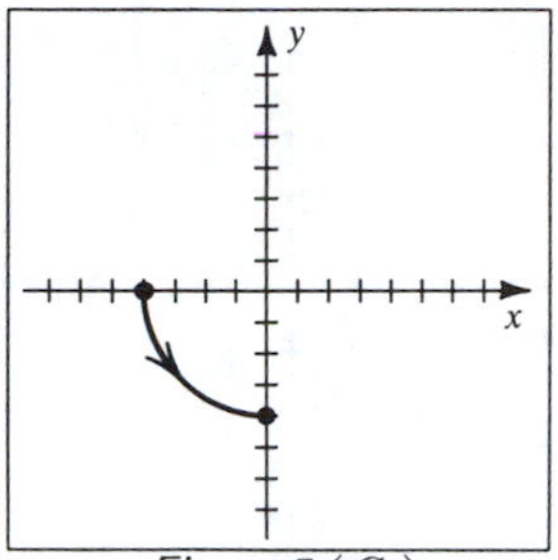

Figure 5 (C_2)

C_2: $x = -\sqrt{16 - t} = -\sqrt{16 - (-\sqrt{t})^2} = -\sqrt{16 - y^2}$.

This is the left half of the circle. Since $y = -\sqrt{t}$, y can only be nonpositive.

Hence we have only the third quadrant portion of the circle.

C_3: $x = 4\cos t$, $y = 4\sin t \Rightarrow \frac{x}{4} = \cos t$, $\frac{y}{4} = \sin t \Rightarrow \frac{x^2}{16} = \cos^2 t$, $\frac{y^2}{16} = \sin^2 t \Rightarrow$

$\frac{x^2}{16} + \frac{y^2}{16} = \cos^2 t + \sin^2 t = 1 \Rightarrow x^2 + y^2 = 16$. This is the entire circle.

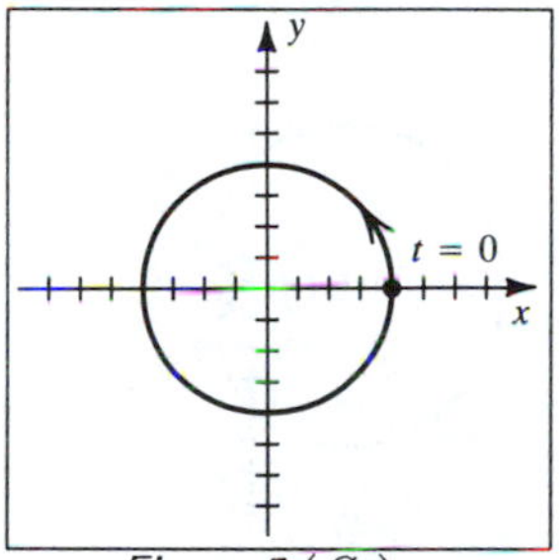

Figure 5 (C_3)

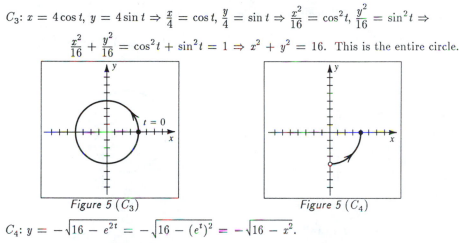

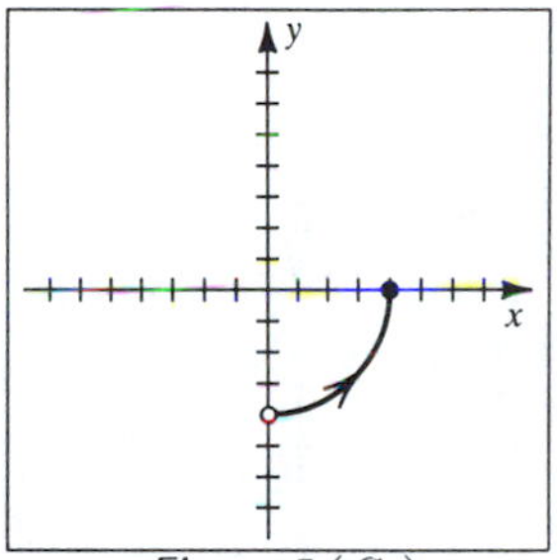

Figure 5 (C_4)

C_4: $y = -\sqrt{16 - e^{2t}} = -\sqrt{16 - (e^t)^2} = -\sqrt{16 - x^2}$.

This is the bottom half of the circle. Since e^t is positive, x takes on all positive

real values. Note that $(0, -4)$ is *not* included on the graph since $x \neq 0$.

7 (a) $\frac{dy}{dx} = \frac{dy/dt}{dx/dt} = \frac{6t^2 + 4}{2t} = \frac{3t^2 + 2}{t}$.

(b) Since $\frac{3t^2 + 2}{t} \neq 0$, there are no horizontal tangents.

There is a vertical tangent at $t = 0$.

(c) $\frac{d^2y}{dx^2} = \frac{dy'/dt}{dx/dt} = \frac{D_t\left[3t + \frac{2}{t}\right]}{2t} = \frac{3 - \frac{2}{t^2}}{2t} = \frac{3t^2 - 2}{2t^3}$.

[9] $r = -4\sin\theta \Rightarrow r^2 = -4r\sin\theta \Rightarrow x^2 + y^2 = -4y \Rightarrow$

$x^2 + y^2 + 4y + \underline{4} = \underline{4} \Rightarrow x^2 + (y + 2)^2 = 4$, a circle of radius 2 and center $(0, -2)$. See Example 1, §13.3 for a similar polar equation.

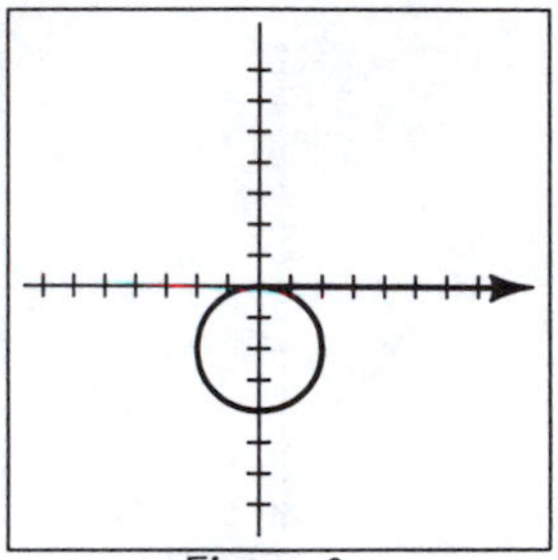

Figure 9

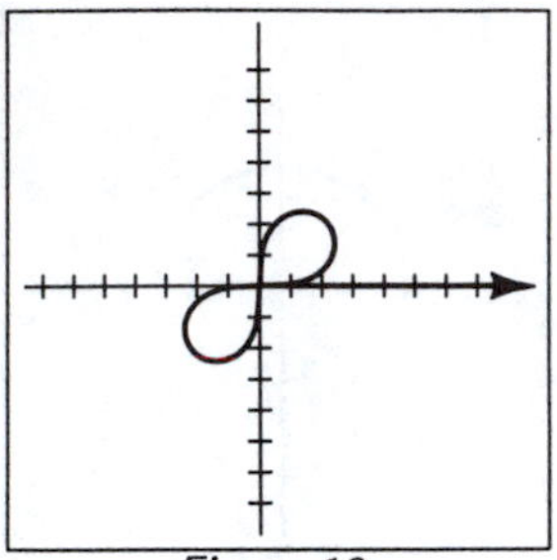

Figure 13

[13] See Exercise 19, §13.3 for a similar polar equation.

The equation is defined when $\sin 2\theta \geq 0$, or when $\theta \in [0, \frac{\pi}{2}] \cup [\pi, \frac{3\pi}{2}]$ for $0 \leq \theta \leq 2\pi$. $r^2 = 9\sin 2\theta$ is a lemniscate with loops in QI and QIII.

[15] $r = 3\sin 5\theta$ is a 5-leafed rose. One leaf is centered on the line $\theta = \frac{\pi}{2}$ and the others are equally spaced 72° apart $\left(\frac{360^\circ}{5} = 72^\circ\right)$.

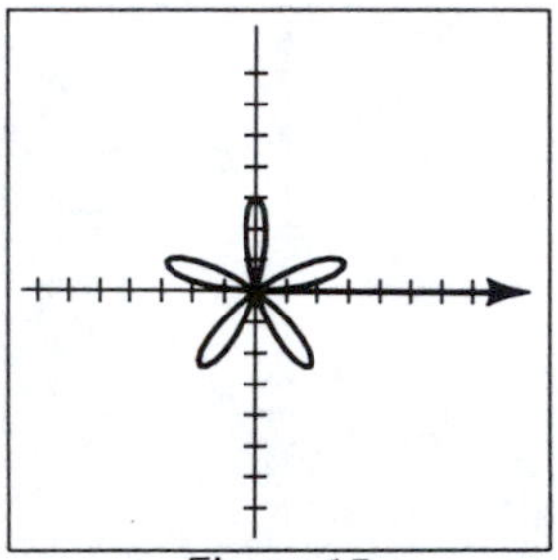

Figure 15

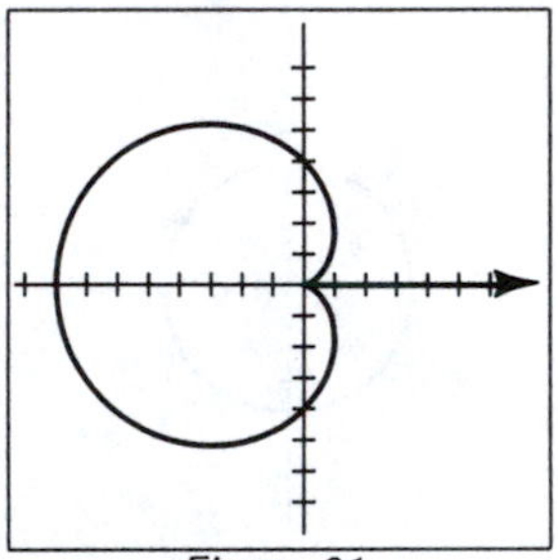

Figure 21

[21] $r = 4 - 4\cos\theta$ is a cardioid.

The "v" shaped portion of the curve occurs when $r = 0$, that is, $4 - 4\cos\theta = 0$, or, $\theta = 0$. See Exercise 7, §13.3, for a similar equation.

[25] This is an equation of a conic section. Dividing each term by 3 to obtain the form in (13.16) gives us

$$r = \frac{8}{3 + \cos\theta} = \frac{\frac{8}{3}}{1 + \frac{1}{3}\cos\theta} \Rightarrow$$

$e = \frac{1}{3} < 1$, an ellipse.

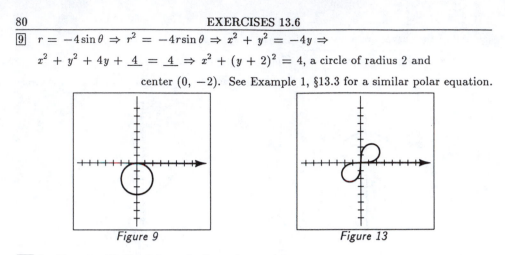

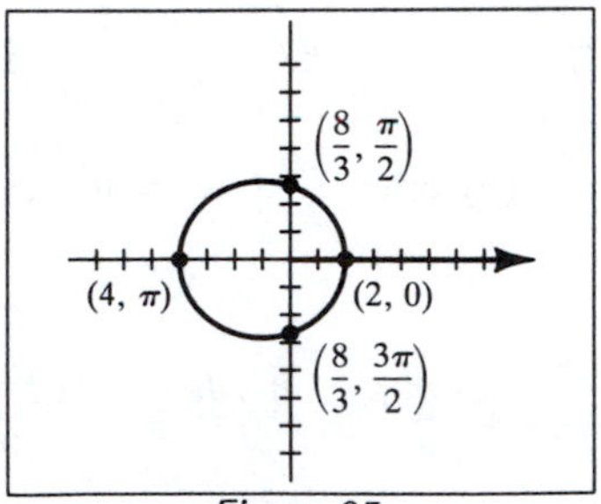

Figure 25

[27] $y^2 = 4x \Rightarrow r^2\sin^2\theta = 4r\cos\theta \Rightarrow r = \dfrac{4r\cos\theta}{r\sin^2\theta} = 4 \cdot \dfrac{\cos\theta}{\sin\theta} \cdot \dfrac{1}{\sin\theta} \Rightarrow r = 4\cot\theta\csc\theta$.

[31] $y^2 = x^2 - 2x \Rightarrow r^2\sin^2\theta = r^2\cos^2\theta - 2r\cos\theta \Rightarrow 2r\cos\theta = r^2(\cos^2\theta - \sin^2\theta) \Rightarrow$

$r(\cos^2\theta - \sin^2\theta) = 2\cos\theta \Rightarrow r\cos 2\theta = 2\cos\theta \Rightarrow r = 2\cos\theta\sec 2\theta.$

[33] $r^2 = \tan\theta \Rightarrow x^2 + y^2 = \frac{y}{x} \Rightarrow x^3 + xy^2 = y.$

[37] $\theta = \sqrt{3} \Rightarrow \tan^{-1}(\frac{y}{x}) = \sqrt{3} \Rightarrow \frac{y}{x} = \tan\sqrt{3} \Rightarrow y = (\tan\sqrt{3})\,x.$

Note that $\tan\sqrt{3} \approx -6.15$. This is a line through the origin making an angle of approximately 99.24° with the positive x-axis. The line is *not* $y = \frac{\pi}{3}x$.

[39] $r' = \dfrac{6\sin\theta}{(2 + 2\cos\theta)^2}$, $r'(\frac{\pi}{2}) = \frac{3}{2}$, $r(\frac{\pi}{2}) = \frac{3}{2}$,

$$m = \frac{(\frac{3}{2})\sin\frac{\pi}{2} + (\frac{3}{2})\cos\frac{\pi}{2}}{(\frac{3}{2})\cos\frac{\pi}{2} - (\frac{3}{2})\sin\frac{\pi}{2}} = -1 \text{ using (13.10).}$$

[43] Using (13.5), $\frac{dx}{dt} = 2\cos t$ and $\frac{dy}{dt} = 2\sin t\cos t \Rightarrow$

$$L = \int_0^{\pi/2}\sqrt{4\cos^2 t + 4\sin^2 t\cos^2 t}\,dt$$

$$= \int_0^{\pi/2}\sqrt{4\cos^2 t(1 + \sin^2 t)}\,dt$$

$$= \int_0^{\pi/2}\sqrt{1 + \sin^2 t}\,(2\cos t)\,dt = 2\int_0^1\sqrt{1 + u^2}\,du\ \{u = \sin t\}$$

$$= 2\left[\tfrac{u}{2}\sqrt{1 + u^2} + \tfrac{1}{2}\ln\left|u + \sqrt{1 + u^2}\right|\right]_0^1\ \{\text{Formula 21}\}$$

$$= \sqrt{2} + \ln(1 + \sqrt{2}) \approx 2.30.$$

[45] Using the formula in blue below (13.7), $\frac{dx}{dt} = 4t$ and $\frac{dy}{dt} = 4 \Rightarrow$

$$S = 2\pi\int_0^1 (2t^2 + 1)\sqrt{16t^2 + 16}\,dt = 16\pi\int_0^1 t^2\sqrt{t^2 + 1}\,dt + 8\pi\int_0^1\sqrt{t^2 + 1}\,dt$$

$$= 16\pi\left[\tfrac{t}{8}(1 + 2t^2)\sqrt{1 + t^2} - \tfrac{1}{8}\ln\left|t + \sqrt{1 + t^2}\right|\right]_0^1\ \{\text{Formula 22}\} +$$

$$8\pi\left[\tfrac{t}{2}\sqrt{1 + t^2} + \tfrac{1}{2}\ln\left|t + \sqrt{1 + t^2}\right|\right]_0^1\ \{\text{Formula 21}\}$$

$$= 16\pi\left[\tfrac{3}{8}\sqrt{2} - \tfrac{1}{8}\ln(1 + \sqrt{2})\right] + 8\pi\left[\tfrac{1}{2}\sqrt{2} + \tfrac{1}{2}\ln(1 + \sqrt{2})\right]$$

$$= 2\pi\left[5\sqrt{2} + \ln(1 + \sqrt{2})\right] \approx 49.97.$$

[47] See Exercise 19, §13.3 for a graph of this form with $a = 2$.

$$r^2 = a^2\cos 2\theta \Rightarrow 2rr' = -2a^2\sin 2\theta \Rightarrow r' = -\frac{a^2\sin 2\theta}{r} = -\frac{a\sin 2\theta}{\sqrt{\cos 2\theta}} \text{ on } [0, \tfrac{\pi}{4}].$$

Using symmetry and (13.14) with $r = a\sqrt{\cos 2\theta}$,

$$S = 2\int_0^{\pi/4} 2\pi\, a\sqrt{\cos 2\theta}\sin\theta\sqrt{a^2\cos 2\theta + \left(-\frac{a\sin 2\theta}{\sqrt{\cos 2\theta}}\right)^2}\,d\theta$$

$$= 4\pi a\int_0^{\pi/4}\sqrt{\cos 2\theta}\sqrt{a^2\cos 2\theta + \frac{a^2\sin^2 2\theta}{\cos 2\theta}}\sin\theta\,d\theta$$

$$= 4\pi a\int_0^{\pi/4}\sqrt{a^2\cos^2 2\theta + a^2\sin^2 2\theta}\sin\theta\,d\theta = 4\pi a^2\int_0^{\pi/4}\sin\theta\,d\theta$$

$$= 4\pi a^2\Big[-\cos\theta\Big]_0^{\pi/4} = -4\pi a^2\left(\tfrac{\sqrt{2}}{2} - 1\right) = 2\pi a^2(2 - \sqrt{2}) \approx 3.68a^2.$$

Chapter 14: Vectors and Surfaces

Exercises 14.1

1 A position vector always has its initial point located at the origin.

Since $\mathbf{a} = <2, 5>$, the associated position vector will have the terminal point located at (2, 5). Draw an arrow from (0, 0) to (2, 5).

By (14.1), $\mathbf{a} = <2, 5> \Rightarrow \|\mathbf{a}\| = \sqrt{2^2 + 5^2} = \sqrt{29}$.

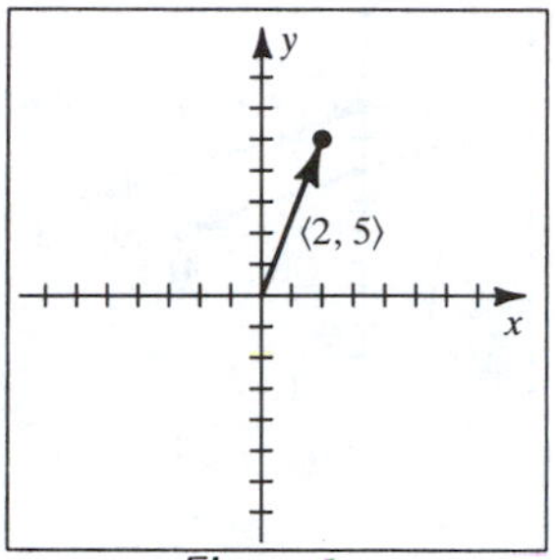

Figure 1

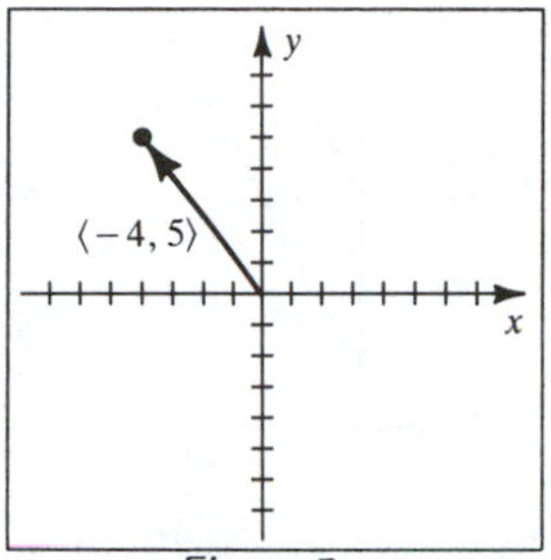

Figure 5

5 $\mathbf{a} = -4\mathbf{i} + 5\mathbf{j}$ is equivalent to $\mathbf{a} = <-4, 5>$. The position vector has initial point (0, 0) and terminal point (−4, 5). $\mathbf{a} = -4\mathbf{i} + 5\mathbf{j} \Rightarrow \|\mathbf{a}\| = \sqrt{(-4)^2 + 5^2} = \sqrt{41}$.

7 Using (14.2), $\mathbf{a} + \mathbf{b} = <2, -3> + <1, 4> = <2 + 1, -3 + 4> = <3, 1>$.

Using (14.6), $\mathbf{a} - \mathbf{b} = <2, -3> - <1, 4> = <2 - 1, -3 - 4> = <1, -7>$.

Using (14.3), $2\mathbf{a} = 2<2, -3> = <2 \cdot 2, 2 \cdot (-3)> = <4, -6>$.

Using (14.3), $-3\mathbf{b} = -3<1, 4> = <-3 \cdot 1, -3 \cdot 4> = <-3, -12>$.

Using (14.3) and (14.6), $4\mathbf{a} - 5\mathbf{b} = 4<2, -3> - 5<1, 4> =$

$$<8, -12> - <5, 20> = <8 - 5, -12 - 20> = <3, -32>.$$

11 Using (14.2), $\mathbf{a} + \mathbf{b} = (3\mathbf{i} + 2\mathbf{j}) + (-\mathbf{i} + 5\mathbf{j}) = [3 + (-1)]\mathbf{i} + (2 + 5)\mathbf{j} = 2\mathbf{i} + 7\mathbf{j}$.

Using (14.6), $\mathbf{a} - \mathbf{b} = (3\mathbf{i} + 2\mathbf{j}) - (-\mathbf{i} + 5\mathbf{j}) = [3 - (-1)]\mathbf{i} + (2 - 5)\mathbf{j} = 4\mathbf{i} - 3\mathbf{j}$.

Using (14.3), $2\mathbf{a} = 2(3\mathbf{i} + 2\mathbf{j}) = 2 \cdot 3\mathbf{i} + 2 \cdot 2\mathbf{j} = 6\mathbf{i} + 4\mathbf{j}$.

Using (14.3), $-3\mathbf{b} = -3(-\mathbf{i} + 5\mathbf{j}) = -3 \cdot (-1)\mathbf{i} + (-3) \cdot 5\mathbf{j} = 3\mathbf{i} - 15\mathbf{j}$.

Using (14.3) and (14.6), $4\mathbf{a} - 5\mathbf{b} = 4(3\mathbf{i} + 2\mathbf{j}) - 5(-\mathbf{i} + 5\mathbf{j}) =$

$$(12\mathbf{i} + 8\mathbf{j}) - (-5\mathbf{i} + 25\mathbf{j}) = (12 + 5)\mathbf{i} + (8 - 25)\mathbf{j} = 17\mathbf{i} - 17\mathbf{j}.$$

13 From the figure, we see that $\mathbf{a} = <2, 0>$ since it has initial point (0, 0) and terminal point (2, 0). Similarly, $\mathbf{b} = <-1, 0>$.

$$\mathbf{a} + \mathbf{b} = <2, 0> + <-1, 0> = <1, 0> = -<-1, 0> = -\mathbf{b}.$$

As an alternate solution, note that the vector $<1, 0>$ has the same direction as $\mathbf{a}$ and has $\frac{1}{2}$ of the magnitude of $\mathbf{a}$. Hence, $\frac{1}{2}\mathbf{a}$ is another vector which is equal to $\mathbf{a} + \mathbf{b}$.

[17] $\mathbf{b} + \mathbf{d} = <-1, 0> + <0, -1> = <-1, -1>$.

Since $<-1, -1>$ has the opposite direction of $\mathbf{e}$ and has $\frac{1}{2}$ the magnitude of $\mathbf{e}$,

we may represent $<-1, -1>$ as $-\frac{1}{2}<2, 2>$, or, equivalently, $-\frac{1}{2}\mathbf{e}$.

[19] By (14.7) with $P(1, -4)$ and $Q(5, 3)$, $\mathbf{a} = <5 - 1, 3 - (-4)> = <4, 7>$.

$\overrightarrow{PQ}$ has initial point $(1, -4)$ and terminal point $(5, 3)$.

$\mathbf{a}$ is the equivalent vector for $\overrightarrow{PQ}$ that has initial point at the origin.

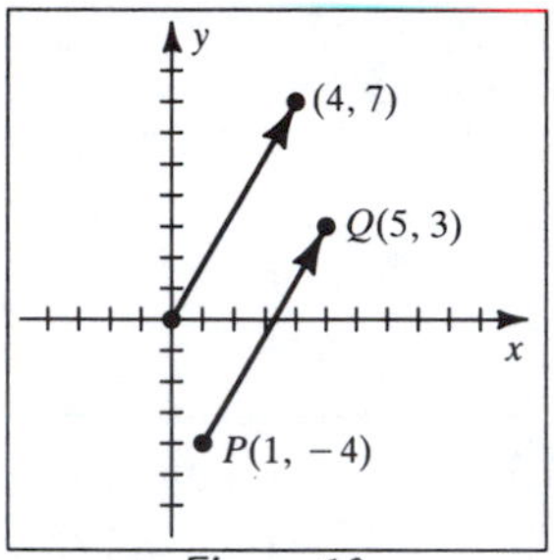

Figure 19

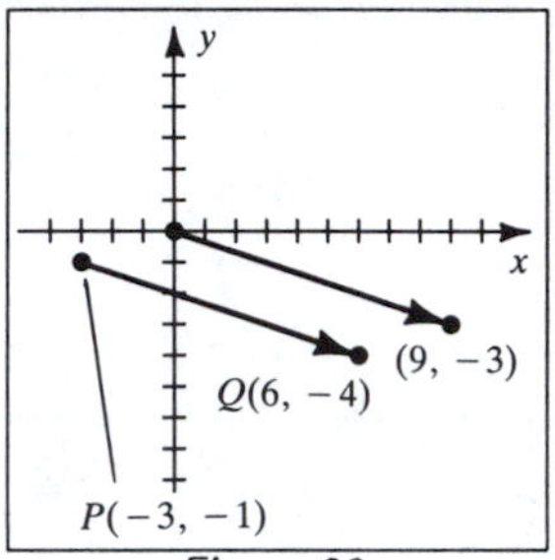

Figure 23

[23] As in Exercise 19 with $P(-3, -1)$ and $Q(6, -4)$,

$$\mathbf{a} = <6 - (-3), -4 - (-1)> = <9, -3>.$$

Note: In Exercises 25–28, let $\mathbf{u}$ denote the unit vector in the direction of $\mathbf{a}$.

[25] (a) By (14.12), the unit vector in the direction of $\mathbf{a}$ is $\frac{1}{\|\mathbf{a}\|}\mathbf{a}$.

$$\mathbf{a} = -8\mathbf{i} + 15\mathbf{j} \Rightarrow \|\mathbf{a}\| = \sqrt{(-8)^2 + 15^2} = \sqrt{64 + 225} = \sqrt{289} = 17.$$

$$\mathbf{u} = \frac{1}{\|\mathbf{a}\|}\mathbf{a} = \tfrac{1}{17}(-8\mathbf{i} + 15\mathbf{j}) = -\tfrac{8}{17}\mathbf{i} + \tfrac{15}{17}\mathbf{j}.$$

(b) The unit vector $-\mathbf{u}$ has the opposite direction of $\mathbf{u}$ and hence,

the opposite direction of $\mathbf{a}$. $-\mathbf{u} = -(-\tfrac{8}{17}\mathbf{i} + \tfrac{15}{17}\mathbf{j}) = \tfrac{8}{17}\mathbf{i} - \tfrac{15}{17}\mathbf{j}$.

[27] (a) As in Exercise 25, $\mathbf{a} = <2, -5> \Rightarrow$

$$\|\mathbf{a}\| = \sqrt{2^2 + (-5)^2} = \sqrt{29} \text{ and } \mathbf{u} = \frac{\mathbf{a}}{\|\mathbf{a}\|} = \left< \frac{2}{\sqrt{29}}, -\frac{5}{\sqrt{29}} \right>.$$

(b) $-\mathbf{u} = -\left(\left< \frac{2}{\sqrt{29}}, -\frac{5}{\sqrt{29}} \right>\right) = \left< -\frac{2}{\sqrt{29}}, \frac{5}{\sqrt{29}} \right>$

[29] (a) By (14.9), $2\mathbf{a}$ has twice the magnitude of $\mathbf{a}$ since $\|2\mathbf{a}\| = 2\|\mathbf{a}\|$.

By (14.8), $2\mathbf{a}$ has the same direction as $\mathbf{a}$. Hence, $2<-6, 3> = <-12, 6>$.

(b) As in part (a), $\frac{1}{2}<-6, 3> = <-3, \frac{3}{2}>$.

[31] The unit vector $\frac{\mathbf{a}}{\|\mathbf{a}\|}$ has the same direction as $\mathbf{a}$. $6\left(\frac{\mathbf{a}}{\|\mathbf{a}\|}\right)$

will have a magnitude of 6 by (14.9). $\mathbf{a} = 4\mathbf{i} - 7\mathbf{j} \Rightarrow \|\mathbf{a}\| = \sqrt{16 + 49} = \sqrt{65}$.

$$6\left(\frac{\mathbf{a}}{\|\mathbf{a}\|}\right) = 6\left(\frac{4}{\sqrt{65}}\mathbf{i} - \frac{7}{\sqrt{65}}\mathbf{j}\right) = \frac{24}{\sqrt{65}}\mathbf{i} - \frac{42}{\sqrt{65}}\mathbf{j}.$$

[33] (a) $\mathbf{a} = 3\mathbf{i} - 4\mathbf{j} \Rightarrow \|\mathbf{a}\| = \sqrt{9 + 16} = 5$. By (14.9), $\|c\mathbf{a}\| = |c|\,\|\mathbf{a}\| = |c|\,(5)$.

$$\|c\mathbf{a}\| = 3 \Rightarrow 5\,|c| = 3 \Rightarrow |c| = \tfrac{3}{5} \Rightarrow c = \pm\tfrac{3}{5}.$$

(b) None, since $\|c\mathbf{a}\| \geq 0$ for all $c\mathbf{a}$.

(c) $\|c\mathbf{a}\| = 0 \Rightarrow c\mathbf{a} = \mathbf{0} \Rightarrow c = 0$, since $\mathbf{a} \neq \mathbf{0}$.

[35] Refer to Figure 14.20 and substitute 35° for 30° and 50 for 200.

Then, the horizontal component $= \mathbf{v}_1 = 50\cos 35° \approx 40.96$ and

the vertical component $= \mathbf{v}_2 = 50\sin 35° \approx 28.68$.

[39] Let **v** denote the plane's air velocity, **r** the plane's ground velocity, and **w** the wind velocity. Refer to Example 7. The true course and ground speed will be given by **v** + **w**. **v** has direction 150° measured clockwise from the positive y-axis or 60° *below* the positive x-axis.

$\mathbf{v} = \langle 300\cos(-60°),\ 300\sin(-60°)\rangle$

$= \langle 150,\ -150\sqrt{3}\rangle$. **w** has direction 60° measured clockwise from the positive y-axis or 30° *above* the positive x-axis. $\mathbf{w} = \langle 30\cos 30°,\ 30\sin 30°\rangle = \langle 15\sqrt{3},\ 15\rangle$.

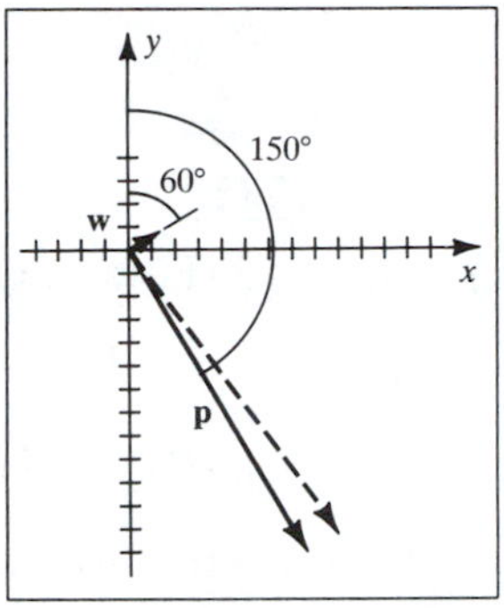

Figure 39

Thus, $\mathbf{r} = \mathbf{v} + \mathbf{w} = \langle 150 + 15\sqrt{3},\ 15 - 150\sqrt{3}\rangle$ and

$\|\mathbf{r}\| = \sqrt{90{,}900} \approx 301.5$ mi/hr. Since **r** has its terminal point in the fourth quadrant,

we compute $\tan^{-1}\left(\dfrac{15 - 150\sqrt{3}}{150 + 15\sqrt{3}}\right) \approx -54.3°$, indicating that **r** makes an angle of 54.3°

below the positive x-axis. Hence, the true course is $90° + 54.3° = 144.3°$.

[43] $p\langle 3, -1\rangle + q\langle 4, 3\rangle = \langle -6, -11\rangle \Rightarrow \langle 3p, -p\rangle + \langle 4q, 3q\rangle = \langle -6, -11\rangle$

$\Rightarrow \langle 3p + 4q, -p + 3q\rangle = \langle -6, -11\rangle \Rightarrow \begin{cases} 3p + 4q = -6 & (E_1) \\ -p + 3q = -11 & (E_2) \end{cases}$

$E_1 + 3E_2 \Rightarrow 13q = -39 \Rightarrow q = -3$ and $p = 2$.

[47] $\|\mathbf{r} - \mathbf{r}_0\| = c \Leftrightarrow \|\mathbf{r} - \mathbf{r}_0\|^2 = c^2 \Leftrightarrow \|\langle x, y\rangle - \langle x_0, y_0\rangle\|^2 = c^2 \Leftrightarrow$

$\|\langle x - x_0, y - y_0\rangle\|^2 = c^2 \Leftrightarrow \left(\sqrt{(x - x_0)^2 + (y - y_0)^2}\right)^2 = c^2 \Leftrightarrow$

$(x - x_0)^2 + (y - y_0)^2 = c^2$. This is a circle with center (x_0, y_0) and radius c.

[49] $\mathbf{a} + (\mathbf{b} + \mathbf{c}) = \langle a_1, a_2\rangle + (\langle b_1, b_2\rangle + \langle c_1, c_2\rangle)$

$= \langle a_1, a_2\rangle + \langle b_1 + c_1, b_2 + c_2\rangle$

$= \langle a_1 + (b_1 + c_1), a_2 + (b_2 + c_2)\rangle$

$= \langle a_1 + b_1 + c_1, a_2 + b_2 + c_2\rangle$

$= \langle (a_1 + b_1) + c_1, (a_2 + b_2) + c_2\rangle$

$= \langle a_1 + b_1, a_2 + b_2\rangle + \langle c_1, c_2\rangle$

$= (\langle a_1, a_2\rangle + \langle b_1, b_2\rangle) + \langle c_1, c_2\rangle = (\mathbf{a} + \mathbf{b}) + \mathbf{c}$

53 $(p + q)\mathbf{a} = (p + q)\langle a_1, a_2\rangle$

$= \langle (p + q)a_1, (p + q)a_2\rangle$

$= \langle pa_1 + qa_1, pa_2 + qa_2\rangle$

$= \langle pa_1, pa_2\rangle + \langle qa_1, qa_2\rangle$

$= p\langle a_1, a_2\rangle + q\langle a_1, a_2\rangle = p\mathbf{a} + q\mathbf{a}$

Exercises 14.2

1 (a) Using (14.13) with $A(2, 4, -5)$ and $B(4, -2, 3)$,

$$d(A, B) = \sqrt{(4 - 2)^2 + (-2 - 4)^2 + (3 + 5)^2} = \sqrt{4 + 36 + 64} = \sqrt{104}.$$

(b) Using (14.14), $M_{AB} = \left(\frac{2 + 4}{2}, \frac{4 + (-2)}{2}, \frac{-5 + 3}{2}\right) = (3, 1, -1)$.

(c) Using the 3-D generalization of (14.7), the vector in V_3 that corresponds to $\overrightarrow{AB}$ is $\langle 4 - 2, -2 - 4, 3 - (-5)\rangle = \langle 2, -6, 8\rangle$.

5 (a) $A(1, 0, 0)$, $B(0, 1, 1) \Rightarrow d(A, B) = \sqrt{(0 - 1)^2 + (1 - 0)^2 + (1 - 0)^2} = \sqrt{3}$.

(b) $M_{AB} = \left(\frac{1 + 0}{2}, \frac{0 + 1}{2}, \frac{0 + 1}{2}\right) = \left(\frac{1}{2}, \frac{1}{2}, \frac{1}{2}\right)$.

(c) $\langle 0 - 1, 1 - 0, 1 - 0\rangle = \langle -1, 1, 1\rangle$.

7 For the following exercises, we use the 3-D generalizations of (14.5)–(14.9).

(a) $\mathbf{a} + \mathbf{b} = \langle -2, 6, 1\rangle + \langle 3, -3, -1\rangle =$

$\langle -2 + 3, 6 + (-3), 1 + (-1)\rangle = \langle 1, 3, 0\rangle$.

(b) $\mathbf{a} - \mathbf{b} = \langle -2, 6, 1\rangle - \langle 3, -3, -1\rangle =$

$\langle -2 - 3, 6 - (-3), 1 - (-1)\rangle = \langle -5, 9, 2\rangle$.

(c) $5\mathbf{a} - 4\mathbf{b} = 5\langle -2, 6, 1\rangle - 4\langle 3, -3, -1\rangle$

$= \langle -10, 30, 5\rangle - \langle 12, -12, -4\rangle$

$= \langle -10 - 12, 30 - (-12), 5 - (-4)\rangle = \langle -22, 42, 9\rangle$.

(d) $\|\mathbf{a}\| = \sqrt{(-2)^2 + 6^2 + 1^2} = \sqrt{4 + 36 + 1} = \sqrt{41}$.

(e) $\|-3\mathbf{a}\| = |-3|\|\mathbf{a}\| = 3\|\mathbf{a}\| = 3\sqrt{41}$.

9 (a) $\mathbf{a} + \mathbf{b} = (3\mathbf{i} - 4\mathbf{j} + 2\mathbf{k}) + (\mathbf{i} + 2\mathbf{j} - 5\mathbf{k}) =$

$(3 + 1)\mathbf{i} + (-4 + 2)\mathbf{j} + (2 - 5)\mathbf{k} = 4\mathbf{i} - 2\mathbf{j} - 3\mathbf{k}$.

(b) $\mathbf{a} - \mathbf{b} = (3\mathbf{i} - 4\mathbf{j} + 2\mathbf{k}) - (\mathbf{i} + 2\mathbf{j} - 5\mathbf{k}) =$

$(3 - 1)\mathbf{i} + (-4 - 2)\mathbf{j} + (2 + 5)\mathbf{k} = 2\mathbf{i} - 6\mathbf{j} + 7\mathbf{k}$.

(c) $5\mathbf{a} - 4\mathbf{b} = 5(3\mathbf{i} - 4\mathbf{j} + 2\mathbf{k}) - 4(\mathbf{i} + 2\mathbf{j} - 5\mathbf{k})$

$= (15\mathbf{i} - 20\mathbf{j} + 10\mathbf{k}) - (4\mathbf{i} + 8\mathbf{j} - 20\mathbf{k})$

$= (15 - 4)\mathbf{i} + (-20 - 8)\mathbf{j} + (10 + 20)\mathbf{k} = 11\mathbf{i} - 28\mathbf{j} + 30\mathbf{k}$.

(d) $\|\mathbf{a}\| = \sqrt{3^2 + (-4)^2 + 2^2} = \sqrt{9 + 16 + 4} = \sqrt{29}$.

(e) $\|-3\mathbf{a}\| = |-3|\|\mathbf{a}\| = 3\|\mathbf{a}\| = 3\sqrt{29}$.

[13] $2\mathbf{a} = 2<2, 3, 4> = <4, 6, 8>$. $\quad -3\mathbf{b} = -3<1, -2, 2> = <-3, 6, -6>$.

$\mathbf{a} + \mathbf{b} = <2, 3, 4> + <1, -2, 2> = <3, 1, 6>$.

$\mathbf{a} - \mathbf{b} = <2, 3, 4> - <1, -2, 2> = <1, 5, 2>$.

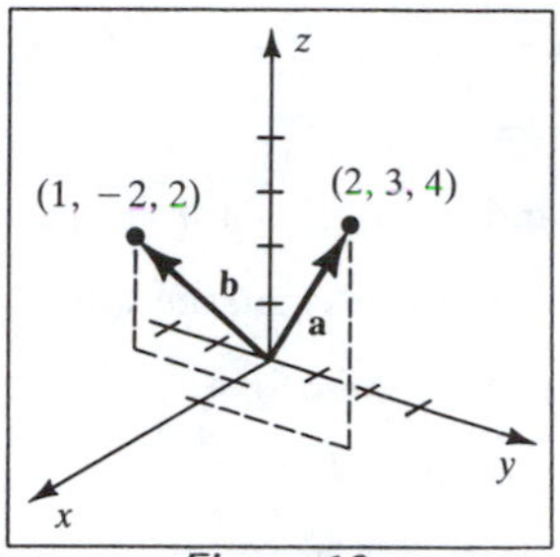

Figure 13a

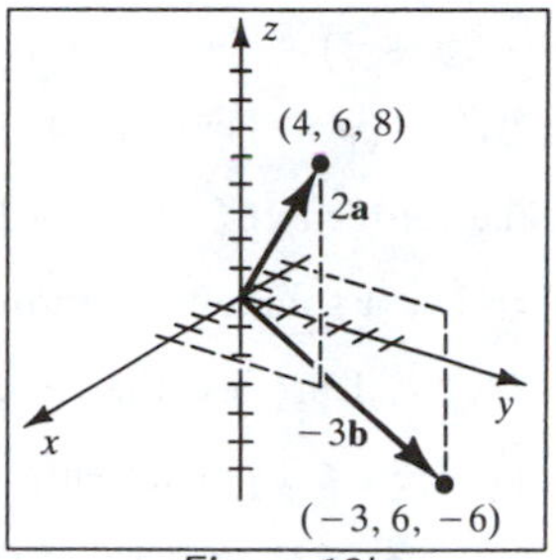

Figure 13b

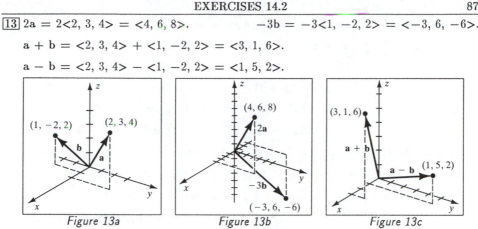

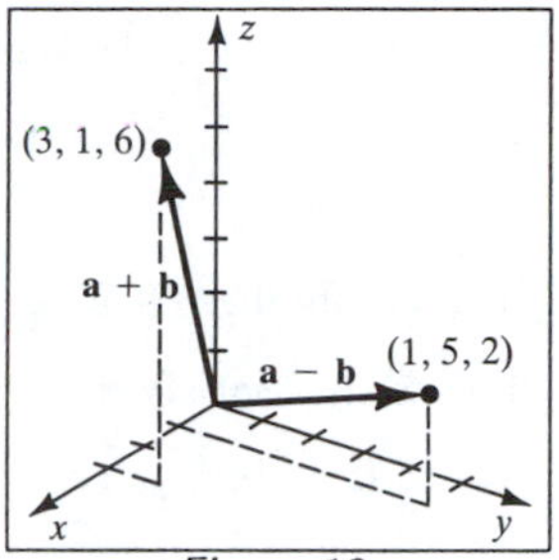

Figure 13c

[15] Using the 3-D generalization of (14.12), $\mathbf{a} = 2<-2, 5, -1> = <-4, 10, -2> \Rightarrow$

$\|\mathbf{a}\| = \sqrt{(-4)^2 + 10^2 + (-2)^2} = \sqrt{16 + 100 + 4} = \sqrt{120} = 2\sqrt{30}$.

$$\mathbf{u} = \frac{\mathbf{a}}{\|\mathbf{a}\|} = \frac{1}{\sqrt{30}}<-2, 5, -1>.$$

[17] (a) Since $2\mathbf{a}$ has the same direction as $\mathbf{a}$ and $\|2\mathbf{a}\| = |2|\|\mathbf{a}\| = 2\|\mathbf{a}\|$,

the required vector is $2\mathbf{a} = 2(14\mathbf{i} - 15\mathbf{j} + 6\mathbf{k}) = 28\mathbf{i} - 30\mathbf{j} + 12\mathbf{k}$.

(b) Since $-\frac{1}{3}\mathbf{a}$ has the opposite direction as $\mathbf{a}$ and $\left\|-\frac{1}{3}\mathbf{a}\right\| = \left|-\frac{1}{3}\right|\|\mathbf{a}\| = \frac{1}{3}\|\mathbf{a}\|$,

the required vector is $-\frac{1}{3}\mathbf{a} = -\frac{1}{3}(14\mathbf{i} - 15\mathbf{j} + 6\mathbf{k}) = -\frac{14}{3}\mathbf{i} + 5\mathbf{j} - 2\mathbf{k}$.

(c) $\frac{\mathbf{a}}{\|\mathbf{a}\|}$ is a unit vector in the direction of $\mathbf{a}$ and $2\frac{\mathbf{a}}{\|\mathbf{a}\|}$ has a magnitude of 2.

Since $\|\mathbf{a}\| = \sqrt{14^2 + (-15)^2 + 6^2} = \sqrt{457}$, $\frac{2\mathbf{a}}{\|\mathbf{a}\|} = \frac{2}{\sqrt{457}}(14\mathbf{i} - 15\mathbf{j} + 6\mathbf{k})$.

[19] Using (14.15), $C(3, -1, 2)$, $r = 3 \Rightarrow (x - 3)^2 + (y + 1)^2 + (z - 2)^2 = 3^2 = 9$.

[23] (a) The point $(-2, 4, 0)$ is in the xy-plane and the point $(-2, 4, -6)$ is 6 units below the xy-plane. If the sphere is to touch the xy-plane at only one point (since it is tangent), it must have a radius of 6.

Thus, an equation is $(x + 2)^2 + (y - 4)^2 + (z + 6)^2 = 6^2 = 36$.

(b) Tangent to xz-plane $\Rightarrow r = 4$; $(x + 2)^2 + (y - 4)^2 + (z + 6)^2 = 4^2 = 16$

(c) Tangent to yz-plane $\Rightarrow r = 2$; $(x + 2)^2 + (y - 4)^2 + (z + 6)^2 = 2^2 = 4$

[25] The center of the sphere is the midpoint of AB. $A(1, 4, -2)$ and $B(-7, 1, 2) \Rightarrow$

$M_{AB} = \left(\frac{1 + (-7)}{2}, \frac{4 + 1}{2}, \frac{-2 + 2}{2}\right) = \left(-3, \frac{5}{2}, 0\right)$. The radius is

$\frac{1}{2} \cdot d(A, B) = \frac{1}{2}\sqrt{(-7 - 1)^2 + (1 - 4)^2 + [2 - (-2)]^2} = \frac{1}{2}\sqrt{64 + 9 + 16} = \frac{1}{2}\sqrt{89}$.

An equation is $(x + 3)^2 + (y - \frac{5}{2})^2 + z^2 = (\frac{1}{2}\sqrt{89})^2 = \frac{89}{4}$.

[27] The sphere must be a distance of 1 unit from each plane.

Therefore, the center of the sphere must be $(1, 1, 1)$,

and an equation is $(x - 1)^2 + (y - 1)^2 + (z - 1)^2 = 1^2 = 1$.

[29] As in Example 7, we will complete the squares for x, y, and z.

$$x^2 + y^2 + z^2 + 4x - 2y + 2z + 2 = 0$$
$$(x^2 + 4x) + (y^2 - 2y) + (z^2 + 2z) = -2$$
$$(x^2 + 4x + 4) + (y^2 - 2y + 1) + (z^2 + 2z + 1) = -2 + 4 + 1 + 1$$
$$(x + 2)^2 + (y - 1)^2 + (z + 1)^2 = 4$$

The center is $C(-2, 1, -1)$ and the radius is $r = \sqrt{4} = 2$.

[35] The graph of $x^2 + y^2 + z^2 = 1$ is a sphere with radius 1 and center at the origin. Points that satisfy $x^2 + y^2 + z^2 < 1$ are less than 1 unit from the origin. Thus, $R = \{(x, y, z): x^2 + y^2 + z^2 \leq 1\}$ represents all points inside or on the sphere of radius 1 with center at the origin.

[39] $x^2 + y^2 = 25$ is a circle of radius 5 in the xy-plane. In 3-D, $x^2 + y^2 = 25$ is a right circular cylinder. Since $|z| \leq 3$, the z coordinates on the cylinder will vary between -3 and 3. Thus, $R = \{(x, y, z): x^2 + y^2 \leq 25, |z| \leq 3\}$ represents all points inside or on a cylindrical region of radius 5 and altitude 6 with center at the origin and axis along the z-axis.

[43] Following the hint, let $A = (x_1, y_1, 0)$, $B = (x_2, y_2, 0)$, $C = (x_3, y_3, 0)$, and $D = (0, 0, z_4)$. Since P_1 is the midpoint of segment AB,

$P_1 = \left(\frac{x_1 + x_2}{2}, \frac{y_1 + y_2}{2}, 0\right)$, Similarly, $P_2 = \left(\frac{x_2}{2}, \frac{y_2}{2}, \frac{z_4}{2}\right)$,

$P_3 = \left(\frac{x_2 + x_3}{2}, \frac{y_2 + y_3}{2}, 0\right)$, $P_1' = \left(\frac{x_3}{2}, \frac{y_3}{2}, \frac{z_4}{2}\right)$, $P_2' = \left(\frac{x_1 + x_3}{2}, \frac{y_1 + y_3}{2}, 0\right)$, and

$P_3' = \left(\frac{x_1}{2}, \frac{y_1}{2}, \frac{z_4}{2}\right)$. P_1P_1', P_2P_2', and P_3P_3' all have the common midpoint,

$P = \left(\frac{x_1 + x_2 + x_3}{4}, \frac{y_1 + y_2 + y_3}{4}, \frac{z_4}{4}\right)$, and hence, are bisected by P.

Exercises 14.3

[1] Using (14.16), $\mathbf{a} \cdot \mathbf{b} = \langle -2, 3, 1 \rangle \cdot \langle 7, 4, 5 \rangle =$

$$(-2)(7) + (3)(4) + (1)(5) = -14 + 12 + 5 = 3.$$

[3] (a) $\mathbf{a} \cdot (\mathbf{b} + \mathbf{c}) = \langle -2, 3, 1 \rangle \cdot (\langle 7, 4, 5 \rangle + \langle 1, -5, 2 \rangle)$

$= \langle -2, 3, 1 \rangle \cdot \langle 8, -1, 7 \rangle$

$= (-2)(8) + (3)(-1) + (1)(7) = -16 - 3 + 7 = -12$

(b) $\mathbf{a} \cdot \mathbf{b} + \mathbf{a} \cdot \mathbf{c} = 3 + (-15) = -12$

[7] Using (14.25) with $\|\mathbf{c}\| = \sqrt{1^2 + (-5)^2 + 2^2} = \sqrt{30}$, $\text{comp}_{\mathbf{c}}\,\mathbf{b} = \mathbf{b} \cdot \frac{1}{\|\mathbf{c}\|}\mathbf{c} =$

$$\langle 7, 4, 5 \rangle \cdot \tfrac{1}{\sqrt{30}}\langle 1, -5, 2 \rangle = 7\left(\tfrac{1}{\sqrt{30}}\right) + 4\left(\tfrac{-5}{\sqrt{30}}\right) + 5\left(\tfrac{2}{\sqrt{30}}\right) = -\tfrac{3}{\sqrt{30}}.$$

[11] Using (14.20), $\cos\theta = \frac{\mathbf{a} \cdot \mathbf{b}}{\|\mathbf{a}\|\|\mathbf{b}\|} = \frac{-3}{\sqrt{89}\sqrt{6}} = \frac{-3}{\sqrt{534}} \Rightarrow \theta = \arccos\frac{-3}{\sqrt{534}} \approx 97.5^\circ$.

[15] Since $\mathbf{a}\cdot\mathbf{b} = (3\mathbf{i} - 2\mathbf{j} + \mathbf{k})\cdot(4\mathbf{i} + 5\mathbf{j} - 2\mathbf{k}) = (3)(4) + (-2)(5) + (1)(-2) = 12 - 10 - 2 = 0$, $\mathbf{a}$ and $\mathbf{b}$ are orthogonal by Theorem (14.21).

[17] By (14.21), $\mathbf{a}$ and $\mathbf{b}$ are orthogonal if and only if $\mathbf{a}\cdot\mathbf{b} = 0$. We must find all values of c such that $\mathbf{a}\cdot\mathbf{b} = 0$. $\mathbf{a}\cdot\mathbf{b} = 0 \Rightarrow \langle c, -2, 3\rangle\cdot\langle c, c, -5\rangle = 0 \Rightarrow$
$c^2 - 2c - 15 = 0 \Rightarrow (c + 3)(c - 5) = 0 \Rightarrow c = -3, 5.$

[19] $P(3, -2, -1)$, $Q(1, 5, 4) \Rightarrow \overrightarrow{PQ} = \langle 1 - 3, 5 - (-2), 4 - (-1)\rangle = \langle -2, 7, 5\rangle$.
$R(2, 0, -6)$, $S(-4, 1, 5) \Rightarrow \overrightarrow{RS} = \langle -4 - 2, 1 - 0, 5 - (-6)\rangle = \langle -6, 1, 11\rangle$.
$\overrightarrow{PQ}\cdot\overrightarrow{RS} = \langle -2, 7, 5\rangle\cdot\langle -6, 1, 11\rangle = (-2)(-6) + (7)(1) + (5)(11) = 74.$

[21] From Exercise 19, $\overrightarrow{PQ} = \langle -2, 7, 5\rangle$, $\overrightarrow{RS} = \langle -6, 1, 11\rangle$, and $\overrightarrow{PQ}\cdot\overrightarrow{RS} = 74$.

$$\text{By (14.20), } \cos\theta = \frac{\overrightarrow{PQ}\cdot\overrightarrow{RS}}{\|\overrightarrow{PQ}\|\|\overrightarrow{RS}\|} = \frac{74}{\sqrt{78}\sqrt{158}} \Rightarrow \theta = \cos^{-1}\frac{37}{\sqrt{3081}} \approx 48.2^\circ.$$

[23] $P(3, -2, -1)$ and $S(-4, 1, 5) \Rightarrow \overrightarrow{PS} = \langle -4 - 3, 1 - (-2), 5 - (-1)\rangle = \langle -7, 3, 6\rangle$. $Q(1, 5, 4)$ and $R(2, 0, -6) \Rightarrow \overrightarrow{QR} = \langle 2 - 1, 0 - 5, -6 - 4\rangle = \langle 1, -5, -10\rangle$. $\|\overrightarrow{QR}\| = \sqrt{1 + 25 + 100} = \sqrt{126}$. By (14.25),

$$\text{comp}_{\overrightarrow{QR}}\,\overrightarrow{PS} = \overrightarrow{PS}\cdot\frac{1}{\|\overrightarrow{QR}\|}\overrightarrow{QR} = \langle -7, 3, 6\rangle\cdot\frac{1}{\sqrt{126}}\langle 1, -5, -10\rangle = \frac{-82}{\sqrt{126}}.$$

[25] We will apply (14.26). The point of application moves along $\overrightarrow{PQ} = \langle 2 - 4, 4 - 0, 0 - (-7)\rangle = \langle -2, 4, 7\rangle$. Since $\mathbf{a} = -\mathbf{i} + 5\mathbf{j} - 3\mathbf{k}$ is the constant force, the work W done is $W = \mathbf{a}\cdot\overrightarrow{PQ} = \langle -1, 5, -3\rangle\cdot\langle -2, 4, 7\rangle = 1.$

[27] The point of application moves along the vector $\langle 0 - 0, -1 - 2, 0 - 0\rangle = \langle 0, -3, 0\rangle$, or, equivalently, $-3\mathbf{j}$. The unit vector in the direction of $\mathbf{a} = \mathbf{i} + \mathbf{j} + \mathbf{k}$ is $\frac{1}{\|\mathbf{a}\|}\mathbf{a} = \frac{1}{\sqrt{3}}(\mathbf{i} + \mathbf{j} + \mathbf{k})$. The force vector is $\frac{4\mathbf{a}}{\|\mathbf{a}\|} = \frac{4}{\sqrt{3}}(\mathbf{i} + \mathbf{j} + \mathbf{k})$.
By (14.26), the work $W = \frac{4}{\sqrt{3}}(\mathbf{i} + \mathbf{j} + \mathbf{k})\cdot(-3\mathbf{j}) = -4\sqrt{3} \approx 6.93$ ft-lb.

[31] Refer to *Figure 31*. Following the hint,
$\mathbf{v}_2 + \overrightarrow{PA} = \mathbf{v}_1 \Rightarrow \overrightarrow{PA} = \mathbf{v}_1 - \mathbf{v}_2$ and
$\mathbf{v}_2 + \overrightarrow{PB} = -\mathbf{v}_1 \Rightarrow \overrightarrow{PB} = -\mathbf{v}_1 - \mathbf{v}_2$.
We must show that $\overrightarrow{PA}$ and $\overrightarrow{PB}$ are orthogonal, or, equivalently, $\overrightarrow{PA}\cdot\overrightarrow{PB} = 0$.

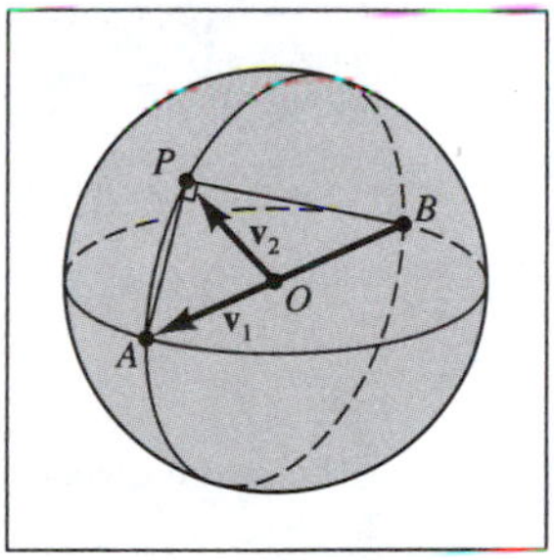

Figure 31

$\overrightarrow{PA}\cdot\overrightarrow{PB} = (\mathbf{v}_1 - \mathbf{v}_2)\cdot(-\mathbf{v}_1 - \mathbf{v}_2)$
$= (\mathbf{v}_1 - \mathbf{v}_2)\cdot(-\mathbf{v}_1) + (\mathbf{v}_1 - \mathbf{v}_2)\cdot(-\mathbf{v}_2)$ {(14.17)(iii)}
$= -\mathbf{v}_1\cdot(\mathbf{v}_1 - \mathbf{v}_2) - \mathbf{v}_2\cdot(\mathbf{v}_1 - \mathbf{v}_2)$ {(14.17)(ii)}
$= -\mathbf{v}_1\cdot\mathbf{v}_1 + \mathbf{v}_1\cdot\mathbf{v}_2 - \mathbf{v}_2\cdot\mathbf{v}_1 + \mathbf{v}_2\cdot\mathbf{v}_2$ {(14.17)(iii)}
$= -\mathbf{v}_1\cdot\mathbf{v}_1 + \mathbf{v}_1\cdot\mathbf{v}_2 - \mathbf{v}_1\cdot\mathbf{v}_2 + \mathbf{v}_2\cdot\mathbf{v}_2$ {(14.17)(ii)} (continued)

$= -(\mathbf{v}_1 \cdot \mathbf{v}_1) + (\mathbf{v}_2 \cdot \mathbf{v}_2)$ {(14.17)(iv)}

$= -\|\mathbf{v}_1\|^2 + \|\mathbf{v}_2\|^2$ {(14.17)(i)}

$= -r^2 + r^2 = 0.$

Note that $\mathbf{v}_1$ and $\mathbf{v}_2$ both correspond to radii of the sphere and therefore, have magnitudes equal to r. Thus, $\overrightarrow{PA}$ and $\overrightarrow{PB}$ are orthogonal and APB is a right Δ.

[35] (a) The angle between $\mathbf{a} = a_1\mathbf{i} + a_2\mathbf{j} + a_3\mathbf{k}$ and $\mathbf{i}$ can be determined using (14.19).

$a_1 = \mathbf{a} \cdot \mathbf{i} = \|\mathbf{a}\|\|\mathbf{i}\|\cos\alpha = \|\mathbf{a}\|(1)\cos\alpha \Rightarrow \cos\alpha = \dfrac{a_1}{\|\mathbf{a}\|}$. Similarly,

$a_2 = \mathbf{a} \cdot \mathbf{j} = \|\mathbf{a}\|\|\mathbf{j}\|\cos\beta = \|\mathbf{a}\|(1)\cos\beta \Rightarrow \cos\beta = \dfrac{a_2}{\|\mathbf{a}\|}$ and

$a_3 = \mathbf{a} \cdot \mathbf{k} = \|\mathbf{a}\|\|\mathbf{k}\|\cos\gamma = \|\mathbf{a}\|(1)\cos\gamma \Rightarrow \cos\gamma = \dfrac{a_3}{\|\mathbf{a}\|}$.

(b) $\cos^2\alpha + \cos^2\beta + \cos^2\gamma = \dfrac{a_1^2 + a_2^2 + a_3^2}{\|\mathbf{a}\|^2} = \dfrac{\|\mathbf{a}\|^2}{\|\mathbf{a}\|^2} = 1.$

[37] $d = (l^2 + m^2 + n^2)^{1/2} = (k^2\cos^2\alpha + k^2\cos^2\beta + k^2\cos^2\gamma)^{1/2}$

$= \left[k^2(\cos^2\alpha + \cos^2\beta + \cos^2\gamma)\right]^{1/2} = [k^2(1)]^{1/2} = k.$

Since $d = k$, $l = k\cos\alpha = d\cos\alpha \Rightarrow \cos\alpha = l/d$.

Similarly, $\cos\beta = m/d$, and $\cos\gamma = n/d$.

[41] $(c\mathbf{a}) \cdot \mathbf{b} = (c\langle a_1, a_2, a_3\rangle) \cdot \langle b_1, b_2, b_3\rangle$

$= \langle ca_1, ca_2, ca_3\rangle \cdot \langle b_1, b_2, b_3\rangle$

$= ca_1b_1 + ca_2b_2 + ca_3b_3$

$= c(a_1b_1 + a_2b_2 + a_3b_3) = c(\mathbf{a} \cdot \mathbf{b})$

$c(\mathbf{a} \cdot \mathbf{b}) = c(\langle a_1, a_2, a_3\rangle \cdot \langle b_1, b_2, b_3\rangle)$

$= c(a_1b_1 + a_2b_2 + a_3b_3)$

$= ca_1b_1 + ca_2b_2 + ca_3b_3$

$= a_1(cb_1) + a_2(cb_2) + a_3(cb_3)$

$= \langle a_1, a_2, a_3\rangle \cdot \langle cb_1, cb_2, cb_3\rangle$

$= \mathbf{a} \cdot (c\langle b_1, b_2, b_3\rangle) = \mathbf{a} \cdot (c\mathbf{b})$

[43] $c\mathbf{a} + d\mathbf{b} = p\mathbf{a} + q\mathbf{b} \Rightarrow c\mathbf{a} - p\mathbf{a} = q\mathbf{b} - d\mathbf{b} \Rightarrow (c - p)\mathbf{a} = (q - d)\mathbf{b}$.

Assume $c \neq p$. Then $\mathbf{a} = \dfrac{q - d}{c - p}\mathbf{b}$. Since $\dfrac{q - d}{c - p}$ is also a scalar, $\mathbf{a}$ and $\mathbf{b}$ must either have the same direction or opposite direction, that is, $\mathbf{a}$ and $\mathbf{b}$ are parallel,

a contradiction. Hence $c = p$. Similarly, $q = d$.

[49] Using the hint, $\mathbf{a} = \mathbf{b} + (\mathbf{a} - \mathbf{b}) \Rightarrow \|\mathbf{a}\| = \|\mathbf{b} + (\mathbf{a} - \mathbf{b})\| \Rightarrow$

$\|\mathbf{a}\| \leq \|\mathbf{b}\| + \|\mathbf{a} - \mathbf{b}\|$ {by (14.23)} $\Rightarrow \|\mathbf{a}\| - \|\mathbf{b}\| \leq \|\mathbf{a} - \mathbf{b}\|$.

Exercises 14.4

1 See (14.27) and Example 1.

$$\mathbf{a} \times \mathbf{b} = \begin{vmatrix} \mathbf{i} & \mathbf{j} & \mathbf{k} \\ 1 & -2 & 3 \\ 2 & 1 & -4 \end{vmatrix} = \begin{vmatrix} -2 & 3 \\ 1 & -4 \end{vmatrix}\mathbf{i} - \begin{vmatrix} 1 & 3 \\ 2 & -4 \end{vmatrix}\mathbf{j} + \begin{vmatrix} 1 & -2 \\ 2 & 1 \end{vmatrix}\mathbf{k} =$$

$$5\mathbf{i} + 10\mathbf{j} + 5\mathbf{k} = \langle 5, 10, 5 \rangle$$

5 $$\mathbf{a} \times \mathbf{b} = \begin{vmatrix} \mathbf{i} & \mathbf{j} & \mathbf{k} \\ 5 & -6 & -1 \\ 3 & 0 & 1 \end{vmatrix} = \begin{vmatrix} -6 & -1 \\ 0 & 1 \end{vmatrix}\mathbf{i} - \begin{vmatrix} 5 & -1 \\ 3 & 1 \end{vmatrix}\mathbf{j} + \begin{vmatrix} 5 & -6 \\ 3 & 0 \end{vmatrix}\mathbf{k} =$$

$$-6\mathbf{i} - 8\mathbf{j} + 18\mathbf{k}$$

11 Since $\mathbf{a} \times \mathbf{b} = \begin{vmatrix} \mathbf{i} & \mathbf{j} & \mathbf{k} \\ -6 & -10 & 4 \\ 3 & 5 & -2 \end{vmatrix} = \begin{vmatrix} -10 & 4 \\ 5 & -2 \end{vmatrix}\mathbf{i} - \begin{vmatrix} -6 & 4 \\ 3 & -2 \end{vmatrix}\mathbf{j} + \begin{vmatrix} -6 & -10 \\ 3 & 5 \end{vmatrix}\mathbf{k} =$

$0\mathbf{i} - 0\mathbf{j} + 0\mathbf{k} = \mathbf{0}$, $\mathbf{a}$ and $\mathbf{b}$ are parallel by Corollary (14.31).

13 $\mathbf{b} \times \mathbf{c} = \langle 4, 12, 5 \rangle$, $\mathbf{a} \times (\mathbf{b} \times \mathbf{c}) = \langle 2, 0, -1 \rangle \times \langle 4, 12, 5 \rangle = \langle 12, -14, 24 \rangle$.

$\mathbf{a} \times \mathbf{b} = \langle 1, 3, 2 \rangle$, $(\mathbf{a} \times \mathbf{b}) \times \mathbf{c} = \langle 1, 3, 2 \rangle \times \langle 1, -2, 4 \rangle = \langle 16, -2, -5 \rangle$.

Note that $\mathbf{a} \times (\mathbf{b} \times \mathbf{c}) \neq (\mathbf{a} \times \mathbf{b}) \times \mathbf{c}$ and hence,

the associative property does not hold for the vector product.

Note: In Exercises 15–18, let c be a nonzero scalar.

15 (a) $P(1, -1, 2)$ and $Q(0, 3, -1) \Rightarrow \overrightarrow{PQ} = \langle -1, 4, -3 \rangle$.

$P(1, -1, 2)$ and $R(3, -4, 1) \Rightarrow \overrightarrow{PR} = \langle 2, -3, -1 \rangle$.

The vectors $\overrightarrow{PQ}$ and $\overrightarrow{PR}$ both lie in the plane determined by P, Q, and R.

By (14.29), $\overrightarrow{PQ} \times \overrightarrow{PR} = \langle 13, 7, 5 \rangle$ is orthogonal to both $\overrightarrow{PQ}$ and $\overrightarrow{PR}$.

Thus, any vector perpendicular to the plane will be a nonzero scalar multiple of $\langle 13, 7, 5 \rangle$, that is, the vectors will have the form $c(\overrightarrow{PR} \times \overrightarrow{PQ}) = c\langle 13, 7, 5 \rangle$.

(b) As in Example 2,

$$A = \tfrac{1}{2}\left\| \overrightarrow{PR} \times \overrightarrow{PQ} \right\| = \tfrac{1}{2}\|\langle 13, 7, 5 \rangle\| = \tfrac{1}{2}\sqrt{13^2 + 7^2 + 5^2} = \tfrac{1}{2}\sqrt{243} = \tfrac{9}{2}\sqrt{3}.$$

19 In this exercise, Q and R lie on a line and P is the point not on the line, whereas in Example 3, P and Q lie on a line and R is a point not on the line. Thus, the formula becomes $d = \dfrac{\left\|\overrightarrow{QR} \times \overrightarrow{QP}\right\|}{\left\|\overrightarrow{QR}\right\|} = \dfrac{\|\langle -3, -1, 1 \rangle \times \langle 1, -4, -3 \rangle\|}{\|\langle -3, -1, 1 \rangle\|} = \dfrac{\|\langle 7, -8, 13 \rangle\|}{\|\langle -3, -1, 1 \rangle\|}$

$$= \frac{\sqrt{282}}{\sqrt{11}} \approx 5.06.$$

[21] $\mathbf{a} \times \mathbf{b} = \langle a_2 b_3 - a_3 b_2,\ b_1 a_3 - a_1 b_3,\ a_1 b_2 - a_2 b_1 \rangle$ and

$$\mathbf{b} \times \mathbf{c} = \langle b_2 c_3 - b_3 c_2,\ c_1 b_3 - b_1 c_3,\ b_1 c_2 - b_2 c_1 \rangle.$$

$$\mathbf{a} \cdot (\mathbf{b} \times \mathbf{c}) = a_1 b_2 c_3 - a_1 b_3 c_2 + a_2 c_1 b_3 - a_2 b_1 c_3 + a_3 b_1 c_2 - a_3 b_2 c_1 \text{ (1) and}$$

$$(\mathbf{a} \times \mathbf{b}) \cdot \mathbf{c} = a_2 b_3 c_1 - a_3 b_2 c_1 + b_1 a_3 c_2 - a_1 b_3 c_2 + a_1 b_2 c_3 - a_2 b_1 c_3 \text{ (2).}$$

Comparing terms, we see that $\mathbf{a} \cdot (\mathbf{b} \times \mathbf{c}) = (\mathbf{a} \times \mathbf{b}) \cdot \mathbf{c}$.

Also, $\begin{vmatrix} a_1 & a_2 & a_3 \\ b_1 & b_2 & b_3 \\ c_1 & c_2 & c_3 \end{vmatrix}$

$$= \begin{vmatrix} b_2 & b_3 \\ c_2 & c_3 \end{vmatrix} a_1 - \begin{vmatrix} b_1 & b_3 \\ c_1 & c_3 \end{vmatrix} a_2 + \begin{vmatrix} b_1 & b_2 \\ c_1 & c_2 \end{vmatrix} a_3$$

$$= (b_2 c_3 - c_2 b_3)\, a_1 - (b_1 c_3 - c_1 b_3)\, a_2 + (b_1 c_2 - c_1 b_2)\, a_3$$

$$= a_1 b_2 c_3 - a_1 c_2 b_3 - a_2 b_1 c_3 + a_2 c_1 b_3 + a_3 b_1 c_2 - a_3 c_1 b_2$$

$$= \mathbf{a} \cdot (\mathbf{b} \times \mathbf{c}) \text{ \{see (1)\}} = (\mathbf{a} \times \mathbf{b}) \cdot \mathbf{c} \text{ \{see (2)\}}.$$

[23] $\overrightarrow{AB} = \langle 1, -1, 3 \rangle$, $\overrightarrow{AC} = \langle 2, -3, 2 \rangle$, and $\overrightarrow{AD} = \langle 3, -4, 1 \rangle$.

Thus, by Example 4 and Exercise 21, the volume of the box is

$$V = \left| (\overrightarrow{AB} \times \overrightarrow{AC}) \cdot \overrightarrow{AD} \right| = \begin{vmatrix} 1 & -1 & 3 \\ 2 & -3 & 2 \\ 3 & -4 & 1 \end{vmatrix} = |1(5) + 1(-4) + 3(1)| = |4| = 4.$$

[25] By (14.29), $\mathbf{a} \times \mathbf{b}$ is orthogonal to $\mathbf{b}$. By (14.21), $(\mathbf{a} \times \mathbf{b}) \cdot \mathbf{b} = 0$.

[29] $(\mathbf{a} + \mathbf{b}) \times \mathbf{c} = \begin{vmatrix} \mathbf{i} & \mathbf{j} & \mathbf{k} \\ a_1 + b_1 & a_2 + b_2 & a_3 + b_3 \\ c_1 & c_2 & c_3 \end{vmatrix} =$

$$\big[(a_2 + b_2)\, c_3 - (a_3 + b_3)\, c_2\big]\mathbf{i} - \big[(a_1 + b_1)\, c_3 - (a_3 + b_3)\, c_1\big]\mathbf{j} + \big[(a_1 + b_1)\, c_2 - (a_2 + b_2)\, c_1\big]\mathbf{k}.$$

$\mathbf{a} \times \mathbf{c} = \begin{vmatrix} \mathbf{i} & \mathbf{j} & \mathbf{k} \\ a_1 & a_2 & a_3 \\ c_1 & c_2 & c_3 \end{vmatrix} = (a_2 c_3 - c_2 a_3)\mathbf{i} - (a_1 c_3 - c_1 a_3)\mathbf{j} + (a_1 c_2 - c_1 a_2)\mathbf{k}.$

$\mathbf{b} \times \mathbf{c} = \begin{vmatrix} \mathbf{i} & \mathbf{j} & \mathbf{k} \\ b_1 & b_2 & b_3 \\ c_1 & c_2 & c_3 \end{vmatrix} = (b_2 c_3 - c_2 b_3)\mathbf{i} - (b_1 c_3 - c_1 b_3)\mathbf{j} + (b_1 c_2 - c_1 b_2)\mathbf{k}.$

Hence, $(\mathbf{a} \times \mathbf{c}) + (\mathbf{b} \times \mathbf{c}) = \big[(a_2 c_3 + b_2 c_3) - (a_3 c_2 + b_3 c_2)\big]\mathbf{i}$

$$- \big[(a_1 c_3 + b_1 c_3) - (a_3 c_1 + b_3 c_1)\big]\mathbf{j} + \big[(a_1 c_2 + b_1 c_2) - (a_2 c_1 + b_2 c_1)\big]\mathbf{k}$$

$$= \big[(a_2 + b_2)c_3 - (a_3 + b_3)c_2\big]\mathbf{i} - \big[(a_1 + b_1)c_3 - (a_3 + b_3)c_1\big]\mathbf{j} + \big[(a_1 + b_1)c_2 - (a_2 + b_2)c_1\big]\mathbf{k} = (\mathbf{a} + \mathbf{b}) \times \mathbf{c}.$$

[33] Using (14.33)(vi), LHS $= \mathbf{a} \times (\mathbf{b} \times \mathbf{c}) + \mathbf{b} \times (\mathbf{c} \times \mathbf{a}) + \mathbf{c} \times (\mathbf{a} \times \mathbf{b}) =$

$(\mathbf{a} \cdot \mathbf{c})\mathbf{b} - (\mathbf{a} \cdot \mathbf{b})\mathbf{c} + (\mathbf{b} \cdot \mathbf{a})\mathbf{c} - (\mathbf{b} \cdot \mathbf{c})\mathbf{a} + (\mathbf{c} \cdot \mathbf{b})\mathbf{a} - (\mathbf{c} \cdot \mathbf{a})\mathbf{b}$. Rearranging terms

gives us $(\mathbf{a} \cdot \mathbf{c} - \mathbf{c} \cdot \mathbf{a})\mathbf{b} + (\mathbf{b} \cdot \mathbf{a} - \mathbf{a} \cdot \mathbf{b})\mathbf{c} + (\mathbf{c} \cdot \mathbf{b} - \mathbf{b} \cdot \mathbf{c})\mathbf{a} = 0\,\mathbf{b} + 0\,\mathbf{c} + 0\,\mathbf{a} = \mathbf{0}$.

37 $(\mathbf{a} \times \mathbf{b}) \cdot (\mathbf{b} \times \mathbf{c}) \times (\mathbf{c} \times \mathbf{a})$

$= (\mathbf{a} \times \mathbf{b}) \cdot [(\mathbf{b} \times \mathbf{c}) \times (\mathbf{c} \times \mathbf{a})]$ {group terms}

$= (\mathbf{a} \times \mathbf{b}) \cdot \{ [(\mathbf{b} \times \mathbf{c}) \cdot \mathbf{a}]\mathbf{c} - [(\mathbf{b} \times \mathbf{c}) \cdot \mathbf{c}]\mathbf{a} \}$ {(14.33)(vi)}

$= (\mathbf{a} \times \mathbf{b}) \cdot \{ [(\mathbf{b} \times \mathbf{c}) \cdot \mathbf{a}]\mathbf{c} - 0 \}$ {since $\mathbf{b} \times \mathbf{c} \cdot \mathbf{c} = 0$ by Exercise 25}

$= (\mathbf{a} \times \mathbf{b}) \cdot [(\mathbf{b} \times \mathbf{c}) \cdot \mathbf{a}]\mathbf{c}$ {simplify}

$= [(\mathbf{b} \times \mathbf{c}) \cdot \mathbf{a}] (\mathbf{a} \times \mathbf{b}) \cdot \mathbf{c}$ {$(\mathbf{b} \times \mathbf{c}) \cdot \mathbf{a}$ is a scalar}

$= (\mathbf{b} \times \mathbf{c} \cdot \mathbf{a})(\mathbf{a} \cdot \mathbf{b} \times \mathbf{c})$ {(14.33)(v)}

$= (\mathbf{a} \cdot \mathbf{b} \times \mathbf{c})(\mathbf{a} \cdot \mathbf{b} \times \mathbf{c}) = (\mathbf{a} \cdot \mathbf{b} \times \mathbf{c})^2$

Note that in the expression $\mathbf{a} \cdot \mathbf{b} \times \mathbf{c}$, the cross product must be performed first since $\mathbf{a} \cdot \mathbf{b}$ is a scalar and we cannot cross a scalar and a vector.

Exercises 14.5

1 By (14.34), a parametric equation for the line through $P(4, 2, -3)$ and parallel to $\mathbf{a} = \langle \frac{1}{3}, 2, \frac{1}{2} \rangle$ is $x = 4 + \frac{1}{3}t$, $y = 2 + 2t$, and $z = -3 + \frac{1}{2}t$.

5 $\mathbf{a} = \overrightarrow{P_1P_2} = \langle 2 - 5, 6 + 2, 1 - 4 \rangle = \langle -3, 8, -3 \rangle$. Thus, by (14.34), parametric equations for the line are $x = 5 - 3t$, $y = -2 + 8t$, and $z = 4 - 3t$. The line intersects the xy-plane when $z = 0$. $z = 4 - 3t = 0 \Rightarrow t = \frac{4}{3}$. Substituting $t = \frac{4}{3}$ in the equation for x and y gives us $x = 5 - 3(\frac{4}{3}) = 1$ and $y = -2 + 8(\frac{4}{3}) = \frac{26}{3}$. $(1, \frac{26}{3}, 0)$ is the point of intersection with the xy-plane. Similarly, $y = 0 \Rightarrow t = \frac{1}{4} \Rightarrow xz$-plane intersection point is $(\frac{17}{4}, 0, \frac{13}{4})$, and $x = 0 \Rightarrow t = \frac{5}{3} \Rightarrow yz$-plane intersection point is $(0, \frac{34}{3}, -1)$.

9 By (14.34), l is parallel to $\mathbf{a} = \langle -3, 1, 9 \rangle$. Using the point $P(-6, 4, -3)$ and a with the parameter s, we have $x = -6 - 3s$, $y = 4 + s$, and $z = -3 + 9s$.

11 Let t_0 and v_0 denote the values of t and v for the intersection point. We will equate x, y, and z values for the two lines and solve for t_0 and v_0.

$$\begin{aligned} 1 + 2t_0 &= 4 - v_0 & 2t_0 + v_0 &= 3 \\ 1 - 4t_0 &= -1 + 6v_0 \Rightarrow & 4t_0 + 6v_0 &= 2 \Rightarrow t_0 = 2,\ v_0 = -1 \\ 5 - t_0 &= 4 + v_0 & t_0 + v_0 &= 1 \end{aligned}$$

Substituting into the equation of the first line gives $x = 1 + 2t_0 = 1 + 2(2) = 5$, $y = 1 - 4t_0 = 1 - 4(2) = -7$, and $z = 5 - t_0 = 5 - 2 = 3$. We also must check to make sure that $t_0 = 2$ and $v_0 = -1$ substituted into the equation of the second line gives the same values for x, y, and z. The lines intersect at $(5, -7, 3)$.

15 By (14.34) the first line is parallel to $\mathbf{a} = \langle -2, 3, 5 \rangle$ and the second line is parallel to $\mathbf{b} = \langle 4, 4, 1 \rangle$. By (14.20) and (14.35),

$$\cos\theta = \frac{\mathbf{a} \cdot \mathbf{b}}{\|\mathbf{a}\|\|\mathbf{b}\|} = \frac{-8 + 12 + 5}{\sqrt{38}\sqrt{33}} \Rightarrow \theta = \cos^{-1}\left(\frac{9}{\sqrt{38}\sqrt{33}}\right) \approx 75^\circ \text{ and } 180^\circ - \theta.$$

Note: The answer in the first printing was for $x = -1 + t$ instead of $x = -1 + 4t$.

[19] (a) All points in a plane parallel to the xy-plane must have a constant z-coordinate value. Since the plane passes through $P(6, -7, 4)$, the equation of the plane must be $z = 4$.

(b) As in part (a), $x = 6$. (c) As in part (a), $y = -7$.

[21] By (14.36), an equation of the plane through $P(-11, 4, -2)$ with normal vector $\mathbf{a} = 6\mathbf{i} - 5\mathbf{j} - \mathbf{k}$ is $6(x + 11) - 5(y - 4) - 1(z + 2) = 0$, or $6x - 5y - z = -84$.

[25] The trace of the plane $x + 4y - 5z = 8$ in the xz-plane is the set of all points that are in the intersection of the plane and the xz-plane. This will occur when the y-coordinates of the points in the given plane are 0. Letting $y = 0$ to find the trace of $x + 4y - 5z = 8$ in the xz-plane gives us $x - 5z = 8$. Thus, the desired plane has equation $x - 5z = 8$ when $y = 0$. Therefore, an equation of the desired plane is $x + ky - 5z = 8$ for some constant k. Substituting $x = -4$, $y = 1$, and $z = 6$ yields $-4 + k(1) - 5(6) = 8$, or, equivalently, $k = 42$. ★ $x + 42y - 5z = 8$

[27] $\overrightarrow{PQ} = \langle -2, 2, -1 \rangle$ and $\overrightarrow{PR} = \langle 0, -2, -1 \rangle$ are two vectors in a plane. We must find a vector $\mathbf{a}$ that is orthogonal to both of these vectors. Using the cross product, $\mathbf{a} = \overrightarrow{PQ} \times \overrightarrow{PR} = \langle -4, -2, 4 \rangle$. $\mathbf{a}$ is normal to the plane, so an equation is $-4(x - 1) - 2(y - 1) + 4(z - 3) = 0$, or $2x + y - 2z = -3$.

[31] The plane has normal vector $\mathbf{a} = \langle 2, 1, 0 \rangle$, and hence, $\mathbf{a}$ lies in the xy-plane. Thus, the plane is orthogonal to the xy-plane and passes through $(3, 0, 0)$ and $(0, 6, 0)$.

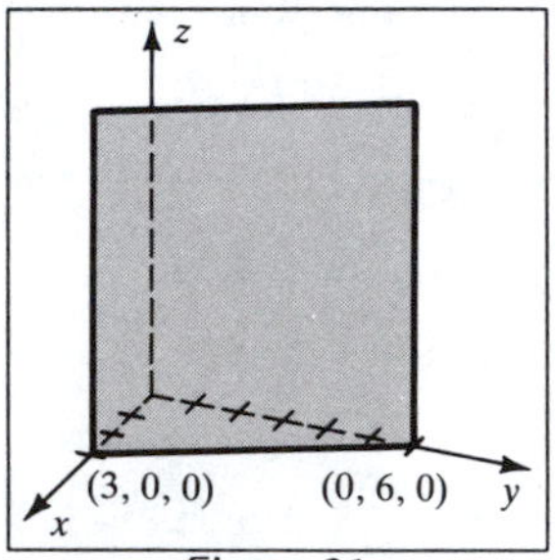

Figure 31

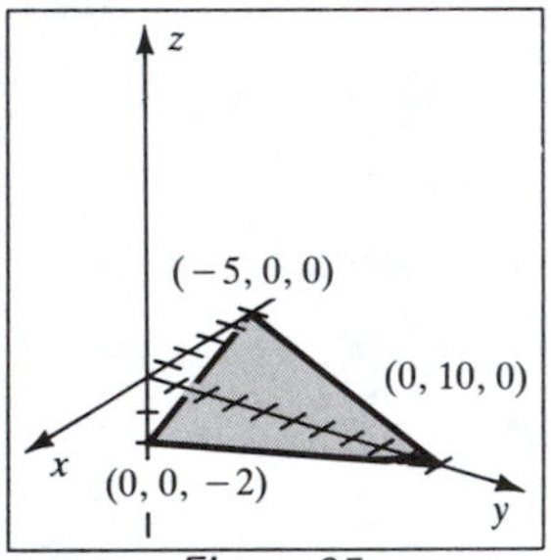

Figure 35

[35] To find the x-intercept, let y and z equal 0. This gives us $2x + 10 = 0$, or $x = -5$. Similarly, the y-intercept is 10 and the z-intercept is -2. Thus, the plane passes through the points $(-5, 0, 0)$, $(0, 10, 0)$, and $(0, 0, -2)$. Plot these points and draw the resulting triangle.

[37] In the xz-plane, the plane passes through the points $(5, 0, 0)$ and $(0, 0, 5)$. We may consider this as a two-dimensional problem with the points $(5, 0)$ and $(0, 5)$ in the xz-plane. Using the point-slope form of a line, we obtain $z - 5 = \frac{5-0}{0-5}(x - 0)$, or $x + z = 5$. Since the plane is orthogonal to the xz-plane, the y-coordinate coefficient is zero.

[41] A normal vector for the plane with equation $4x - y + 3z - 7 = 0$ is $<4, -1, 3>$.

By (14.36), an equation for the plane through $P(1, 2, -3)$ is

$$4(x - 1) - 1(y - 2) + 3(z + 3) = 0, \text{ or } 4x - y + 3z + 7 = 0.$$

[43] $P_1(5, -2, 4)$, $P_2(2, 6, 1) \Rightarrow \overrightarrow{P_1P_2} = <-3, 8, -3>$ and the line is parallel to

$\mathbf{a} = <-3, 8, -3>$. By (14.39), $\frac{x-5}{-3} = \frac{y+2}{8} = \frac{z-4}{-3}$ is a symmetric form for

the line through P_1 and P_2. We could also use the point P_2 and write the equation

of the line as $\frac{x-2}{-3} = \frac{y-6}{8} = \frac{z-1}{-3}$.

[47] As in Example 12, we must determine if the two planes intersect.

Note that two planes do not intersect at a unique point but rather in a line.

$x + 2y = 7 + 9z\ (\mathrm{E}_1)$; $\quad 2x - 3y = -17z\ (\mathrm{E}_2)$

$2\,\mathrm{E}_1 - \mathrm{E}_2 \Rightarrow 7y = 14 + 35z \Rightarrow y = 2 + 5z$

$3\,\mathrm{E}_1 + 2\,\mathrm{E}_2 \Rightarrow 7x = 21 - 7z \Rightarrow x = 3 - z$

For simplicity, we let $z = t$. ★ $x = 3 - t$, $y = 2 + 5t$, $z = t$

[51] $P(1, -1, 2) \Rightarrow x_0 = 1$, $y_0 = -1$, and $z_0 = 2$.

$3x - 7y + z - 5 = 0 \Rightarrow a = 3$, $b = -7$, $c = 1$, and $d = -5$.

$$h = \frac{|ax_0 + by_0 + cz_0 + d|}{\sqrt{a^2 + b^2 + c^2}} = \frac{|3(1) - 7(-1) + 1(2) - 5|}{\sqrt{3^2 + (-7)^2 + 1^2}} = \frac{7}{\sqrt{59}} \approx 0.91.$$

[53] The normal vectors for the planes $4x - 2y + 6z = 3$ and $-6x + 3y - 9z = 4$ are

$\mathbf{a} = <4, -2, 6>$ and $\mathbf{b} = <-6, 3, -9>$, respectively. They are parallel since

$\mathbf{b} = -\frac{3}{2}\mathbf{a}$ and hence, the planes are parallel. Since $(0, 0, \frac{1}{2})$ is on the first plane, its

distance from the second plane is $h = \frac{|-6(0) + 3(0) - 9(\frac{1}{2}) - 4|}{\sqrt{(-6)^2 + 3^2 + (-9)^2}} = \frac{17}{6\sqrt{14}} \approx 0.76$ by

Example 13. Note that we could have picked any point on the plane since the two

planes are parallel. Also, we could have picked any point on the second plane and

found its distance from the first plane using the same formula.

[55] l_1 is parallel to $\mathbf{a} = \overrightarrow{AB} = <1, 2, 2>$ and l_2 is parallel to $\mathbf{b} = \overrightarrow{CD} = <-6, 2, 5>$.

$\mathbf{a} \times \mathbf{b} = <6, -17, 14>$ is orthogonal to l_1 and l_2 and $\|\mathbf{a} \times \mathbf{b}\| = \sqrt{521}$.

Also, $\mathbf{c} = \overrightarrow{AC} = <3, 3, -4>$ is a vector between the two lines.

Thus, $d = \frac{|(\mathbf{a} \times \mathbf{b}) \cdot \mathbf{c}|}{\|\mathbf{a} \times \mathbf{b}\|} = \frac{89}{\sqrt{521}} \approx 3.90.$

[57] Since three points determine a plane, we will first find two points that lie on the line, and then proceed as in Exercise 27, using those two points and P.

By letting $t = 0$ and $t = 1$ in the equations $x = 3t + 1$, $y = -2t + 4$, and $z = t - 3$, we obtain the points $Q(1, 4, -3)$ and $R(4, 2, -2)$, respectively.

Q and R lie on the line and P, Q, and R determine a plane.

$\overrightarrow{QP} = \langle 4, -4, 5\rangle$ and $\overrightarrow{QR} = \langle 3, -2, 1\rangle$. $\overrightarrow{QP} \times \overrightarrow{QR} = \langle 6, 11, 4\rangle$.

The plane is $6(x - 5) + 11(y - 0) + 4(z - 2) = 0$, or $6x + 11y + 4z = 38$.

[59] We could proceed as in Example 3 of Section 14.2. However, this uses a cross product rather than a dot product to determine the distance. Refer to *Figure 59*. Let D denote the projection of A onto the line through B and C. Let $\mathbf{a} = \overrightarrow{BA} = \langle -1, -10, 3\rangle$ and $\mathbf{c} = \overrightarrow{BC} = \langle 4, -5, 7\rangle$. $\|\mathbf{a}\|$, $\|\overrightarrow{BD}\|$, and $\overline{AD}$ form a right triangle. $\|\mathbf{a}\| = \sqrt{110}$ and $\|\overrightarrow{BD}\| =$

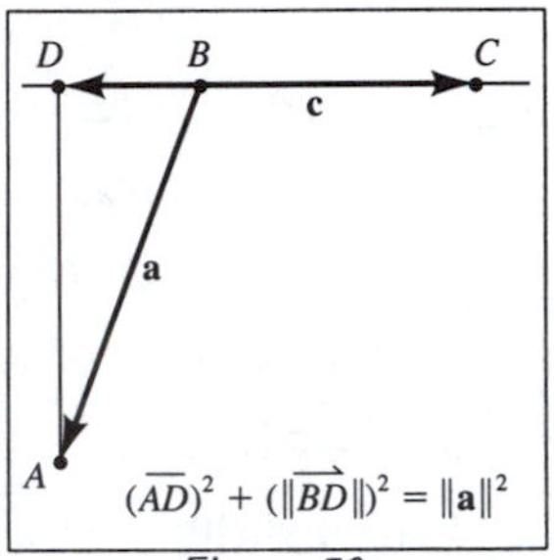

Figure 59

$\text{comp}_{\mathbf{c}}\,\mathbf{a} = \frac{\mathbf{a}\cdot\mathbf{c}}{\|\mathbf{c}\|} = \frac{67}{\sqrt{90}}$. Thus, $(\overline{AD})^2 + (\|\overrightarrow{BD}\|)^2 = \|\mathbf{a}\|^2 \Rightarrow$

$$\overline{AD} = \sqrt{\|\mathbf{a}\|^2 - (\|\overrightarrow{BD}\|)^2} = \sqrt{(\sqrt{110})^2 - (67/\sqrt{90})^2} = \sqrt{5411/90} \approx 7.75.$$

[61] Refer to Example 3 of Section 14.4. To obtain two points on the line, let $t = 0$ and $t = 1$ to get $A(3, -4, 1)$ and $B(1, -1, 3)$. $\mathbf{a} = \overrightarrow{AB} = \langle -2, 3, 2\rangle$ and

$$\mathbf{b} = \overrightarrow{AP} = \langle -1, 5, -3\rangle.\quad d = \frac{\|\mathbf{a}\times\mathbf{b}\|}{\|\mathbf{a}\|} = \frac{\|\langle -19, -8, -7\rangle\|}{\sqrt{17}} = \sqrt{\frac{474}{17}} \approx 5.28.$$

[63] To find the intercept form, divide the equation $ax + by + cz = d$ by d, obtaining a "1" on the right side. $(10x - 15y + 6z = 30)\cdot\frac{1}{30} \Leftrightarrow \frac{x}{3} + \frac{y}{-2} + \frac{z}{5} = 1$. An alternative method would be to let $y = z = 0$, obtaining $x = 3$ and hence, the x-intercept is 3. Similarly, $x = z = 0 \Rightarrow y = b = -2$ and $x = y = 0 \Rightarrow z = c = 5$.

[65] The x-intercept is 2, the y-intercept is 3, and the z-intercept is 4.

Using the intercept form, an equation is $\frac{x}{2} + \frac{y}{3} + \frac{z}{4} = 1$.

Multiplying by the lcd, 12, we obtain an equivalent form $6x + 4y + 3z = 12$.

Exercises 14.6

1 The graph of $x^2 + y^2 = 9$ is a circle of radius 3 in the xy-plane. Since z does not occur in the equation, there are no restrictions on the z-coordinates of the cylinder. Thus, this is a circular cylinder with directrix $x^2 + y^2 = 9$ and rulings parallel to the z-axis.

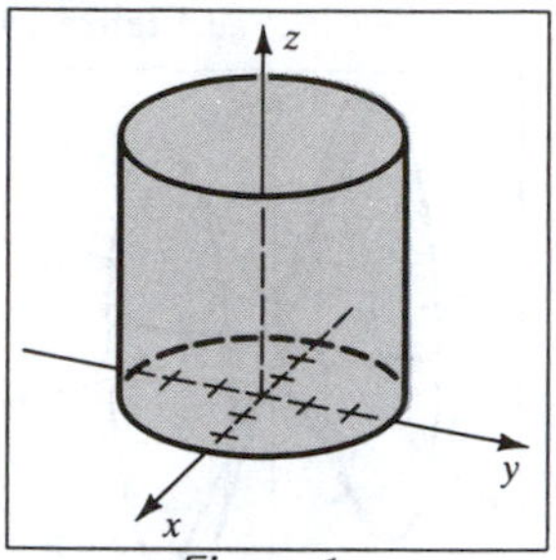

Figure 1

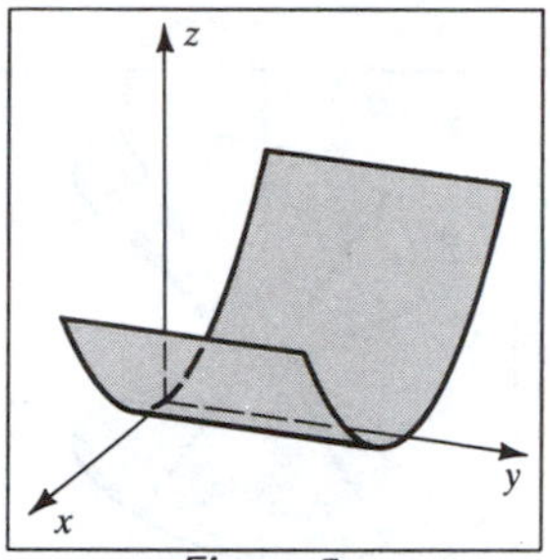

Figure 5

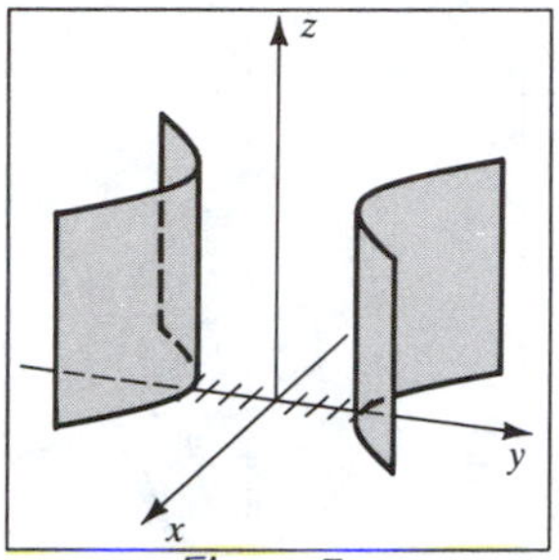

Figure 7

5 The graph of $x^2 = 9z$ is a parabola in the xz-plane. There are no restrictions on the y-coordinates. Thus, this is a parabolic cylinder with directrix $x^2 = 9z$ and rulings parallel to the y-axis.

7 The graph of $y^2 - x^2 = 16$ is a hyperbola in the xy-plane. There are no restrictions on the z-coordinates. Thus, this is a hyperbolic cylinder with directrix $y^2 - x^2 = 16$ and rulings parallel to the z-axis.

9 The traces in the xz-plane and the yz-plane are parabolas.
The trace in the plane $z = k > 0$ is an ellipse.
The surface is an elliptic paraboloid with z positive—choice K. See (14.45).

11 The traces in the xy-plane and the xz-plane are hyperbolas. The trace in the yz-plane is a circle. The surface is a hyperboloid of one sheet with the x^2 term being negative—choice C. See (14.42). A mnemonic device that may help to keep (14.42) and (14.43) straight is "one sheet—one negative sign" and "two sheets—two negative signs." Of course, this is true only if the equations are in the form "expression = 1."

13 The traces in the xy-plane and yz-plane are hyperbolas. There are two sheets. The surface is a hyperboloid of two sheets with the y^2 term being positive—choice Q. See (14.43). The positive term is the one that corresponds to the axis of the hyperboloid.

15 The surface is a cone opening in the positive and negative y direction and having the form $y^2 = Ax^2 + Bz^2$—choice P. See (14.44). If the cone opens in the positive and negative x direction, then it can be written in the form "$x^2 =$." If the cone opens in the positive and negative z direction, then it can be written in the form "$z^2 =$."

[19] The graph has a saddle shape. The surface is a hyperbolic paraboloid with z positive in the xz-plane, z negative in the yz-plane, and "facing along the x-axis"—choice E. Figure 14.68 is similar to this figure, except it has been rotated 90° in the xy-plane.

[21] $\frac{x^2}{4} + \frac{y^2}{9} + \frac{z^2}{16} = 1$ is an ellipsoid with x-, y-, and z-intercepts at ± 2, ± 3, and ± 4.

See Figure 14.63.

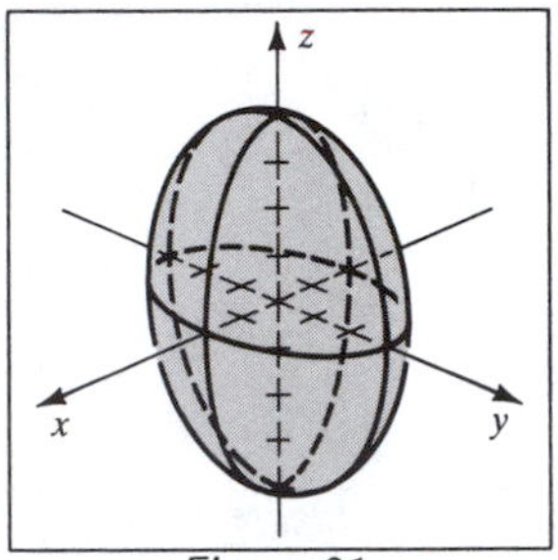

Figure 21

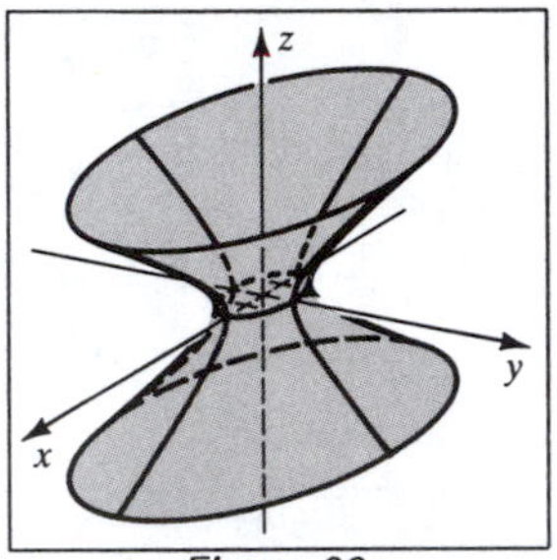

Figure 23a

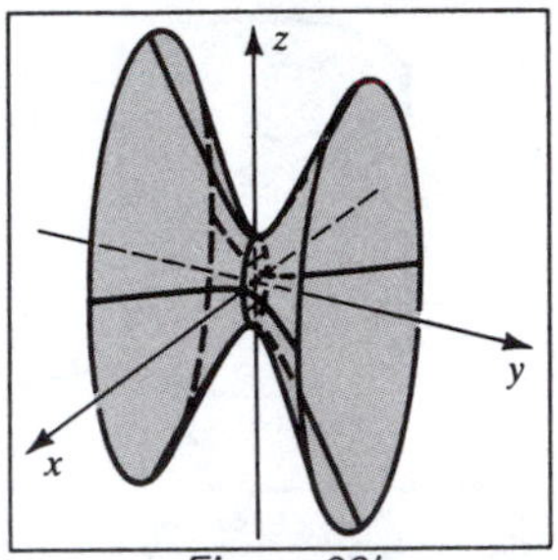

Figure 23b

[23] (a) See Figure 14.65. The traces in the xz-plane and yz-plane are hyperbolas. Since the z^2 term is negative, we know that the axis of the hyperboloid is the z-axis.

(b) See Figure 14.69. The traces in the xy-plane and yz-plane are hyperbolas. Since the y^2 term is negative, we know that the axis of the hyperboloid is the y-axis.

[27] (a) See Figure 14.67. Since the equation is of the form $z^2 = Ax^2 + By^2$,

it is a cone opening in the positive and negative z direction.

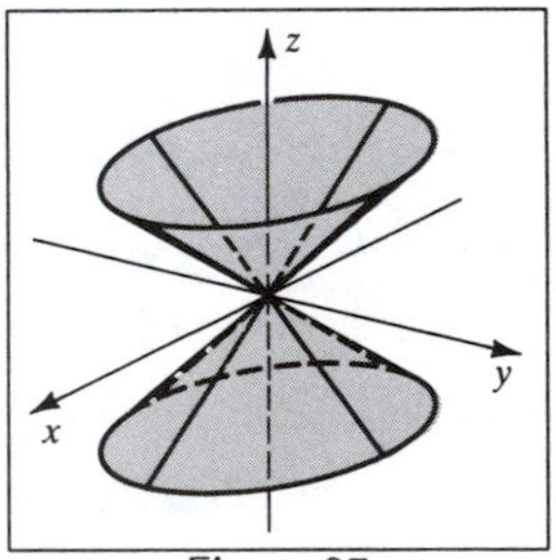

Figure 27a

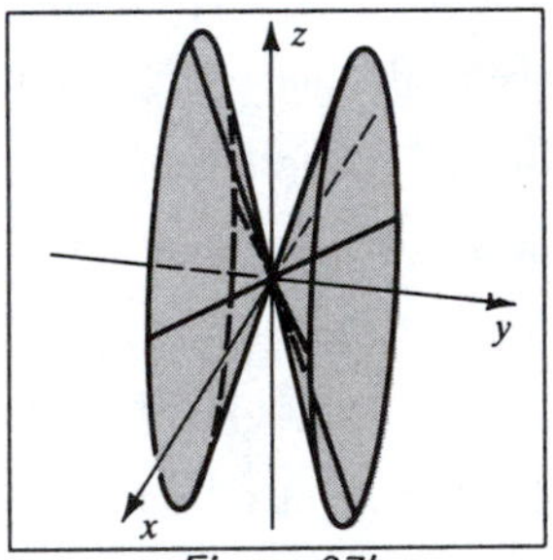

Figure 27b

(b) The graph will be similar to the figure in Exercise 15.

Since the equation can be written in the form $y^2 = Ax^2 + Bz^2$,

it is a cone opening in the positive and negative z direction.

[31] (a) The graph will be similar to the figure in Exercise 19

since z is positive in the xz-plane and negative in the yz-plane.

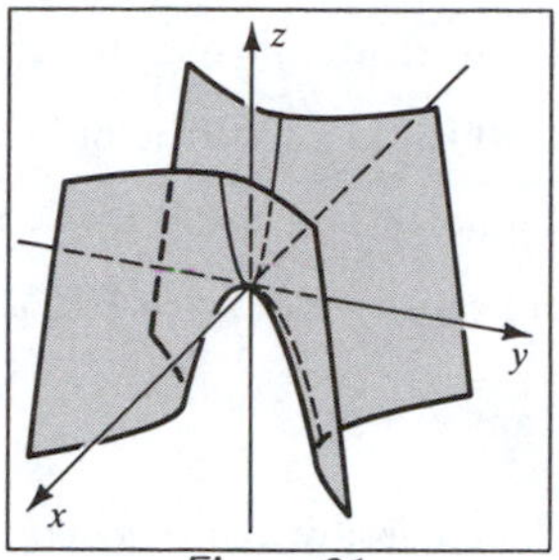

Figure 31a

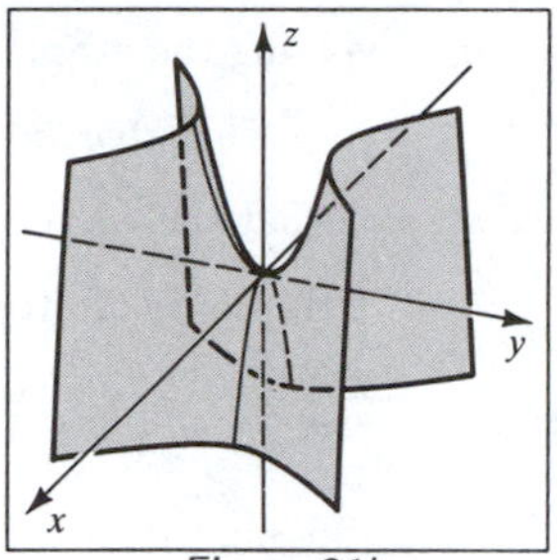

Figure 31b

(b) See Figure 14.68.

[35] $36x = 9y^2 + z^2 \Leftrightarrow x = \frac{y^2}{4} + \frac{z^2}{36}$; a paraboloid with axis along the x-axis.

See Figure 14.70.

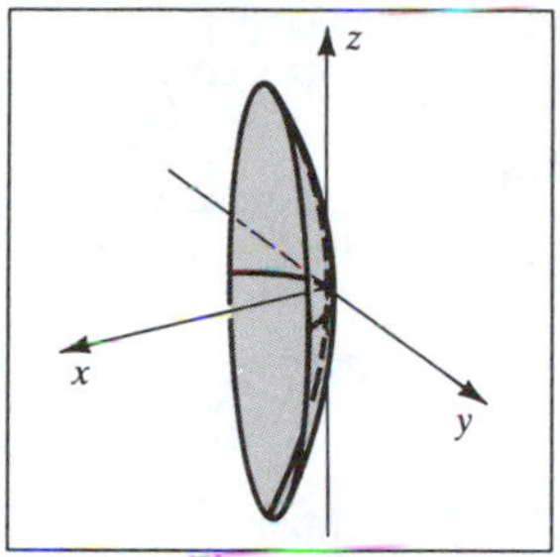

Figure 35

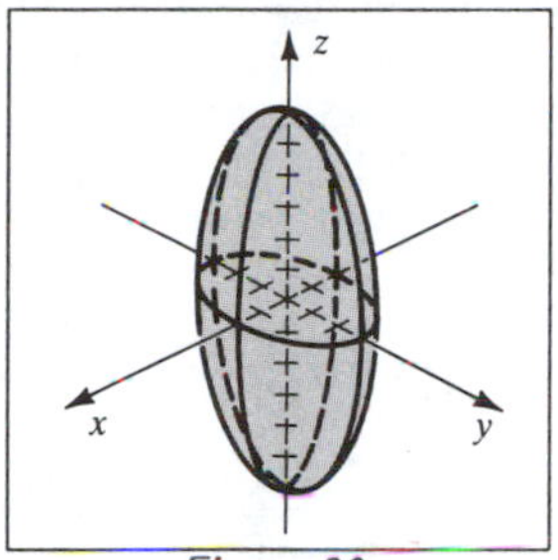

Figure 39

[39] $9x^2 + 4y^2 + z^2 = 36 \Leftrightarrow \frac{x^2}{4} + \frac{y^2}{9} + \frac{z^2}{36} = 1$ {divide by 36};

an ellipsoid with x-, y-, and z-intercepts at ± 2, ± 3, and ± 6.

[41] In the yz-plane the trace will be the graph of $z = e^y$. Since x does not occur in the equation, there are no restrictions on the x-coordinates. The graph will be orthogonal to the yz-plane and is a cylinder with directrix $z = e^y$ and rulings parallel to the x-axis.

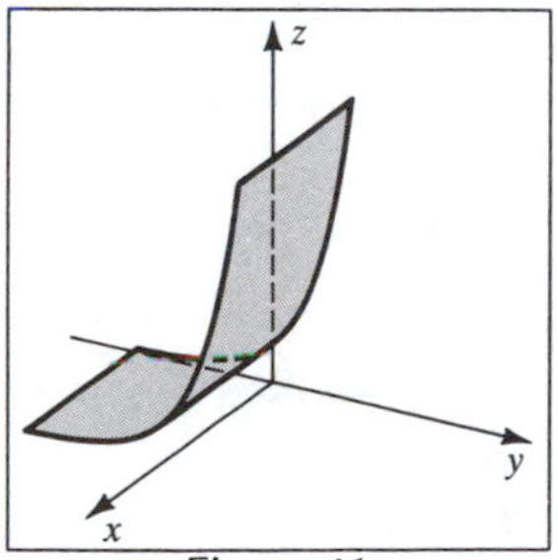

Figure 41

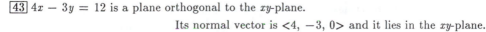

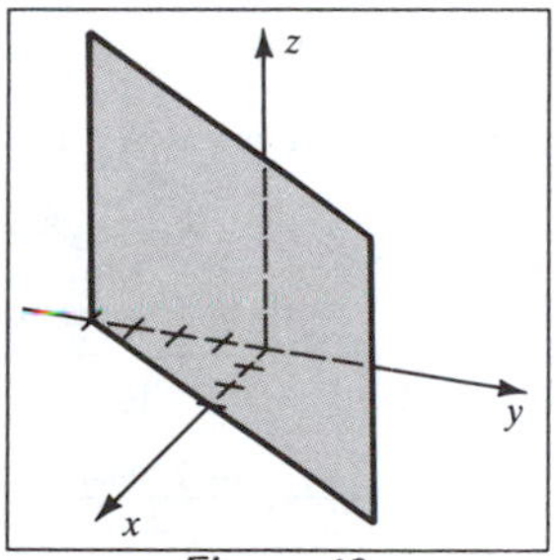

Figure 43

[43] $4x - 3y = 12$ is a plane orthogonal to the xy-plane.

Its normal vector is $<4, -3, 0>$ and it lies in the xy-plane.

[51] As in Example 6, we substitute $x^2 + z^2$ for x^2 and obtain $x^2 + z^2 + 4y^2 = 16$, which is an ellipsoid.

[55] Since we are revolving the graph about the x-axis, substitute $y^2 + z^2$ for z^2.

This gives $y^2 + z^2 - x^2 = 1$, which is a hyperboloid of one sheet.

[57] (a) If we assume that the z-axis passes through the north and south poles, the Clarke ellipsoid is flatter at the north and south poles since $c < a$ and $c < b$.

(b) $z = k \Rightarrow \frac{x^2}{a^2} + \frac{y^2}{b^2} = 1 - \frac{k^2}{c^2}$, which is a constant.

Since $a = b$, these curves will be circles, or more generally, ellipses.

(c) $y = mx \Rightarrow \frac{x^2}{a^2} + \frac{m^2x^2}{b^2} + \frac{z^2}{c^2} = 1 \Rightarrow \left(\frac{1 + m^2}{a^2}\right)x^2 + \frac{z^2}{c^2} = 1$ {since $a = b$},

which are ellipses.

14.7 Review Exercises

[1] $3\mathbf{a} - 2\mathbf{b} = 3(3\mathbf{i} - \mathbf{j} - 4\mathbf{k}) - 2(2\mathbf{i} + 5\mathbf{j} - 2\mathbf{k}) =$

$(9\mathbf{i} - 3\mathbf{j} - 12\mathbf{k}) - (4\mathbf{i} + 10\mathbf{j} - 4\mathbf{k}) = 5\mathbf{i} - 13\mathbf{j} - 8\mathbf{k}$

[3] $\|-3\mathbf{b}\| = \|-3(2\mathbf{i} + 5\mathbf{j} - 2\mathbf{k})\| = \|-6\mathbf{i} - 15\mathbf{j} + 6\mathbf{k}\| =$

$\sqrt{(-6)^2 + (-15)^2 + 6^2} = \sqrt{297} = 3\sqrt{33}$

[7] $\cos\theta = \frac{\mathbf{a}\cdot\mathbf{c}}{\|\mathbf{a}\|\|\mathbf{c}\|} = \frac{(3\mathbf{i} - \mathbf{j} - 4\mathbf{k})\cdot(-\mathbf{i} + 6\mathbf{k})}{\|3\mathbf{i} - \mathbf{j} - 4\mathbf{k}\|\|-\mathbf{i} + 6\mathbf{k}\|} =$

$\frac{3(-1) - 1(0) - 4(6)}{\sqrt{3^2 + (-1)^2 + (-4)^2}\sqrt{(-1)^2 + 0^2 + 6^2}} = \frac{-27}{\sqrt{26}\sqrt{37}} = \frac{-27}{\sqrt{962}} \Rightarrow$

$\theta = \arccos\frac{-27}{\sqrt{962}} \approx 150.52°.$

[11] $\mathbf{a} \times \mathbf{b} = (3\mathbf{i} - \mathbf{j} - 4\mathbf{k}) \times (2\mathbf{i} + 5\mathbf{j} - 2\mathbf{k}) =$

$$\begin{vmatrix} \mathbf{i} & \mathbf{j} & \mathbf{k} \\ 3 & -1 & -4 \\ 2 & 5 & -2 \end{vmatrix} = \begin{vmatrix} -1 & -4 \\ 5 & -2 \end{vmatrix}\mathbf{i} - \begin{vmatrix} 3 & -4 \\ 2 & -2 \end{vmatrix}\mathbf{j} + \begin{vmatrix} 3 & -1 \\ 2 & 5 \end{vmatrix}\mathbf{k} =$$

$(2 + 20)\mathbf{i} - (-6 + 8)\mathbf{j} + (15 + 2)\mathbf{k} = 22\mathbf{i} - 2\mathbf{j} + 17\mathbf{k}.$

[13] $\text{comp}_{\mathbf{b}}\,\mathbf{a} = \mathbf{a}\cdot\frac{1}{\|\mathbf{b}\|}\mathbf{b} = \langle 3, -1, -4\rangle \cdot \frac{1}{\sqrt{33}}\langle 2, 5, -2\rangle = \frac{9}{\sqrt{33}} \approx 1.57$

[15] $(2\mathbf{a})\cdot(3\mathbf{a}) = (2\cdot 3)(\mathbf{a}\cdot\mathbf{a}) = 6\left[(3^2 + (-1)^2 + (-4)^2\right] = 6(26) = 156$

[19] Using Example 4 of Section 14.4,

$V = |(\mathbf{a} \times \mathbf{b})\cdot\mathbf{c}| = |\langle 22, -2, 17\rangle\cdot\langle -1, 0, 6\rangle| = |80| = 80.$

[23] (a) $A(5, -3, 2)$ and $B(-1, -4, 3) \Rightarrow d(A, B) =$

$\sqrt{(-1 - 5)^2 + (-4 + 3)^2 + (3 - 2)^2} = \sqrt{(-6)^2 + (-1)^2 + 1^2} = \sqrt{38}.$

(b) $M_{AB} = \left(\frac{5 - 1}{2}, \frac{-3 - 4}{2}, \frac{2 + 3}{2}\right) = \left(2, -\frac{7}{2}, \frac{5}{2}\right)$

(c) The point B is a distance of 4 from the xz-plane since its y-coordinate is -4.

The sphere must have a radius of 4 and an equation is

$(x + 1)^2 + (y + 4)^2 + (z - 3)^2 = 4^2 = 16.$

(d) Points on a plane that is parallel to the xz-plane must have a constant y-coordinate. Since the plane passes through $B(-1, -4, 3)$,

an equation of the plane is $y = -4$.

(e) Using the point $A(5, -3, 2)$ and $\overrightarrow{BA} = <6, 1, -1>$, parametric equations for the line through A and B are $x = 5 + 6t$, $y = -3 + t$, $z + 2 - t$.

(f) Using the point $A(5, -3, 2)$ and $\overrightarrow{AB} = <-6, -1, 1>$, an equation of the plane through A with normal vector $\overrightarrow{AB}$ is $-6(x - 5) - 1(y + 3) + 1(z - 2) = 0$,

or $6x + y - z = 25$.

[25] Using the intercept form in Exercises 63–64 of Section 14.5 and multiplying it by 30,

we have $\left(\frac{x}{5} + \frac{y}{-2} + \frac{z}{6} = 1\right) \cdot 30$, which simplifies to $6x - 15y + 5z = 30$.

[27] The second equation should be $x + 3y - z = -1$.

[29] The yz-trace of $2x + 3y - 4z = 11$ {let $x = 0$} is $3y - 4z = 11$.

Thus, the plane's equation has the form $ax + 3y - 4z = 11$.

Substituting $x = 4$, $y = 1$, and $z = 2$ yields $4a = 16$, or $a = 4$.

[33] (a) $P(2, -1, 1)$, $Q(-3, 2, 0)$, and $R(4, -5, 3) \Rightarrow$

$\overrightarrow{PQ} = <-5, 3, -1>$ and $\overrightarrow{PR} = <2, -4, 2>$.

$\overrightarrow{PQ} \times \overrightarrow{PR} = <2, 8, 14> \Rightarrow \mathbf{u} = \frac{1}{2\sqrt{66}}<2, 8, 14> = \frac{1}{\sqrt{66}}<1, 4, 7>$.

(b) Using $P(2, -1, 1)$ and the vector $<1, 4, 7>$ from part (a) gives us

$1(x - 2) + 4(y + 1) + 7(z - 1) = 0$, or $x + 4y + 7z = 5$.

(c) $\overrightarrow{QR} = <7, -7, 3>$; $x = 2 + 7t$, $y = -1 - 7t$, $z = 1 + 3t$

(d) $\overrightarrow{QP} \cdot \overrightarrow{QR} = <5, -3, 1> \cdot <7, -7, 3> = 59$

(e) $\cos\theta = \dfrac{\overrightarrow{QP} \cdot \overrightarrow{QR}}{\|\overrightarrow{QP}\|\|\overrightarrow{QR}\|} = \dfrac{59}{\sqrt{35}\sqrt{107}} \Rightarrow \theta = \arccos\dfrac{59}{\sqrt{3745}} \approx 15.40^\circ$

(f) As in Example 2 of §14.4, area $= \frac{1}{2}\|\overrightarrow{PQ} \times \overrightarrow{PR}\| = \frac{1}{2}(2\sqrt{66}) = \sqrt{66} \approx 8.12$.

(g) As in Example 3 of §14.4, $d = \dfrac{\|\overrightarrow{PQ} \times \overrightarrow{PR}\|}{\|\overrightarrow{PQ}\|} = \dfrac{2\sqrt{66}}{\sqrt{35}} \approx 2.75$.

[35] We set each part equal to t and solve for x, y, and z. $t = \frac{x - 3}{2} \Rightarrow x = 3 + 2t$;

$t = \frac{y + 1}{-4} \Rightarrow y = -1 - 4t$; $t = \ = \frac{z - 5}{8} \Rightarrow z = 5 + 8t$. Similarly, solving for

x, y, and z with the second line gives $x = -1 + 7t$, $y = 6 - 2t$, $z = -\frac{7}{2} - 2t$.

[37] Direction vectors for the given lines are $\mathbf{a}_1 = <1, 1, 7>$ and $\mathbf{a}_2 = <5, -2, -4>$.

$\cos\theta = \dfrac{\mathbf{a}_1 \cdot \mathbf{a}_2}{\|\mathbf{a}_1\|\|\mathbf{a}_2\|} = \dfrac{-25}{\sqrt{51}\sqrt{45}} \Rightarrow \theta = \arccos\dfrac{-25}{\sqrt{2295}} \approx 121.46^\circ$ and $180^\circ - \theta$

[39] Completing the squares on x, y, and z gives us:

$$x^2 + y^2 + z^2 - 14x + 6y - 8z + 10 = 0$$
$$(x^2 - 14x) + (y^2 + 6y) + (z^2 - 8z) = -10$$
$$(x^2 - 14x + 49) + (y^2 + 6y + 9) + (z^2 - 8z + 16) = -10 + 49 + 9 + 16$$
$$(x - 7)^2 + (y + 3)^2 + (z - 4)^2 = 64 = 8^2$$

This is a sphere with center $(7, -3, 4)$ and radius 8.

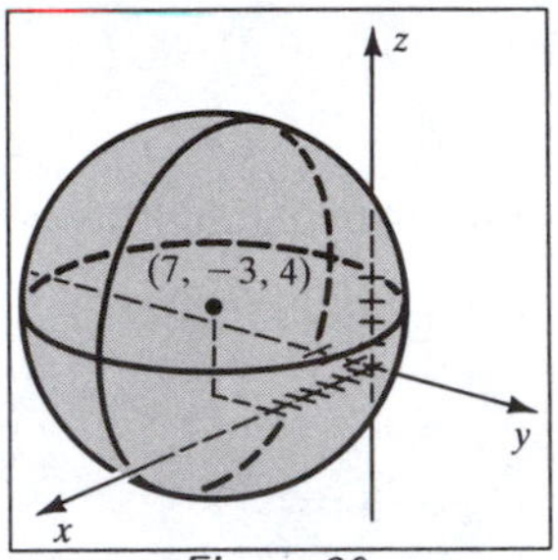

Figure 39

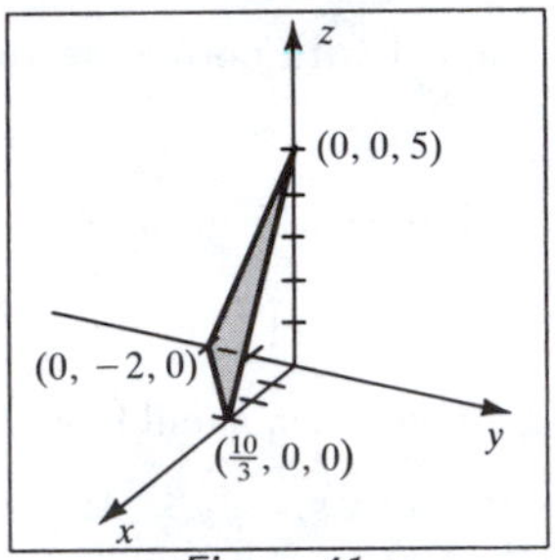

Figure 41

[41] Writing $3x - 5y + 2z = 10$ in the intercept form gives us $\frac{x}{10/3} + \frac{y}{-2} + \frac{z}{5} = 1$.

Hence, this is a plane passing through the points $(\frac{10}{3}, 0, 0)$, $(0, -2, 0)$, and $(0, 0, 5)$.

[45] $z^2 - 4x^2 = 9 - 4y^2 \Leftrightarrow \frac{z^2}{9} + \frac{y^2}{9/4} - \frac{x^2}{9/4} = 1$;

a hyperboloid of one sheet with axis along the x-axis.

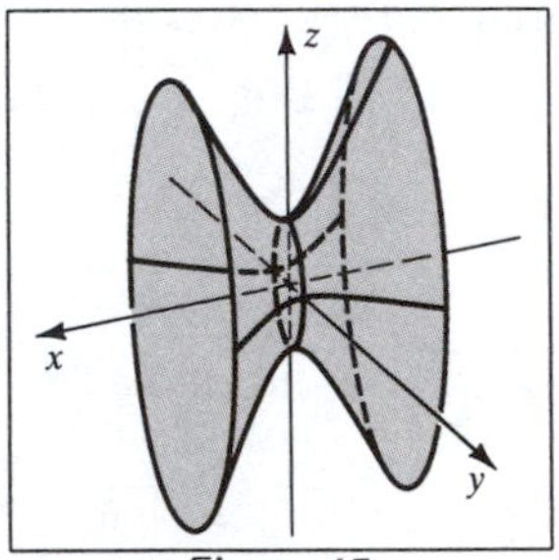

Figure 45

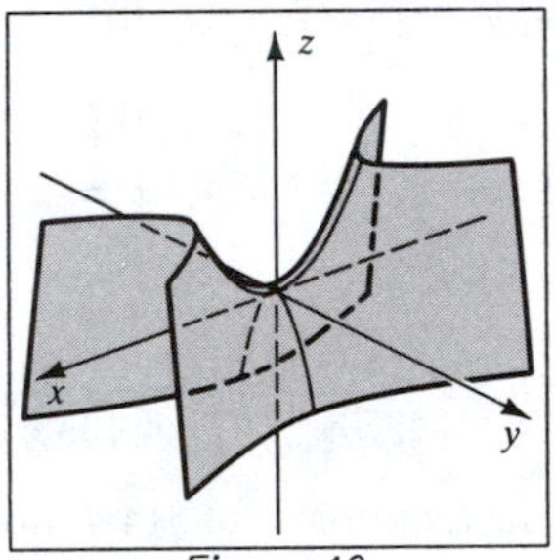

Figure 49

[49] $x^2 - 4y^2 = 4z \Leftrightarrow z = \frac{x^2}{4} - y^2$; a hyperbolic paraboloid with z positive in the

xz-plane, z negative in the yz-plane, and "facing along the x-axis".

Chapter 15: Vector-Valued Functions

Exercises 15.1

[1] $\mathbf{r}(t) = 3t\mathbf{i} + (1 - 9t^2)\mathbf{j} \Rightarrow \mathbf{r}(0) = 3(0)\mathbf{i} + (1 - 9 \cdot 0^2)\mathbf{j} = 0\mathbf{i} + 1\mathbf{j} = \langle 0, 1 \rangle$.
$\mathbf{r}(1) = 3(1)\mathbf{i} + (1 - 9 \cdot 1^2)\mathbf{j} = \langle 3, -8 \rangle$. To graph $\mathbf{r}(0)$, draw a vector with its initial point at the origin and its terminal point at $(0, 1)$. $\mathbf{r}(1)$ is drawn in similar fashion. $x = 3t$ and $y = 1 - 9t^2 \Rightarrow y = 1 - (3t)^2 = 1 - x^2$. This is a parabola. To determine the orientation, examine $\mathbf{r}(t)$ for increasing values of t. From this we can see that the $\mathbf{i}$-component of C always increases while the $\mathbf{j}$-component of C increases to a maximum of 1 and then decreases. The orientation of C is indicated by the arrows in *Figure 1*. See Example 5 and Figure 15.7.

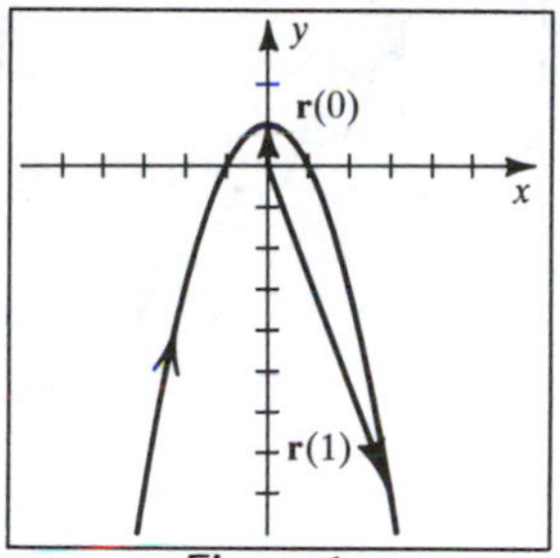

Figure 1

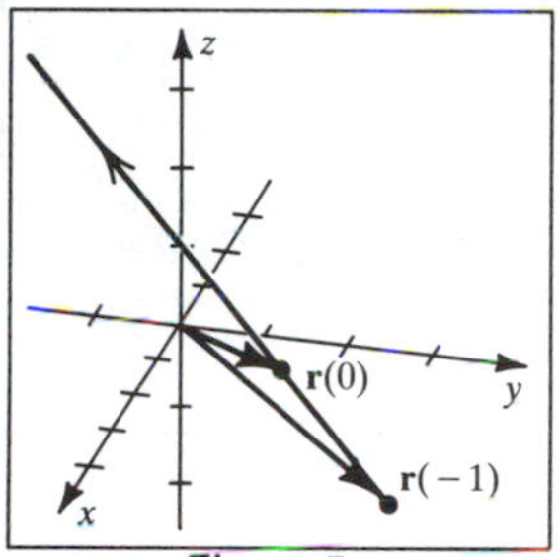

Figure 5

[5] $\mathbf{r}(-1) = \langle 2, 3, -1 \rangle$. $\mathbf{r}(0) = \langle 3, 2, 1 \rangle$. $x = 3 + t$, $y = 2 - t$, and
$z = 1 + 2t \Rightarrow \dfrac{x - 3}{1} = \dfrac{2 - y}{1} = \dfrac{z - 1}{2}$, the symmetric form of a half-line with end point $(2, 3, -1)$ $\{t = -1\}$. As t increases from $t = -1$, the curve moves away from the point $(2, 3, -1)$ in the direction of $\mathbf{r}(0) - \mathbf{r}(-1) = \langle 1, -1, 2 \rangle$.

[7] $\mathbf{r}(0) = \langle 0, 4, 0 \rangle$. $\mathbf{r}(\frac{\pi}{2}) = \langle \frac{\pi}{2}, 0, 9 \rangle$. $y = 4\cos t$ and $z = 9\sin t \Rightarrow \dfrac{y^2}{16} + \dfrac{z^2}{81} = 1$.
C is an elliptic helix along the positive x-axis with end point $(0, 4, 0)$.
The curve will start at $(0, 4, 0)$ and spiral around the x-axis in an elliptical orbit.
See Example 4 and Figure 15.6.

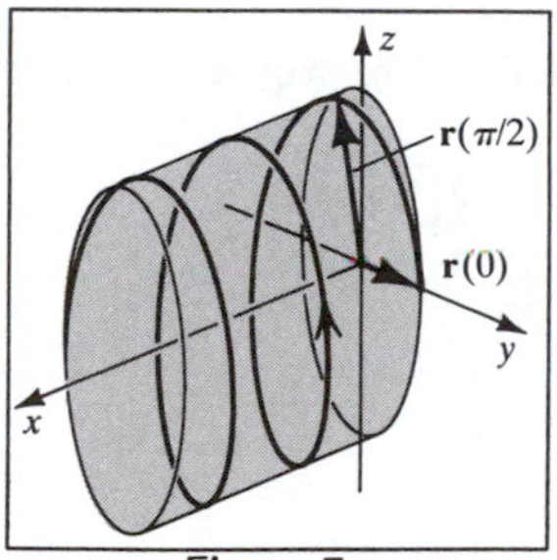

Figure 7

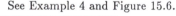

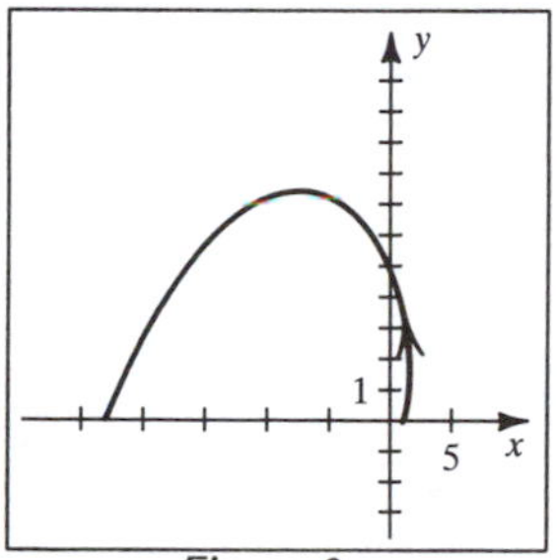

Figure 9

[9] $x = e^t \cos t$ and $y = e^t \sin t \Rightarrow$

$$\begin{aligned} x^2 + y^2 &= (e^t \cos t)^2 + (e^t \sin t)^2 \\ &= e^{2t} \cos^2 t + e^{2t} \sin^2 t \\ &= e^{2t}(\cos^2 t + \sin^2 t) = e^{2t} \Rightarrow r^2 = e^{2t} \Rightarrow r = e^t. \end{aligned}$$

Since $0 \le t \le \pi$, $y = e^t \sin t \ge 0$ and C is in quadrants I and II. Because $e^t > 0$ and $\cos t$ varies from 1 to -1 as t varies from 0 to π, $x = e^t \cos t$ decreases from positive to negative values. The orientation of C is indicated by the arrows in *Figure 9.*

[13] Since $z = 3$, all points lie in the plane $z = 3$. $x = t^2 + 1$ and $y = t \Rightarrow x = y^2 + 1$. To determine the orientation, we can evaluate the two vectors $\mathbf{r}(0) = \langle 1, 0, 3 \rangle$ and $\mathbf{r}(1) = \langle 2, 1, 3 \rangle$. The orientation of C is indicated by the arrows in *Figure 13.*

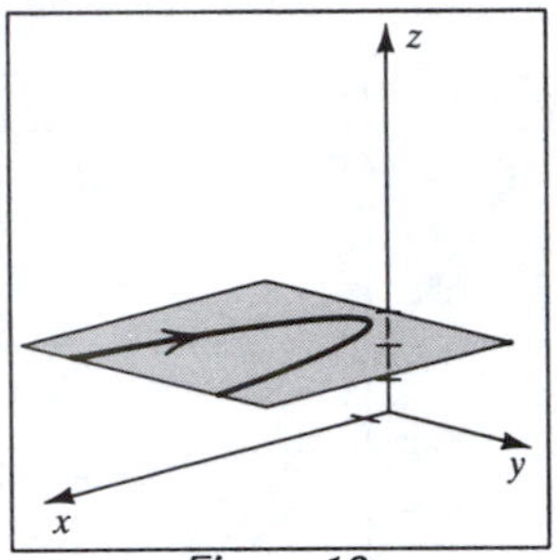

Figure 13

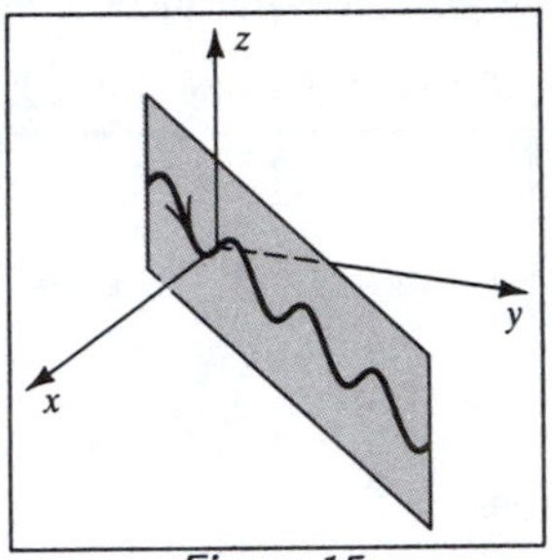

Figure 15

[15] $x = t$, $y = t$, and $z = \sin t \Rightarrow y = x$ and C is a sine curve along the line $y = x$, orthogonal to the xy-plane. The orientation of C is in the direction of increasing values of x and y.

[19] From (15.3), $L = \int_0^2 \sqrt{\left(\frac{dx}{dt}\right)^2 + \left(\frac{dy}{dt}\right)^2 + \left(\frac{dz}{dt}\right)^2}\, dt$

$$= \int_0^2 \sqrt{(5)^2 + (8t)^2 + (6t)^2}\, dt = \int_0^2 \sqrt{25 + 64t^2 + 36t^2}\, dt$$

$$= \int_0^2 \sqrt{100t^2 + 25}\, dt = \int_0^2 \sqrt{25(4t^2 + 1)}\, dt$$

$$= 5\int_0^2 \sqrt{4t^2 + 1}\, dt = \tfrac{5}{2}\int_0^4 \sqrt{1 + u^2}\, du \;\{u = 2t,\ \tfrac{1}{2}\, du = dt\}$$

$$= \tfrac{5}{2}\left[\tfrac{u}{2}\sqrt{1 + u^2} + \tfrac{1}{2}\ln\left|u + \sqrt{1 + u^2}\right|\right]_0^4 \;\{\text{Formula 21 or trig substitution}\} =$$

$$\tfrac{5}{4}\left[4\sqrt{17} + \ln(4 + \sqrt{17})\right] \approx 23.23$$

[21] $L = \int_0^{2\pi} \sqrt{(-e^t \sin t + e^t \cos t)^2 + (e^t)^2 + (e^t \cos t + e^t \sin t)^2}\, dt$

$= \int_0^{2\pi} \sqrt{\mathrm{A} + (e^{2t}) + \mathrm{B}}\, dt$, where $\mathrm{A} = e^{2t} \sin^2 t - 2e^{2t} \sin t \cos t + e^{2t} \cos^2 t$ and

$\mathrm{B} = e^{2t} \cos^2 t + 2e^{2t} \cos t \sin t + e^{2t} \sin^2 t$

$= \int_0^{2\pi} \sqrt{e^{2t}(\sin^2 t + \cos^2 t) + (e^{2t}) + e^{2t}(\cos^2 t + \sin^2 t)}\, dt$

$= \int_0^{2\pi} \sqrt{3e^{2t}}\, dt = \sqrt{3}\int_0^{2\pi} \sqrt{(e^t)^2}\, dt = \sqrt{3}\int_0^{2\pi} e^t\, dt = \sqrt{3}(e^{2\pi} - 1) \approx 925.77$

[25] (a) $b^2(x^2 + y^2) = b^2(a^2 e^{2\mu t} \cos^2 t + a^2 e^{2\mu t} \sin^2 t)$

$= b^2 a^2 e^{2\mu t}(\cos^2 t + \sin^2 t)$

$= a^2(b^2 e^{2\mu t}) = a^2 z^2$

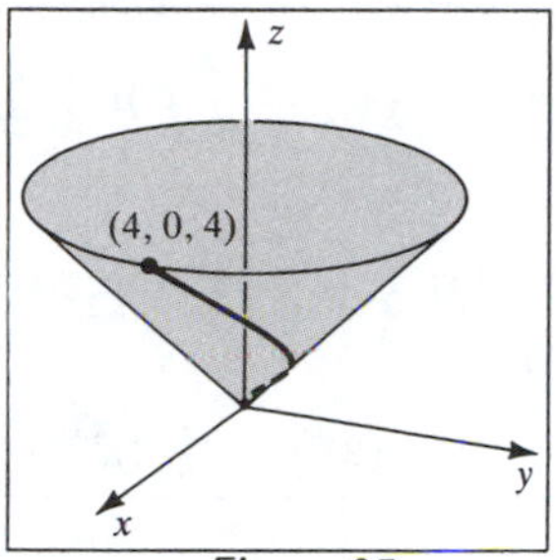

Figure 25

(b) The curve C starts at the point (4, 0, 4) {when $t = 0$} and "spirals" on the cone $z^2 = x^2 + y^2$ toward the vertex of the cone since $z = 4e^{-t} \to 0$ as $t \to \infty$.

(c) $x = 4e^{-t} \cos t$, $y = 4e^{-t} \sin t$, and $z = 4e^{-t} \Rightarrow \sqrt{\left(\frac{dx}{dt}\right)^2 + \left(\frac{dy}{dt}\right)^2 + \left(\frac{dz}{dt}\right)^2} =$

$\sqrt{\left[-4e^{-t}(\sin t + \cos t)\right]^2 + \left[4e^{-t}(\cos t - \sin t)\right]^2 + \left[-4e^{-t}\right]^2} = \sqrt{48e^{-2t}} =$

$4\sqrt{3}\, e^{-t}$. Thus, $L = \int_0^{\infty} 4\sqrt{3}\, e^{-t}\, dt = \lim_{s \to \infty} \int_0^s 4\sqrt{3}\, e^{-t}\, dt = 4\sqrt{3} \lim_{s \to \infty} \left[-e^{-t}\right]_0^s =$

$-4\sqrt{3} \lim_{s \to \infty} (e^{-s} - 1) = -4\sqrt{3}(0 - 1) = 4\sqrt{3}$.

[27] (a) Let $Ax + By + Cz = D$ be the equation of an arbitrary plane.

Then $x = at$, $y = bt^2$, and $z = ct^3 \Rightarrow A(at) + B(bt^2) + C(ct^3) = D$,

which is a cubic equation in t. This equation has at most 3 distinct roots.

(b) $L = \int_0^1 \sqrt{(6)^2 + (6t)^2 + (3t^2)^2}\, dt = \int_0^1 \sqrt{36 + 36t^2 + 9t^4}\, dt$

$= \int_0^1 \sqrt{9(t^4 + 4t^2 + 4)}\, dt = 3\int_0^1 \sqrt{(t^2 + 2)^2}\, dt$

$= 3\int_0^1 (t^2 + 2)\, dt = 3\left[\frac{1}{3}t^3 + 2t\right]_0^1 = 3(\frac{7}{3}) = 7$

Exercises 15.2

1 (a) r will be defined at all values of t for which each component is defined.

$t - 1 \geq 0 \Rightarrow t \geq 1$ and $2 - t \geq 0 \Rightarrow t \leq 2$. Domain $D = [1, 2]$.

(b) Using (15.7) with $h(t) = 0$, $\mathbf{r}(t) = \sqrt{t-1}\,\mathbf{i} + \sqrt{2-t}\,\mathbf{j} \Rightarrow$

$$\mathbf{r}'(t) = \frac{d}{dt}\Big[(t-1)^{1/2}\Big]\mathbf{i} + \frac{d}{dt}\Big[(2-t)^{1/2}\Big]\mathbf{j} = \tfrac{1}{2}(t-1)^{-1/2}\mathbf{i} - \tfrac{1}{2}(2-t)^{-1/2}\mathbf{j}.$$

$$\mathbf{r}''(t) = -\tfrac{1}{4}(t-1)^{-3/2}\mathbf{i} - \tfrac{1}{4}(2-t)^{-3/2}\mathbf{j}.$$

5 (a) r will be defined whenever $\tan t$ is defined. Thus, $D = \{t: t \neq \frac{\pi}{2} + \pi n\}$.

(b) $\mathbf{r}(t) = t^2\mathbf{i} + \tan t\,\mathbf{j} + 3\mathbf{k} \Rightarrow$

$$\mathbf{r}'(t) = \frac{d}{dt}(t^2)\mathbf{i} + \frac{d}{dt}(\tan t)\mathbf{j} + \frac{d}{dt}(3)\mathbf{k} = 2t\mathbf{i} + \sec^2 t\,\mathbf{j}.$$

$$\mathbf{r}''(t) = 2\mathbf{i} + 2(\sec t)(\sec t\tan t)\mathbf{j} = 2\mathbf{i} + 2\sec^2 t\tan t\,\mathbf{j}.$$

9 (a) $x = -\frac{1}{4}t^4$ and $y = t^2 \Rightarrow x = -\frac{1}{4}y^2$, a parabola.

Since $y = t^2$, we must have $y \geq 0$.

(b) $\mathbf{r}(2) = -\frac{1}{4}(2)^4\mathbf{i} + (2)^2\mathbf{j} = -4\mathbf{i} + 4\mathbf{j}$. $\mathbf{r}'(t) = -t^3\mathbf{i} + 2t\mathbf{j}$; $\mathbf{r}'(2) = -8\mathbf{i} + 4\mathbf{j}$.

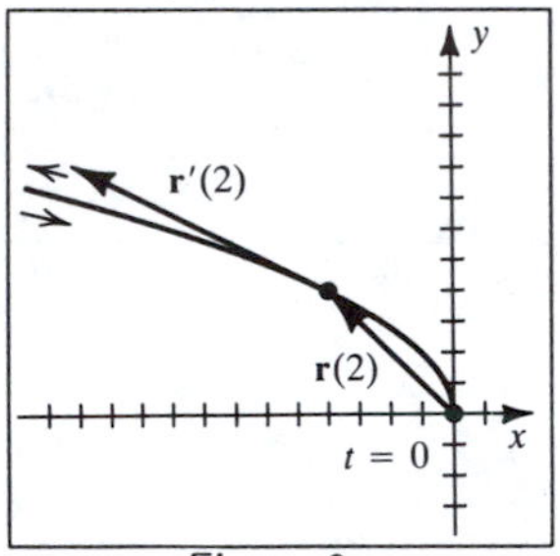

Figure 9

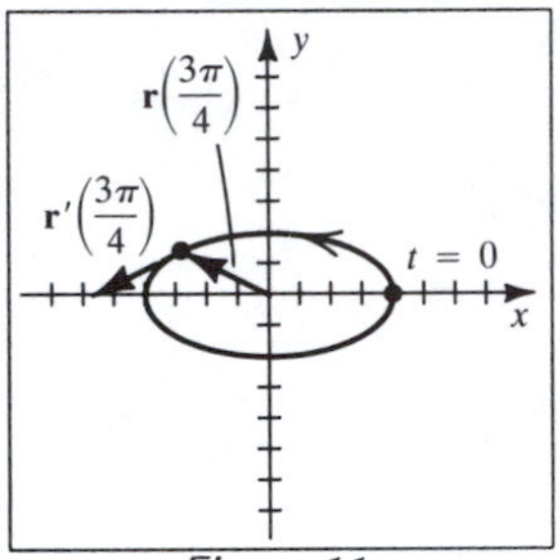

Figure 11

11 (a) $x = 4\cos t$ and $y = 2\sin t \Rightarrow \frac{x}{4} = \cos t$ and $\frac{y}{2} = \sin t \Rightarrow$

$$\left(\frac{x}{4}\right)^2 = \cos^2 t \text{ and } \left(\frac{y}{2}\right)^2 = \sin^2 t \Rightarrow \frac{x^2}{16} + \frac{y^2}{4} = \cos^2 t + \sin^2 t = 1.$$

(b) $\mathbf{r}(\frac{3\pi}{4}) = 4\cos(\frac{3\pi}{4})\mathbf{i} + 2\sin(\frac{3\pi}{4})\mathbf{j} = 4(-\frac{\sqrt{2}}{2})\mathbf{i} + 2(\frac{\sqrt{2}}{2})\mathbf{j} = -2\sqrt{2}\,\mathbf{i} + \sqrt{2}\,\mathbf{j}$.

$$\mathbf{r}'(t) = -4\sin t\,\mathbf{i} + 2\cos t\,\mathbf{j};\ \mathbf{r}'(\tfrac{3\pi}{4}) = -2\sqrt{2}\,\mathbf{i} - \sqrt{2}\,\mathbf{j}.$$

17 x determines the i-component, y the j-component, and z the k-component.

$\mathbf{r}(t) = (2t^3 - 1)\mathbf{i} + (-5t^2 + 3)\mathbf{j} + (8t + 2)\mathbf{k} \Rightarrow \mathbf{r}'(t) = 6t^2\mathbf{i} - 10t\mathbf{j} + 8\mathbf{k}$.

$P(1, -2, 10)$ occurs when $z = 8t + 2 = 10$, or $t = 1$. We could have used $2t^3 - 1 = 1$ or $-5t^2 + 3 = -2$, but $8t + 2 = 10$ is the easiest equation to solve. Thus, $\mathbf{r}'(1) = 6\mathbf{i} - 10\mathbf{j} + 8\mathbf{k}$. A parametric equation for the tangent line to C at P, parallel to $\mathbf{r}'(1)$, is $x = 1 + 6t$, $y = -2 - 10t$, $z = 10 + 8t$.

21 $\mathbf{r}(t) = e^{2t}\mathbf{i} + e^{-t}\mathbf{j} + (t^2 + 4)\mathbf{k} \Rightarrow \mathbf{r}'(t) = 2e^{2t}\mathbf{i} - e^{-t}\mathbf{j} + 2t\mathbf{k}$.

$P(1, 1, 4)$ occurs when $x = 1 \Rightarrow e^{2t} = 1 \Rightarrow 2t = 0 \Rightarrow t = 0$. $\mathbf{r}'(0) = 2\mathbf{i} - \mathbf{j}$ and $\|\mathbf{r}'(0)\| = \sqrt{5}$. Two unit tangent vectors are $\pm\dfrac{\mathbf{r}'(0)}{\|\mathbf{r}'(0)\|} = \pm\frac{1}{\sqrt{5}}(2\mathbf{i} - \mathbf{j})$.

[23] Let θ be the angle between $\mathbf{r}'$ and $\mathbf{k}$. We will show that $\cos\theta$ is constant, and hence, θ is constant. $\mathbf{r}(t) = (ae^{\mu t}\cos t)\mathbf{i} + (ae^{\mu t}\sin t)\mathbf{j} + (be^{\mu t})\mathbf{k} \Rightarrow$

$\mathbf{r}'(t) = (a\mu e^{\mu t}\cos t - ae^{\mu t}\sin t)\mathbf{i} + (a\mu e^{\mu t}\sin t + ae^{\mu t}\cos t)\mathbf{j} + (b\mu e^{\mu t})\mathbf{k} \Rightarrow$

$\|\mathbf{r}'(t)\| = e^{\mu t}\sqrt{a^2\mu^2 + a^2 + b^2\mu^2}$. Using (14.20),

$$\cos\theta = \frac{\mathbf{r}'(t)\cdot\mathbf{k}}{\|\mathbf{r}'(t)\|\|\mathbf{k}\|}\ \{\|\mathbf{k}\| = 1\} = \frac{b\mu e^{\mu t}}{e^{\mu t}\sqrt{a^2\mu^2 + a^2 + b^2\mu^2}} = \frac{b\mu}{\sqrt{a^2\mu^2 + a^2 + b^2\mu^2}},$$

which is a constant.

[27] By (15.10) and (15.11), $\int_0^2 (6t^2\mathbf{i} - 4t\mathbf{j} + 3\mathbf{k})\,dt = \Big[2t^3\mathbf{i} - 2t^2\mathbf{j} + 3t\mathbf{k}\Big]_0^2 =$

$16\mathbf{i} - 8\mathbf{j} + 6\mathbf{k}$.

[29] $\int_0^{\pi/4} (\sin t\mathbf{i} - \cos t\mathbf{j} + \tan t\mathbf{k})\,dt = \Big[-\cos t\mathbf{i} - \sin t\mathbf{j} + \ln|\sec t|\mathbf{k}\Big]_0^{\pi/4} =$

$$\Big[-\tfrac{1}{\sqrt{2}}\mathbf{i} - \tfrac{1}{\sqrt{2}}\mathbf{j} + \ln(\sqrt{2})\mathbf{k}\Big] - \Big[-1\mathbf{i} - 0\mathbf{j} + 0\mathbf{k}\Big] = \Big(1 - \tfrac{1}{\sqrt{2}}\Big)\mathbf{i} - \tfrac{1}{\sqrt{2}}\mathbf{j} + (\ln\sqrt{2})\mathbf{k}$$

[31] $\mathbf{r}'(t) = t^2\mathbf{i} + (6t + 1)\mathbf{j} + 8t^3\mathbf{k} \Rightarrow$

$$\begin{aligned}
\mathbf{r}(t) &= \int\Big[t^2\mathbf{i} + (6t + 1)\mathbf{j} + 8t^3\mathbf{k}\Big]dt \\
&= \Big[\int t^2\,dt\Big]\mathbf{i} + \Big[\int (6t + 1)\,dt\Big]\mathbf{j} + \Big[\int 8t^3\,dt\Big]\mathbf{k} \\
&= (\tfrac{1}{3}t^3 + c_1)\mathbf{i} + (3t^2 + t + c_2)\mathbf{j} + (2t^4 + c_3)\mathbf{k} \\
&= \tfrac{1}{3}t^3\mathbf{i} + (3t^2 + t)\mathbf{j} + 2t^4\mathbf{k} + (c_1\mathbf{i} + c_2\mathbf{j} + c_3\mathbf{k}) \\
&= \tfrac{1}{3}t^3\mathbf{i} + (3t^2 + t)\mathbf{j} + 2t^4\mathbf{k} + \mathbf{c}.
\end{aligned}$$

c_1, c_2, and c_3 are constants and $\mathbf{c} = c_1\mathbf{i} + c_2\mathbf{j} + c_3\mathbf{k}$ is a constant vector. In the future, when evaluating an indefinite integral of a vector-valued function, we will simply include " + c", where c is some constant vector. Now, $\mathbf{r}(0) = 0\mathbf{i} + 0\mathbf{j} + 0\mathbf{k} + \mathbf{c} = 2\mathbf{i} - 3\mathbf{j} + \mathbf{k} \Rightarrow$

$\mathbf{c} = 2\mathbf{i} - 3\mathbf{j} + \mathbf{k}$. Thus, $\mathbf{r}(t) = (\tfrac{1}{3}t^3 + 2)\mathbf{i} + (3t^2 + t - 3)\mathbf{j} + (2t^4 + 1)\mathbf{k}$.

[33] $\mathbf{r}''(t) = 6t\mathbf{i} - 12t^2\mathbf{j} + \mathbf{k} \Rightarrow$

$$\mathbf{r}'(t) = \int\Big[6t\mathbf{i} - 12t^2\mathbf{j} + \mathbf{k}\Big]dt = 3t^2\mathbf{i} - 4t^3\mathbf{j} + t\mathbf{k} + \mathbf{c}_1.$$

$\mathbf{r}'(0) = 0\mathbf{i} + 0\mathbf{j} + 0\mathbf{k} + \mathbf{c}_1$ and $\mathbf{r}'(0) = \mathbf{i} + 2\mathbf{j} - 3\mathbf{k} \Rightarrow \mathbf{c}_1 = \mathbf{i} + 2\mathbf{j} - 3\mathbf{k}$.

Hence, $\mathbf{r}'(t) = (3t^2 + 1)\mathbf{i} + (2 - 4t^3)\mathbf{j} + (t - 3)\mathbf{k}$ and

$$\begin{aligned}
\mathbf{r}(t) &= \int\Big[(3t^2 + 1)\mathbf{i} + (2 - 4t^3)\mathbf{j} + (t - 3)\mathbf{k}\Big]dt \\
&= (t^3 + t)\mathbf{i} + (2t - t^4)\mathbf{j} + (\tfrac{1}{2}t^2 - 3t)\mathbf{k} + \mathbf{c}_2.
\end{aligned}$$

$\mathbf{r}(0) = 0\mathbf{i} + 0\mathbf{j} + 0\mathbf{k} + \mathbf{c}_2$ and $\mathbf{r}(0) = 7\mathbf{i} + \mathbf{k} \Rightarrow \mathbf{c}_2 = 7\mathbf{i} + \mathbf{k}$.

Thus, $\mathbf{r}(t) = (t^3 + t + 7)\mathbf{i} + (2t - t^4)\mathbf{j} + (\tfrac{1}{2}t^2 - 3t + 1)\mathbf{k}$.

[35] $\mathbf{r}(t) = e^t\mathbf{i} + te^t\mathbf{j} + (t^2 + 4)\mathbf{k} \Rightarrow \mathbf{r}'(t) = e^t\mathbf{i} + (t + 1)e^t\mathbf{j} + 2t\mathbf{k}$.

$P(1, 0, 4)$ occurs when $x = e^t = 1$, or $t = 0$. $\mathbf{r}'(0) = \mathbf{i} + \mathbf{j}$.

A normal vector to the plane is $\mathbf{a} = \mathbf{i} + \mathbf{j}$.

An equation of the plane is $1(x - 1) + 1(y - 0) + 0(z - 4) = 0$, or $x + y = 1$.

[37] (1) $\mathbf{u}(t) \cdot \mathbf{v}(t) = t \sin t + t^2 \cos t + 2t^3 \sin t \Rightarrow$

$$D_t\left[\mathbf{u}(t) \cdot \mathbf{v}(t)\right] = t \cos t + \sin t - t^2 \sin t + 2t \cos t + 2t^3 \cos t + 6t^2 \sin t$$
$$= (1 + 5t^2)\sin t + (2t^3 + 3t)\cos t.$$

(2) $\mathbf{u}(t) \times \mathbf{v}(t) = (2t^2 \sin t - t^3 \cos t)\mathbf{i} - (2t \sin t - t^3 \sin t)\mathbf{j} + (t \cos t - t^2 \sin t)\mathbf{k}$

$$\Rightarrow D_t\left[\mathbf{u}(t) \times \mathbf{v}(t)\right] = (2t^2 \cos t + 4t \sin t + t^3 \sin t - 3t^2 \cos t)\mathbf{i}$$
$$- (2t \cos t + 2\sin t - t^3 \cos t - 3t^2 \sin t)\mathbf{j}$$
$$+ (-t \sin t + \cos t - t^2 \cos t - 2t \sin t)\mathbf{k} =$$
$$\left[(t^3 + 4t)\sin t - t^2 \cos t\right]\mathbf{i} + \left[(3t^2 - 2)\sin t + (t^3 - 2t)\cos t\right]\mathbf{j} +$$
$$\left[-3t \sin t + (1 - t^2)\cos t\right]\mathbf{k}.$$

Note: For some remaining exercises, we will use the notation

$\mathbf{u}(t) = \langle u_1, u_2, u_3 \rangle$, where $u_n = f_n(t)$ for $n = 1, 2, 3$.

Similarly, $\mathbf{v}(t) = \langle v_1, v_2, v_3 \rangle$, where $v_n = g_n(t)$ for $n = 1, 2, 3$.

[39] (a) $\lim_{t\to a}\left[\mathbf{u}(t) + \mathbf{v}(t)\right] = \lim_{t\to a} \langle u_1 + v_1, u_2 + v_2, u_3 + v_3 \rangle$

$$= \langle \lim_{t\to a} u_1 + \lim_{t\to a} v_1, \lim_{t\to a} u_2 + \lim_{t\to a} v_2, \lim_{t\to a} u_3 + \lim_{t\to a} v_3 \rangle$$

$$= \langle \lim_{t\to a} u_1, \lim_{t\to a} u_2, \lim_{t\to a} u_3 \rangle + \langle \lim_{t\to a} v_1, \lim_{t\to a} v_2, \lim_{t\to a} v_3 \rangle$$

$$= \lim_{t\to a} \langle u_1, u_2, u_3 \rangle + \lim_{t\to a} \langle v_1, v_2, v_3 \rangle = \lim_{t\to a} \mathbf{u}(t) + \lim_{t\to a} \mathbf{v}(t).$$

(b) $\lim_{t\to a}\left[\mathbf{u}(t) \cdot \mathbf{v}(t)\right] = \lim_{t\to a}\left[u_1 v_1 + u_2 v_2 + u_3 v_3\right]$

$$= \lim_{t\to a} u_1 \lim_{t\to a} v_1 + \lim_{t\to a} u_2 \lim_{t\to a} v_2 + \lim_{t\to a} u_3 \lim_{t\to a} v_3$$

$$= \langle \lim_{t\to a} u_1, \lim_{t\to a} u_2, \lim_{t\to a} u_3 \rangle \cdot \langle \lim_{t\to a} v_1, \lim_{t\to a} v_2, \lim_{t\to a} v_3 \rangle$$

$$= \lim_{t\to a} \mathbf{u}(t) \cdot \lim_{t\to a} \mathbf{v}(t)$$

(c) $\lim_{t\to a} c\mathbf{u}(t) = \lim_{t\to a} \langle cu_1, cu_2, cu_3 \rangle = \langle c \lim_{t\to a} u_1, c \lim_{t\to a} u_2, c \lim_{t\to a} u_3 \rangle =$

$$c \lim_{t\to a} \langle u_1, u_2, u_3 \rangle = c \lim_{t\to a} \mathbf{u}(t)$$

43 $D_t\left[\mathbf{u}(t) + \mathbf{v}(t)\right] = D_t < u_1 + v_1,\ u_2 + v_2,\ u_3 + v_3 >$

$= < u_1' + v_1',\ u_2' + v_2',\ u_3' + v_3' >$

$= < u_1',\ u_2',\ u_3' > + < v_1',\ v_2',\ v_3' > = \mathbf{u}'(t) + \mathbf{v}'(t).$

47 $D_t\left[\mathbf{u}(t) \cdot \mathbf{v}(t) \times \mathbf{w}(t)\right]$

$= D_t\left[\mathbf{u}(t) \cdot [\mathbf{v}(t) \times \mathbf{w}(t)]\right]$ {consider $\mathbf{v}(t) \times \mathbf{w}(t)$ to be a single vector}

$= \mathbf{u}(t) \cdot D_t\left[\mathbf{v}(t) \times \mathbf{w}(t)\right] + \mathbf{u}'(t) \cdot \left[\mathbf{v}(t) \times \mathbf{w}(t)\right]$ {(15.8)(iii)}

$= \mathbf{u}(t) \cdot \left[\mathbf{v}(t) \times \mathbf{w}'(t) + \mathbf{v}'(t) \times \mathbf{w}(t)\right] + \mathbf{u}'(t) \cdot \left[\mathbf{v}(t) \times \mathbf{w}(t)\right]$ {(15.8)(iv)}

$= \left[\mathbf{u}(t) \cdot \mathbf{v}(t) \times \mathbf{w}'(t)\right] + \left[\mathbf{u}(t) \cdot \mathbf{v}'(t) \times \mathbf{w}(t)\right] + \left[\mathbf{u}'(t) \cdot \mathbf{v}(t) \times \mathbf{w}(t)\right]$ {(14.17)(iii)}

49 (a) $\int_a^b \left[\mathbf{u}(t) + \mathbf{v}(t)\right] dt$

$$= \left(\int_a^b (u_1 + v_1)\, dt\right)\mathbf{i} + \left(\int_a^b (u_2 + v_2)\, dt\right)\mathbf{j} + \left(\int_a^b (u_3 + v_3)\, dt\right)\mathbf{k}$$

$$= \int_a^b (u_1\mathbf{i} + u_2\mathbf{j} + u_3\mathbf{k})\, dt + \int_a^b (v_1\mathbf{i} + v_2\mathbf{j} + v_3\mathbf{k})\, dt = \int_a^b \mathbf{u}(t)\, dt + \int_a^b \mathbf{v}(t)\, dt$$

(b) $\int_a^b c\mathbf{u}(t)\, dt = \left(c\int_a^b u_1\, dt\right)\mathbf{i} + \left(c\int_a^b u_2\, dt\right)\mathbf{j} + \left(c\int_a^b u_3\, dt\right)\mathbf{k}$

$$= c\left[\left(\int_a^b u_1\, dt\right)\mathbf{i} + \left(\int_a^b u_2\, dt\right)\mathbf{j} + \left(\int_a^b u_3\, dt\right)\mathbf{k}\right] = c\int_a^b \mathbf{u}(t)\, dt$$

Exercises 15.3

1 $x = 2t$ and $y = 4t^2 + 1 \Rightarrow y = (2t)^2 + 1 = x^2 + 1$, a parabola.

$\mathbf{r}(t) = 2t\mathbf{i} + (4t^2 + 1)\mathbf{j} \Rightarrow \mathbf{v}(t) = \mathbf{r}'(t) = 2\mathbf{i} + 8t\mathbf{j}$; $\mathbf{v}(1) = 2\mathbf{i} + 8\mathbf{j}$.

$\mathbf{a}(t) = \mathbf{v}'(t) = \mathbf{r}''(t) = 8\mathbf{j}$; $\mathbf{a}(1) = 8\mathbf{j}$.

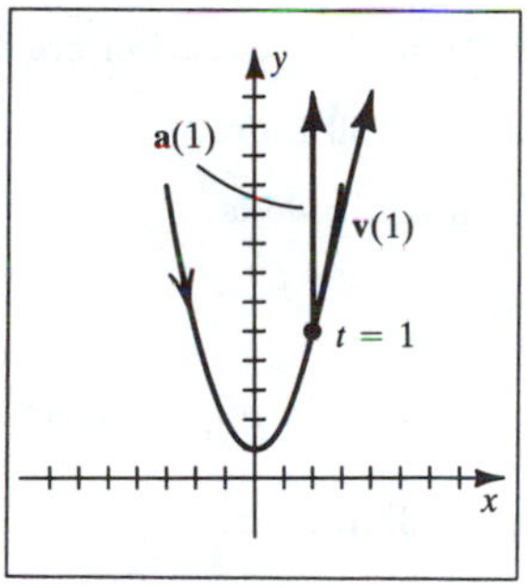

Figure 1

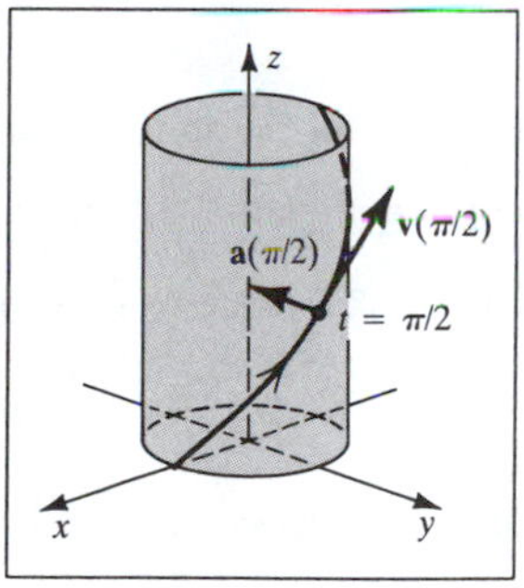

Figure 5

5 This is a circular helix along the z-axis. See Figure 15.6.

$t = 0$ corresponds to $(1, 0, 0)$ since $\mathbf{r}(0) = 1\mathbf{i} + 0\mathbf{j} + 0\mathbf{k}$.

$\mathbf{r}(t) = \cos t\,\mathbf{i} + \sin t\,\mathbf{j} + t\mathbf{k} \Rightarrow \mathbf{v}(t) = \mathbf{r}'(t) = -\sin t\,\mathbf{i} + \cos t\,\mathbf{j} + \mathbf{k}$; $\mathbf{v}(\frac{\pi}{2}) = -\mathbf{i} + \mathbf{k}$.

$\mathbf{a}(t) = \mathbf{v}'(t) = \mathbf{r}''(t) = -\cos t\,\mathbf{i} - \sin t\,\mathbf{j}$; $\mathbf{a}(\frac{\pi}{2}) = -\mathbf{j}$.

[7] $x = t^2$, $y = t$, and $z = 2t \Rightarrow$

$y^2 = t^2$, $z^2 = 4t^2 \Rightarrow y^2 + z^2 = 5t^2 = 5x$.

The curve lies on the circular paraboloid $5x = y^2 + z^2$.

See Figure 14.70 for a similar paraboloid.

$\mathbf{r}(t) = t^2\mathbf{i} + t\mathbf{j} + 2t\mathbf{k} \Rightarrow$

$\mathbf{v}(t) = 2t\mathbf{i} + \mathbf{j} + 2\mathbf{k}$; $\mathbf{v}(1) = 2\mathbf{i} + \mathbf{j} + 2\mathbf{k}$.

$\mathbf{a}(t) = 2\mathbf{i}$; $\mathbf{a}(1) = 2\mathbf{i}$.

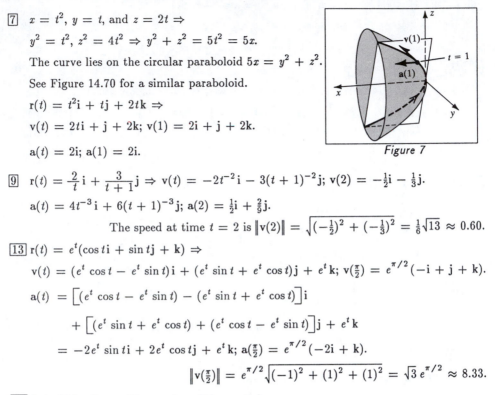

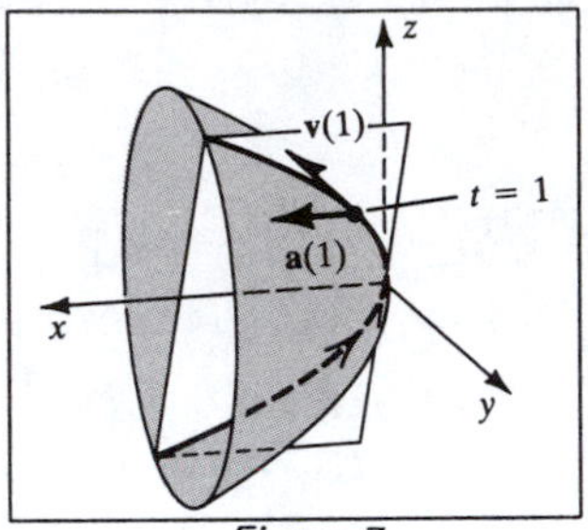

Figure 7

[9] $\mathbf{r}(t) = \frac{2}{t}\mathbf{i} + \frac{3}{t+1}\mathbf{j} \Rightarrow \mathbf{v}(t) = -2t^{-2}\mathbf{i} - 3(t+1)^{-2}\mathbf{j}$; $\mathbf{v}(2) = -\frac{1}{2}\mathbf{i} - \frac{1}{3}\mathbf{j}$.

$\mathbf{a}(t) = 4t^{-3}\mathbf{i} + 6(t+1)^{-3}\mathbf{j}$; $\mathbf{a}(2) = \frac{1}{2}\mathbf{i} + \frac{2}{9}\mathbf{j}$.

The speed at time $t = 2$ is $\|\mathbf{v}(2)\| = \sqrt{(-\frac{1}{2})^2 + (-\frac{1}{3})^2} = \frac{1}{6}\sqrt{13} \approx 0.60$.

[13] $\mathbf{r}(t) = e^t(\cos t\,\mathbf{i} + \sin t\,\mathbf{j} + \mathbf{k}) \Rightarrow$

$\mathbf{v}(t) = (e^t\cos t - e^t\sin t)\mathbf{i} + (e^t\sin t + e^t\cos t)\mathbf{j} + e^t\mathbf{k}$; $\mathbf{v}(\frac{\pi}{2}) = e^{\pi/2}(-\mathbf{i} + \mathbf{j} + \mathbf{k})$.

$$\begin{aligned}\mathbf{a}(t) &= \left[(e^t\cos t - e^t\sin t) - (e^t\sin t + e^t\cos t)\right]\mathbf{i}\\ &\quad + \left[(e^t\sin t + e^t\cos t) + (e^t\cos t - e^t\sin t)\right]\mathbf{j} + e^t\mathbf{k}\\ &= -2e^t\sin t\,\mathbf{i} + 2e^t\cos t\,\mathbf{j} + e^t\mathbf{k};\ \mathbf{a}(\tfrac{\pi}{2}) = e^{\pi/2}(-2\mathbf{i} + \mathbf{k}).\end{aligned}$$

$$\|\mathbf{v}(\tfrac{\pi}{2})\| = e^{\pi/2}\sqrt{(-1)^2 + (1)^2 + (1)^2} = \sqrt{3}\,e^{\pi/2} \approx 8.33.$$

[17] Let $\mathbf{r}(t)$ be the position vector of the particle.

By Theorem (15.9), §15.2, $\|\mathbf{r}'(t)\|$ constant $\Rightarrow \mathbf{r}''(t)$ is orthogonal to $\mathbf{r}'(t)$ for every t;

that is, the velocity and acceleration are orthogonal.

[19] (a) The radius of the shuttle's orbit is $4000 + 150 = 4150$ miles.

The force of gravity is the acceleration, and its magnitude is

$\|\mathbf{a}(t)\| = 32\text{ ft/sec}^2 = \frac{32}{5280}\text{ mi/sec}^2$. From Example 2,

$$\|\mathbf{a}(t)\| = \frac{v^2}{k} \Rightarrow \frac{32}{5280} = \frac{v^2}{4150} \Rightarrow v \approx 5.015\text{ mi/sec} \approx 18{,}054\text{ mi/hr}.$$

(b) The angular speed is $\omega = \frac{v}{k} \approx \dfrac{18{,}054\text{ mi/hr}}{4150\text{ mi}} \approx 4.35$ rad/hr.

Since one complete revolution is 2π radians,

the time required for a complete revolution is $\frac{2\pi}{\omega} \approx 1.4443$ hr ≈ 86.7 min.

[21] $\mathbf{r}(t) = t(v_0\cos\alpha)\mathbf{i} + (-\frac{1}{2}gt^2 + tv_0\sin\alpha + h_0)\mathbf{j}$

{ $h_0 = 0$ is the initial height of the projectile }

$= (1500\cos 30^\circ)\,t\mathbf{i} + (-\frac{1}{2}gt^2 + 1500\sin 30^\circ\,t)\mathbf{j}$

$= 750\sqrt{3}\,t\mathbf{i} + (-\frac{1}{2}gt^2 + 750t)\mathbf{j}$

(a) $\mathbf{v}(t) = \mathbf{r}'(t) = 750\sqrt{3}\,\mathbf{i} + (-gt + 750)\mathbf{j}$

(b) $h = \dfrac{v_0^2 \sin^2\alpha}{2g} = \dfrac{(1500)^2 \sin^2 30^\circ}{2g} = \dfrac{(1500)^2}{8g} \approx 8789$ ft. { $g \approx 32$ }

(c) $d = \dfrac{v_0^2 \sin 2\alpha}{g} = \dfrac{(1500)^2 \sin 60^\circ}{g} = \dfrac{(1500)^2\sqrt{3}}{2g} \approx 60{,}892$ ft.

(d) $t = \dfrac{2v_0 \sin\alpha}{g} = \dfrac{2(1500)\sin 30^\circ}{g} = \dfrac{1500}{g}$.

$\left\|\mathbf{v}(\frac{1500}{g})\right\| \approx \sqrt{(750\sqrt{3})^2 + (-750)^2} = 1500$ ft/sec, which is the initial velocity.

[25] From Example 2, $\|\mathbf{a}(t)\| = \omega^2 k \Rightarrow \omega^2 = \dfrac{\|\mathbf{a}(t)\|}{k}$.

Since we want an acceleration that is 8 times that of gravity { where $g = \|\mathbf{g}\| = 32$ },

$\omega^2 = \dfrac{8g}{k} \Rightarrow \omega \approx \sqrt{\frac{256}{30}} \approx 2.92$ rad/sec ≈ 0.46 rev/sec.

[29] In this problem, the angle α is the angle between the horizontal and the direction of the release of the ball. Since the ball is dropping, α will be negative.

$\mathbf{r}(t) = t(v_0 \cos\alpha)\,\mathbf{i} + (-\frac{1}{2}gt^2 + t\,v_0 \sin\alpha + h_0)\,\mathbf{j}$,

where $v_0 = 100$ mi/hr $= 146\frac{2}{3}$ ft/sec, $h_0 = 6$, and $\tan\alpha = -\frac{2}{58} \Rightarrow \alpha \approx -1.975^\circ$.

The time that the ball travels will be determined when the magnitude of the i-component is 58. Now, $t\,v_0 \cos\alpha = 58 \Rightarrow t = \dfrac{58}{v_0 \cos\alpha} \approx 0.3957$ sec.

The drop caused by gravity is given by $d = -\frac{1}{2}gt^2 \approx (-16)(0.3957)^2 \approx -2.51$ ft.

[31] $D_h\,\mathbf{v} = \frac{d}{dh}\left[(12 + 0.006h^{3/2})\mathbf{i} + (10 + 0.005h^{3/2})\mathbf{j}\right] = 0.009\sqrt{h}\,\mathbf{i} + 0.0075\sqrt{h}\,\mathbf{j}$.

When $h = 150$, $\|D_h\,\mathbf{v}\| = \sqrt{(0.009)^2(150) + (0.0075)^2(150)} \approx 0.14$ (mi/hr)/ft.

Exercises 15.4

[1] (a) $\mathbf{r}(t) = t\mathbf{i} - \frac{1}{2}t^2\mathbf{j} \Rightarrow \mathbf{r}'(t) = \mathbf{i} - t\mathbf{j} \Rightarrow \|\mathbf{r}'(t)\| = (1 + t^2)^{1/2}$.

$$\mathbf{T}(t) = \frac{1}{\|\mathbf{r}'(t)\|}\mathbf{r}'(t) = \frac{1}{(1 + t^2)^{1/2}}\mathbf{i} - \frac{t}{(1 + t^2)^{1/2}}\mathbf{j} \Rightarrow$$

$$\mathbf{T}'(t) = \frac{d}{dt}\left[(1 + t^2)^{-1/2}\right]\mathbf{i} - \frac{d}{dt}\left[t/(1 + t^2)^{1/2}\right]\mathbf{j}$$

$$= -\frac{t}{(1 + t^2)^{3/2}}\mathbf{i} - \frac{1}{(1 + t^2)^{3/2}}\mathbf{j} \Rightarrow$$

$$\|\mathbf{T}'(t)\| = \left[\left(-\frac{t}{(1 + t^2)^{3/2}}\right)^2 + \left(-\frac{1}{(1 + t^2)^{3/2}}\right)^2\right]^{1/2}$$

$$= \left[\frac{t^2 + 1}{(1 + t^2)^3}\right]^{1/2} = \left[\frac{1}{(1 + t^2)^2}\right]^{1/2} = \frac{1}{1 + t^2}.$$

$$\mathbf{N}(t) = \frac{1}{\|\mathbf{T}'(t)\|}\mathbf{T}'(t) = (t^2 + 1)\,\mathbf{T}'(t) = -\frac{t}{(1 + t^2)^{1/2}}\mathbf{i} - \frac{1}{(1 + t^2)^{1/2}}\mathbf{j}.$$

(b) $x = t$ and $y = -\frac{1}{2}t^2 \Rightarrow y = -\frac{1}{2}x^2$.

$\mathbf{T}(1) = \frac{1}{\sqrt{2}}\mathbf{i} - \frac{1}{\sqrt{2}}\mathbf{j}$ and $\mathbf{N}(1) = -\frac{1}{\sqrt{2}}\mathbf{i} - \frac{1}{\sqrt{2}}\mathbf{j}$.

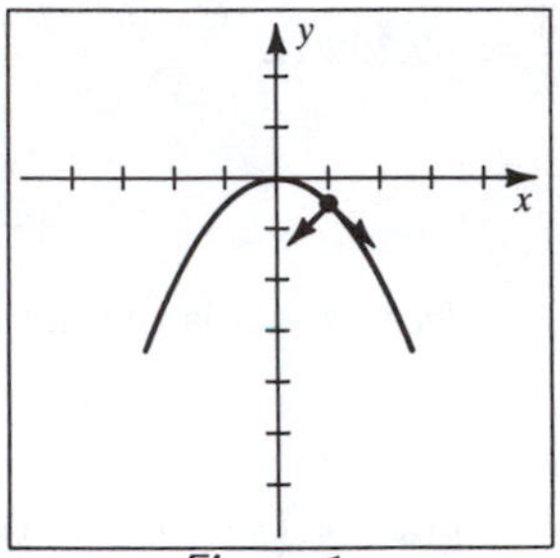

Figure 1

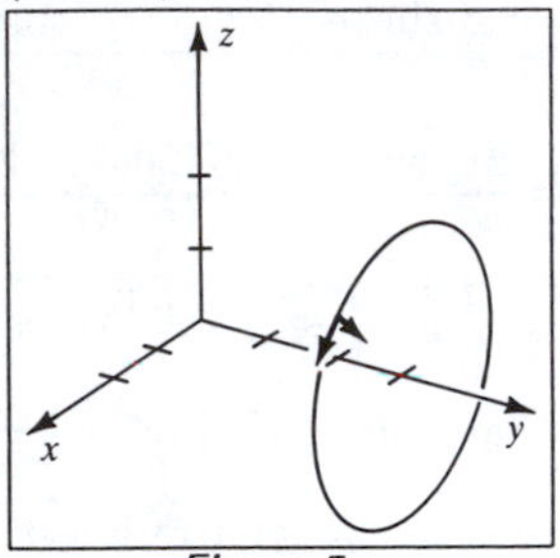

Figure 5

[5] (a) $\mathbf{r}(t) = 2\sin t\,\mathbf{i} + 3\mathbf{j} + 2\cos t\,\mathbf{k} \Rightarrow \mathbf{r}'(t) = 2\cos t\,\mathbf{i} - 2\sin t\,\mathbf{k} \Rightarrow$

$$\|\mathbf{r}'(t)\| = \sqrt{(2\cos t)^2 + (-2\sin t)^2} = \sqrt{4(\cos^2 t + \sin^2 t)} = 2.$$

$$\mathbf{T}(t) = \frac{1}{\|\mathbf{r}'(t)\|}\mathbf{r}'(t) = \cos t\,\mathbf{i} - \sin t\,\mathbf{k} \Rightarrow \mathbf{T}'(t) = -\sin t\,\mathbf{i} - \cos t\,\mathbf{k} \Rightarrow \|\mathbf{T}'(t)\| = 1.$$

$$\mathbf{N}(t) = \frac{1}{\|\mathbf{T}'(t)\|}\mathbf{T}'(t) = -\sin t\,\mathbf{i} - \cos t\,\mathbf{k}.$$

(b) $x = 2\sin t$ and $z = 2\cos t \Rightarrow x^2 + z^2 = 4$ in the plane $y = 3$.

$\mathbf{T}(\frac{\pi}{4}) = \frac{1}{\sqrt{2}}\mathbf{i} - \frac{1}{\sqrt{2}}\mathbf{k}$ and $\mathbf{N}(\frac{\pi}{4}) = -\frac{1}{\sqrt{2}}\mathbf{i} - \frac{1}{\sqrt{2}}\mathbf{k}$.

[7] $y = 2 - x^3 \Rightarrow y' = -3x^2$ and $y'' = -6x$. At $P(1, 1)$, $x = 1$, $y' = -3$, and

$y'' = -6$. Using (15.18), $K = \dfrac{|y''|}{\left[1 + (y')^2\right]^{3/2}} = \dfrac{6}{\left[1 + (-3)^2\right]^{3/2}} = \dfrac{6}{10^{3/2}} \approx 0.19.$

[11] $y = \cos 2x \Rightarrow y' = -2\sin 2x$ and $y'' = -4\cos 2x$. At $P(0, 1)$, $x = 0$, $y' = 0$, and

$y'' = -4$. Using (15.18), $K = \dfrac{|y''|}{\left[1 + (y')^2\right]^{3/2}} = \dfrac{4}{\left[1 + (0)^2\right]^{3/2}} = 4.$

[17] In this exercise, the curve is described parametrically by $x = f(t) = 2\sin t$ and $y = g(t) = 3\cos t$. $f'(t) = 2\cos t$, $f''(t) = -2\sin t$, $g'(t) = -3\sin t$, and $g''(t) = -3\cos t$. At $P(1, \frac{3}{2}\sqrt{3})$, $x = 1 \Rightarrow 2\sin t = 1 \Rightarrow t = \sin^{-1}\frac{1}{2} = \frac{\pi}{6}$.

Using (15.19), $K = \dfrac{|f'(t)g''(t) - g'(t)f''(t)|}{\left[(f'(t))^2 + (g'(t))^2\right]^{3/2}} = \dfrac{\left|(\sqrt{3})(-\frac{3}{2}\sqrt{3}) - (-\frac{3}{2})(-1)\right|}{\left[(\sqrt{3})^2 + (-\frac{3}{2})^2\right]^{3/2}} =$

$48/21^{3/2} \approx 0.50.$

19 (a) $y = \sin x \Rightarrow y' = \cos x$ and $y'' = -\sin x$. At $P(\frac{\pi}{2}, 1)$, $x = \frac{\pi}{2}$, $y' = 0$, and

$$y'' = -1. \quad K = \frac{|y''|}{\left[1 + (y')^2\right]^{3/2}} = \frac{1}{\left[1 + 0^2\right]^{3/2}} = 1 \Rightarrow \rho = \frac{1}{K} = 1.$$

(b) Since $y' = 0$ and the graph of $y = \sin x$ is concave down at $x = \frac{\pi}{2}$, it follows that the center of curvature is on a vertical line 1 unit below P, that is, $(\frac{\pi}{2}, 0)$.

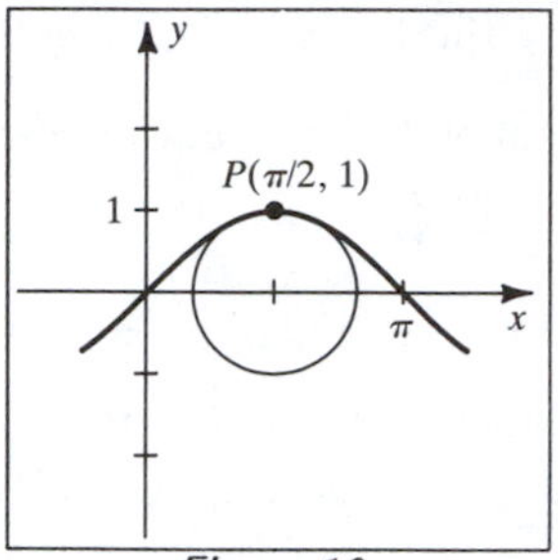

Figure 19

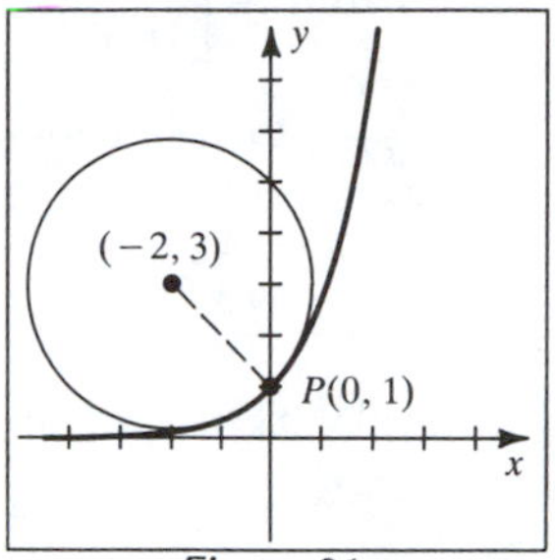

Figure 21

21 (a) $y = e^x \Rightarrow y' = e^x$ and $y'' = e^x$. At $P(0, 1)$, $x = 0$, $y' = 1$, and $y'' = 1$.

$$K = \frac{|y''|}{\left[1 + (y')^2\right]^{3/2}} = \frac{1}{\left[1 + 1^2\right]^{3/2}} = \frac{1}{2^{3/2}} = \frac{1}{2\sqrt{2}} \Rightarrow \rho = \frac{1}{K} = 2\sqrt{2}.$$

(b) Since $y' = 1$ at $x = 0$, it follows that the center of curvature is on the line with slope -1 passing through $(0, 1)$, or, $y = -x + 1$. The center C is a distance of $2\sqrt{2}$ from $(0, 1)$ on the concave side. If a is the side of the isosceles right triangle with hypotenuse $CP = 2\sqrt{2}$, then $a^2 + a^2 = (2\sqrt{2})^2 \Rightarrow 2a^2 = 8 \Rightarrow a^2 = 4 \Rightarrow a = 2$. Hence, the center of curvature is 2 units left and 2 units up from $(0, 1)$—namely, $(-2, 3)$.

23 $y = e^{-x} \Rightarrow y' = -e^{-x}$ and $y'' = e^{-x}$.

Thus, $K(x) = \dfrac{|y''|}{\left[1 + (y')^2\right]^{3/2}} = \dfrac{e^{-x}}{\left[1 + e^{-2x}\right]^{3/2}}$.

To find the maximum value of K, we will differentiate K and solve $K'(x) = 0$.

$$K'(x) = \frac{(1 + e^{-2x})^{3/2}(-e^{-x}) - (e^{-x})(\frac{3}{2})(1 + e^{-2x})^{1/2}(-2e^{-2x})}{(1 + e^{-2x})^3} = 0.$$

The numerator must be equal to 0 for $K'(x)$ to equal 0.

Hence, $(-e^{-x})(1 + e^{-2x})^{1/2}\left[(1 + e^{-2x}) - 3e^{-2x}\right] = 0$.

Since $-e^{-x} \neq 0$ and $1 + e^{-2x} \neq 0$ for any x, $1 - 2e^{-2x} = 0 \Rightarrow e^{-2x} = \frac{1}{2} \Rightarrow (e^{-x})^2 = \frac{1}{2} \Rightarrow e^{-x} = \frac{1}{\sqrt{2}} \Rightarrow e^x = \sqrt{2} \Rightarrow x = \ln\sqrt{2}$.

Since $K'(x) > 0$ if $x < \ln\sqrt{2}$ and $K'(x) < 0$ if $x > \ln\sqrt{2}$, $(\ln\sqrt{2}, 1/\sqrt{2})$ is the point of maximum curvature by the first derivative test.

[25] $9x^2 + 4y^2 = 36 \Rightarrow \frac{x^2}{4} + \frac{y^2}{9} = 1 \Rightarrow \left(\frac{x}{2}\right)^2 + \left(\frac{y}{3}\right)^2 = 1.$

Parametric equations for the ellipse are $x = f(t) = 2\cos t$ and $y = g(t) = 3\sin t$.

$f'(t) = -2\sin t$, $f''(t) = -2\cos t$, $g'(t) = 3\cos t$, and $g''(t) = -3\sin t$.

Using (15.19), $K(t) = \dfrac{|6\sin^2 t + 6\cos^2 t|}{\left[4\sin^2 t + 9\cos^2 t\right]^{3/2}} = \dfrac{|6(\sin^2 t + \cos^2 t)|}{\left[4(\sin^2 t + \cos^2 t) + 5\cos^2 t\right]^{3/2}} =$

$\dfrac{6}{\left[4 + 5\cos^2 t\right]^{3/2}}$. We can determine the maximum of $K(t)$ without differentiation.

$K(t)$ is a maximum when $\cos^2 t = 0$, i.e., when $t = \frac{\pi}{2}, \frac{3\pi}{2}$.

$(0, \pm 3)$ are the points of maximum curvature.

Note: In Exercises 29-32, the curvature is 0 only when $y'' = 0$ from (15.18).

[29] $y = x^4 - 12x^2 \Rightarrow y' = 4x^3 - 24x \Rightarrow y'' = 12x^2 - 24$. $y'' = 0 \Rightarrow x = \pm\sqrt{2}$.

$x = \pm\sqrt{2} \Rightarrow y = (\pm\sqrt{2})^4 - 12(\pm\sqrt{2})^2 = 4 - 24 = -20.$

The curvature is 0 at the points $(\pm\sqrt{2}, -20)$.

[33] Since $r = f(\theta)$, if we let $x = r\cos\theta$ and $y = r\sin\theta$, then x and y are also functions of θ. $x' = -r\sin\theta + r'\cos\theta$. $(x')^2 = r^2\sin^2\theta - 2rr'\sin\theta\cos\theta + (r')^2\cos^2\theta$.

$x'' = \left[-\frac{d}{d\theta}(\sin\theta)r + \frac{d}{d\theta}(-r)\sin\theta\right] + \left[\frac{d}{d\theta}(\cos\theta)r' + \frac{d}{d\theta}(r')\cos\theta\right]$

$= -r\cos\theta - r'\sin\theta - r'\sin\theta + r''\cos\theta.$

Similarly, $y' = r\cos\theta + r'\sin\theta$. $(y')^2 = r^2\cos^2\theta + 2rr'\sin\theta\cos\theta + (r')^2\sin^2\theta$.

$y'' = -r\sin\theta + r'\cos\theta + r'\cos\theta + r''\sin\theta.$

$x'y'' = r^2\sin^2\theta - 2rr'\cos\theta - rr''\sin^2\theta - 3rr'\sin\theta\cos\theta +$

$2(r')^2\cos^2\theta + r'r''\sin\theta\cos\theta$

and $y'x'' = -r^2\cos^2\theta - 3rr'\sin\theta\cos\theta + rr''\cos^2\theta - 2(r')^2\sin^2\theta +$

$r'r''\sin\theta\cos\theta$. Simplifying $K = \dfrac{|x'y'' - y'x''|}{\left[(x')^2 + (y')^2\right]^{3/2}}$ gives us $\dfrac{|2(r')^2 - rr'' + r^2|}{\left[(r')^2 + r^2\right]^{3/2}}$.

[35] $r = \sin 2\theta \Rightarrow r' = 2\cos 2\theta$ and $r'' = -4\sin 2\theta$.

$$K = \frac{|2(r')^2 - rr'' + r^2|}{\left[(r')^2 + r^2\right]^{3/2}} = \frac{|2(2\cos 2\theta)^2 - (\sin 2\theta)(-4\sin 2\theta) + (\sin 2\theta)^2|}{\left[(2\cos 2\theta)^2 + (\sin 2\theta)^2\right]^{3/2}}$$

$$= \frac{|8\cos^2 2\theta + 4\sin^2 2\theta + \sin^2 2\theta|}{\left[4\cos^2 2\theta + \sin^2 2\theta\right]^{3/2}} = \frac{8(\cos^2 2\theta + \sin^2 2\theta) - 3\sin^2\theta}{\left[(\cos^2 2\theta + \sin^2 2\theta) + 3\cos^2 2\theta\right]^{3/2}}$$

$$= \frac{8 - 3\sin^2 2\theta}{(1 + 3\cos^2 2\theta)^{3/2}}.$$

[39] $y = x^4 \Rightarrow y' = 4x^3$ and $y'' = 12x^2$. At $P(1, 1)$, $x = 1$, $y' = 4$, and $y'' = 12$.

From Exercise 37, $h = x - \dfrac{y'[1 + (y')^2]}{y''} = 1 - \dfrac{4[1 + 16]}{12} = -\frac{14}{3}$ and

$k = y + \dfrac{[1 + (y')^2]}{y''} = 1 + \dfrac{[1 + 16]}{12} = \frac{29}{12}.$

[43] The circular arc and the exit ramp must have the same curvature at $P(3, -1)$.

Thus, we need to find the center of curvature for $y = -\frac{1}{27}x^3$ at $P(3, -1)$.

If the circular arc has this as its center, the curvature will be continuous at $x = 3$.

$y' = -\frac{1}{9}x^2$ and $y'' = -\frac{2}{9}x$. When $x = 3$, $y' = -1$, and $y'' = -\frac{2}{3}$. From

Exercise 37, $h = 3 - \dfrac{(-1)\left[1 + (-1)^2\right]}{-\frac{2}{3}} = 0$ and $k = -1 + \dfrac{\left[1 + (-1)^2\right]}{-\frac{2}{3}} = -4$.

[45] Without loss of generality, let $y = ax^2 + bx + c$. Thus, $y' = 2ax + b$ and $y'' = 2a$.

$K(x) = \dfrac{|2a|}{\left[1 + (2ax + b)^2\right]^{3/2}}$ is maximum when the denominator is a minimum.

Since $(2ax + b)^2$ is a minimum when $2ax + b = 0$, or $x = -\dfrac{b}{2a}$, we conclude

that $K(x)$ is a maximum when x is the x-coordinate of the vertex of the parabola.

[47] Without loss of generality, let $\dfrac{x^2}{a^2} - \dfrac{y^2}{b^2} = 1$ with $a, b > 0$. Parametric equations for

the hyperbola are $x = f(t) = a \cosh t$ and $y = g(t) = b \sinh t$.

See Exercise 17, Section 13.1 for similar parametric equations.

Hence $f'(t) = a \sinh t$, $f''(t) = a \cosh t$, $g'(t) = b \cosh t$, $g''(t) = b \sinh t$, and using

(15.19), $K(t) = \dfrac{|ab \sinh^2 t - ab \cosh^2 t|}{\left[a^2 \sinh^2 t + b^2 \cosh^2 t\right]^{3/2}} = \dfrac{|ab(\sinh^2 t - \cosh^2 t)|}{\left[a^2 \sinh^2 t + b^2(1 + \sinh^2 t)\right]^{3/2}} =$

$\dfrac{|ab(1)|}{\left[a^2 \sinh^2 t + b^2 + b^2 \sinh^2 t\right]^{3/2}} = \dfrac{ab}{\left[(a^2 + b^2)\sinh^2 t + b^2\right]^{3/2}}$.

$K(t)$ is maximum when $\sinh^2 t = 0$, or $t = 0$. Thus, maximum curvature occurs at

$(a, 0)$ and by symmetry at $(-a, 0)$, that is, at the ends of the transverse axis.

[49] $x = f(t) = 4t - 3 \Rightarrow f'(t) = 4$ and $y = g(t) = 3t + 5 \Rightarrow g'(t) = 3$.

$s = \displaystyle\int_a^t \sqrt{[f'(u)]^2 + [g'(u)]^2}\, du = \int_0^t \sqrt{(4)^2 + (3)^2}\, du = 5\int_0^t du = 5\Big[\, u \,\Big]_0^t = 5t \Rightarrow$

$t = \frac{1}{5}s$. Thus, $x = \frac{4}{5}s - 3$ and $y = \frac{3}{5}s + 5$; $t \geq 0 \Rightarrow s \geq 0$.

[51] $x = f(t) = 4\cos t \Rightarrow f'(t) = -4\sin t$ and $y = g(t) = 4\sin t \Rightarrow g'(t) = 4\cos t$.

$s = \displaystyle\int_0^t \sqrt{16\sin^2 u + 16\cos^2 u}\, du = \int_0^t 4\, du = 4t \Rightarrow t = \tfrac{1}{4}s$.

Thus, $x = 4\cos\frac{1}{4}s$ and $y = 4\sin\frac{1}{4}s$; $0 \leq t \leq 2\pi$ and $t = \frac{1}{4}s \Rightarrow 0 \leq s \leq 8\pi$.

Exercises 15.5

Note: In Exercises 1-8, we use (15.22) for a_{T}, (15.23) for a_{N}, and (15.25) for K.

You could also use (15.24) for finding a_{N}.

[1] $\mathbf{r}(t) = t^2\mathbf{i} + (3t + 2)\mathbf{j}$, $\mathbf{r}'(t) = 2t\mathbf{i} + 3\mathbf{j}$, $\mathbf{r}''(t) = 2\mathbf{i}$, and $\|\mathbf{r}'(t)\| = \sqrt{4t^2 + 9}$.

Using (15.22), $a_{\mathrm{T}} = \dfrac{\mathbf{r}'(t)\cdot\mathbf{r}''(t)}{\|\mathbf{r}'(t)\|} = \dfrac{4t}{(4t^2+9)^{1/2}}$.

Using (15.23), $a_{\mathrm{N}} = \dfrac{\|\mathbf{r}'(t)\times\mathbf{r}''(t)\|}{\|\mathbf{r}'(t)\|} = \dfrac{\|-6\mathbf{k}\|}{(4t^2+9)^{1/2}} = \dfrac{6}{(4t^2+9)^{1/2}}$.

Using (15.25), $K = a_{\mathrm{N}}\dfrac{1}{\|\mathbf{r}'(t)\|^2} = \dfrac{6}{(4t^2+9)^{1/2}}\cdot\dfrac{1}{4t^2+9} = \dfrac{6}{(4t^2+9)^{3/2}}$.

[7] $\mathbf{r}(t) = 4\cos t\,\mathbf{i} + 9\sin t\,\mathbf{j} + t\mathbf{k}$, $\mathbf{r}'(t) = -4\sin t\,\mathbf{i} + 9\cos t\,\mathbf{j} + \mathbf{k}$,

$\mathbf{r}''(t) = -4\cos t\,\mathbf{i} - 9\sin t\,\mathbf{j}$, and $\|\mathbf{r}'(t)\| = \sqrt{16\sin^2 t + 81\cos^2 t + 1}$.

$$a_{\mathrm{T}} = \frac{\mathbf{r}'(t)\cdot\mathbf{r}''(t)}{\|\mathbf{r}'(t)\|} = \frac{-65\sin t\cos t}{(16\sin^2 t + 81\cos^2 t + 1)^{1/2}}.$$

$$a_{\mathrm{N}} = \frac{\|\mathbf{r}'(t)\times\mathbf{r}''(t)\|}{\|\mathbf{r}'(t)\|} = \frac{\|9\sin t\,\mathbf{i} - 4\cos t\,\mathbf{j} + 36\mathbf{k}\|}{(16\sin^2 t + 81\cos^2 t + 1)^{1/2}} =$$

$$\frac{(81\sin^2 t + 16\cos^2 t + 1296)^{1/2}}{(16\sin^2 t + 81\cos^2 t + 1)^{1/2}}.$$

$$K = a_{\mathrm{N}}\cdot\frac{1}{16\sin^2 t + 81\cos^2 t + 1} = \frac{(81\sin^2 t + 16\cos^2 t + 1296)^{1/2}}{(16\sin^2 t + 81\cos^2 t + 1)^{3/2}}.$$

[9] Let $\mathbf{r}(t) = f(t)\,\mathbf{i} + g(t)\,\mathbf{j}$, where $g(t) = [f(t)]^2$ $\{y = x^2\}$. The velocity of the point is $\mathbf{r}'(t) = f'(t)\,\mathbf{i} + g'(t)\,\mathbf{j}$. Since the horizontal component of the velocity is 3, $f'(t) = 3$. Thus, $f(t) = \int 3\,dt = 3t + C$. Since $f(1) = 1$, $1 = 3 + C$, and $C = -2$.
So $f(t) = 3t - 2$ and $g(t) = (3t-2)^2$. Hence, $\mathbf{r}(t) = (3t-2)\mathbf{i} + (3t-2)^2\mathbf{j}$,
$\mathbf{r}'(t) = 3\mathbf{i} + 6(3t-2)\mathbf{j}$, $\mathbf{r}''(t) = 18\mathbf{j}$, and $\|\mathbf{r}'(t)\| = \sqrt{9 + 36(3t-2)^2}$. Let
$t = 1$ since $P(1, 1) = (f(1), g(1))$. $\mathbf{r}'(1) = 3\mathbf{i} + 6\mathbf{j}$, $\mathbf{r}''(1) = 18\mathbf{j}$, and $\|\mathbf{r}'(1)\| = \sqrt{45}$.

$$a_{\mathrm{T}} = \frac{\mathbf{r}'(1)\cdot\mathbf{r}''(1)}{\|\mathbf{r}'(1)\|} = \frac{108}{\sqrt{45}} = \frac{36}{\sqrt{5}} \approx 16.10.$$

$$a_{\mathrm{N}} = \frac{\|\mathbf{r}'(1)\times\mathbf{r}''(1)\|}{\|\mathbf{r}'(1)\|} = \frac{\|54\,\mathbf{k}\|}{\sqrt{45}} = \frac{54}{\sqrt{45}} = \frac{18}{\sqrt{5}} \approx 8.05.$$

[11] Since $v = \|\mathbf{v}(t)\| = \frac{ds}{dt}$ is a constant, $a_{\mathrm{T}} = \dfrac{d^2s}{dt^2} = 0$.

But a_{T} is also equal to $\dfrac{\mathbf{r}'(t)}{\|\mathbf{r}'(t)\|}\cdot\mathbf{r}''(t)$, which is $\mathbf{T}(t)\cdot\mathbf{r}''(t)$. Thus, the acceleration $\{\mathbf{r}''(t)\}$ is normal to the unit tangent vector $\{\mathbf{T}(t)\}$, and therefore normal to C.

[13] Since the speed is constant, $\dfrac{dv}{dt} = \dfrac{d^2s}{dt^2} = 0$. Thus, the formula for the acceleration given near the bottom of page 780, $\mathbf{a}(t) = \frac{dv}{dt}\mathbf{T}(s) + Kv^2\,\mathbf{N}(s)$, reduces to $Kv^2\,\mathbf{N}(s)$.
Hence, the magnitude of the acceleration is given by $\|\mathbf{a}(t)\| = Kv^2 = cK$,
where $c = v^2$ is a constant and $\|\mathbf{N}(s)\| = 1$.

[17] $\mathbf{r}(t) = x\mathbf{i} + y\mathbf{j} + z\mathbf{k} = a\cos t\mathbf{i} + a\sin t\mathbf{j} + bt\mathbf{k} \Rightarrow$

$\mathbf{r}'(t) = -a\sin t\mathbf{i} + a\cos t\mathbf{j} + b\mathbf{k} \Rightarrow \mathbf{r}''(t) = -a\cos t\mathbf{i} - a\sin t\mathbf{j}.$

$\|\mathbf{r}'(t)\| = \sqrt{a^2 + b^2}$ and $\mathbf{r}'(t) \times \mathbf{r}''(t) = ab\sin t\mathbf{i} - ab\cos t\mathbf{j} + a^2\mathbf{k}.$

Thus, $K = \dfrac{\|\mathbf{r}'(t) \times \mathbf{r}''(t)\|}{\|\mathbf{r}'(t)\|^3} = \dfrac{\left[a^2b^2(\sin^2 t + \cos^2 t) + a^4\right]^{1/2}}{(a^2 + b^2)^{3/2}} =$

$$\frac{\left[a^2(b^2 + a^2)\right]^{1/2}}{(a^2 + b^2)^{3/2}} = \frac{a(b^2 + a^2)^{1/2}}{(a^2 + b^2)^{3/2}} = \frac{a}{a^2 + b^2}.$$

15.7 Review Exercises

[1] $\mathbf{r}(t) = \tan t\mathbf{i} - \sec t\mathbf{j} \Rightarrow x = \tan t$ and $y = -\sec t$.

Because $y^2 - x^2 = \sec^2 t - \tan^2 t = 1$, the point P will move on the hyperbola $y^2 - x^2 = 1$.

Since $-\frac{\pi}{2} < t < \frac{\pi}{2}$, $y = -\sec t < 0$ and $-\infty < x < \infty$.

Thus, C is the lower branch of the hyperbola.

$\mathbf{v}(t) = \mathbf{r}'(t) = \sec^2 t\mathbf{i} - \sec t\tan t\mathbf{j}$; $\mathbf{v}(\frac{\pi}{4}) = 2\mathbf{i} - \sqrt{2}\mathbf{j}$.

$\mathbf{a}(t) = \mathbf{r}''(t) = 2\sec^2 t\tan t\mathbf{i} - (\sec^3 t + \tan^2 t\sec t)\mathbf{j}$;

$\mathbf{a}(\frac{\pi}{4}) = 4\mathbf{i} - 3\sqrt{2}\mathbf{j}$.

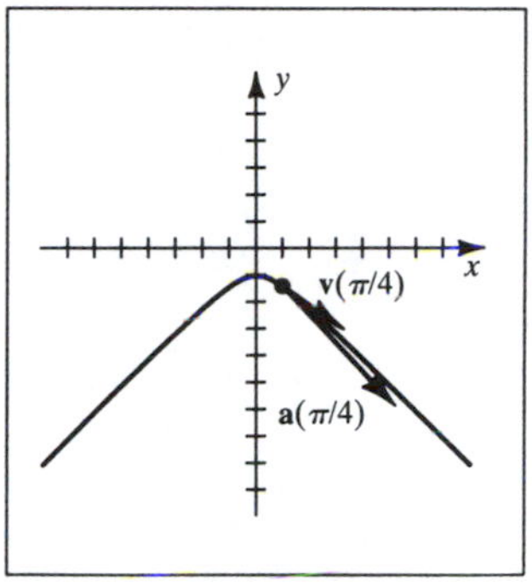

Figure 1

[3] $\mathbf{r}(t) = t^2\mathbf{i} + (4t^2 - t^4)\mathbf{j} \Rightarrow \mathbf{r}'(t) = 2t\mathbf{i} + (8t - 4t^3)\mathbf{j}$, $\mathbf{r}''(t) = 2\mathbf{i} + (8 - 12t^2)\mathbf{j}$,

and $\|\mathbf{r}'(t)\| = \sqrt{(2t)^2 + (8t - 4t^3)^2} = \sqrt{4t^2 + 64t^2 - 64t^4 + 16t^6} =$

$$\sqrt{4t^2(17 - 16t^2 + 4t^4)} = 2|t|\sqrt{17 - 16t^2 + 4t^4}.$$

[7] $\mathbf{v}(t) = t\mathbf{i} - 5t\mathbf{j} + 4t^2\mathbf{k} \Rightarrow \mathbf{v}'(t) = \mathbf{i} - 5\mathbf{j} + 8t\mathbf{k}$. $\mathbf{u}(t)$ and $\mathbf{v}'(t)$ will be orthogonal if their dot product is zero. Their dot product is $\mathbf{u}(t)\cdot\mathbf{v}'(t) =$

$(t^2\mathbf{i} + 6t\mathbf{j} + t\mathbf{k})\cdot(\mathbf{i} - 5\mathbf{j} + 8t\mathbf{k}) = t^2 - 30t + 8t^2 = 9t^2 - 30t = 3t(3t - 10)$.

Hence, $\mathbf{u}(t)\cdot\mathbf{v}'(t) = 0 \Leftrightarrow 3t(3t - 10) = 0 \Leftrightarrow t = 0, \frac{10}{3}$.

[9] In the answer in the first printing, the first $\sqrt{34}$ should be $\sqrt{9 + 9t^4 + 16t^6}$.

[11] $\displaystyle\int_0^1 (4t\mathbf{i} + t^3\mathbf{j} - \mathbf{k})\,dt = \left[2t^2\mathbf{i} + \tfrac{1}{4}t^4\mathbf{j} - t\mathbf{k}\right]_0^1 = 2\mathbf{i} + \tfrac{1}{4}\mathbf{j} - \mathbf{k}$

[13] $\mathbf{a}(t) = 12t\mathbf{j} + 5\mathbf{k} \Rightarrow \mathbf{v}(t) = \int \mathbf{a}(t)\,dt = 6t^2\mathbf{j} + 5t\mathbf{k} + \mathbf{c}$.

$\mathbf{v}(1) = 6\mathbf{j} + 5\mathbf{k} + \mathbf{c}$ and $\mathbf{v}(1) = 3\mathbf{i} - 2\mathbf{j} + 4\mathbf{k} \Rightarrow$

$6\mathbf{j} + 5\mathbf{k} + \mathbf{c} = 3\mathbf{i} - 2\mathbf{j} + 4\mathbf{k} \Rightarrow \mathbf{c} = 3\mathbf{i} + (-2 - 6)\mathbf{j} + (4 - 5)\mathbf{k} = 3\mathbf{i} - 8\mathbf{j} - \mathbf{k}$

and hence $\mathbf{v}(t) = 3\mathbf{i} + (6t^2 - 8)\mathbf{j} + (5t - 1)\mathbf{k}$.

$\mathbf{r}(t) = \int \mathbf{v}(t)\,dt = 3t\mathbf{i} + (2t^3 - 8t)\mathbf{j} + (\frac{5}{2}t^2 - t)\mathbf{k} + \mathbf{c}$.

$\mathbf{r}(1) = 3\mathbf{i} - 6\mathbf{j} + \frac{3}{2}\mathbf{k} + \mathbf{c}$ and $\mathbf{r}(1) = -\mathbf{i} + 3\mathbf{j} + \frac{3}{2}\mathbf{k} \Rightarrow$

$3\mathbf{i} - 6\mathbf{j} + \frac{3}{2}\mathbf{k} + \mathbf{c} = -\mathbf{i} + 3\mathbf{j} + \frac{3}{2}\mathbf{k} \Rightarrow \mathbf{c} = (-1 - 3)\mathbf{i} + [3 - (-6)]\mathbf{j} =$

$-4\mathbf{i} + 9\mathbf{j}$. Thus, $\mathbf{r}(t) = (3t - 4)\mathbf{i} + (2t^3 - 8t + 9)\mathbf{j} + (\frac{5}{2}t^2 - t)\mathbf{k}$.

[15] Using (15.8)(iii) with $\mathbf{v}(t) = \mathbf{u}'(t) \times \mathbf{u}''(t)$,

$$D_t\big[\mathbf{u}(t) \cdot \mathbf{u}'(t) \times \mathbf{u}''(t)\big] = \mathbf{u}(t) \cdot D_t\big[\mathbf{u}'(t) \times \mathbf{u}''(t)\big] + \mathbf{u}'(t) \cdot \big[\mathbf{u}'(t) \times \mathbf{u}''(t)\big].$$

Since $\mathbf{u}'(t) \times \mathbf{u}''(t)$ is orthogonal to $\mathbf{u}'(t)$, the second expression equals 0.

By (15.8)(iv), $\mathbf{u}(t) \cdot D_t\big[\mathbf{u}'(t) \times \mathbf{u}''(t)\big] = \mathbf{u}(t) \cdot \big[\mathbf{u}'(t) \times \mathbf{u}'''(t) + \mathbf{u}''(t) \times \mathbf{u}''(t)\big]$.

Since $\mathbf{u}''(t) \times \mathbf{u}''(t) = \mathbf{0}$, the last expression is $= \mathbf{u}(t) \cdot \mathbf{u}'(t) \times \mathbf{u}'''(t)$.

[17] Let $x = f(t) = \dfrac{1}{1+t} = (1+t)^{-1}$ and $y = g(t) = \dfrac{1}{1-t} = (1-t)^{-1}$.

$f'(t) = -(1+t)^{-2}$, $f''(t) = 2(1+t)^{-3}$, $g'(t) = (1-t)^{-2}$, $g''(t) = 2(1-t)^{-3}$.

At $P(\frac{2}{3}, 2)$, $y = 2 \Rightarrow \dfrac{1}{1-t} = 2 \Rightarrow 1 = 2 - 2t \Rightarrow t = \frac{1}{2}$.

Using (15.19), $K = \dfrac{\left|(-\frac{4}{9})(16) - (4)(\frac{16}{27})\right|}{\left[(-\frac{4}{9})^2 + (4)^2\right]^{3/2}} = \dfrac{108}{82^{3/2}} \approx 0.15$.

[19] $y = x^3 - 3x \Rightarrow y' = 3x^2 - 3$ and $y'' = 6x$. Using (15.18),

$K(x) = \dfrac{|6x|}{\left[1 + (3x^2 - 3)^2\right]^{3/2}} = \dfrac{|6x|}{(9x^4 - 18x^2 + 10)^{3/2}}$. If $x > 0$, then $|6x| = 6x$ and

$K'(x) = \dfrac{(9x^4 - 18x^2 + 10)^{3/2}(6) - (6x)(\frac{3}{2})(9x^4 - 18x^2 + 10)^{1/2}(36x^3 - 36x)}{(9x^4 - 18x^2 + 10)^3}$ and

$K'(x) = 0 \Rightarrow (9x^4 - 18x^2 + 10)(6) - (9x)(36x^3 - 36x) = 0 \Rightarrow$

$45x^4 - 36x^2 - 10 = 0 \Rightarrow x^2 = \dfrac{36 \pm \sqrt{3096}}{90} \Rightarrow x \approx 1.009$ since $x > 0$.

This gives a maximum curvature since $K'(x) > 0$ on $[0, 1.009)$ and $K'(x) < 0$ on $(1.009, \infty)$. Since the graph of $y = x^3 - 3x$ is symmetric with respect to the origin, another maximum curvature exists at $x \approx -1.009$.

[23] $\mathbf{r}(t) = \sin 2t\,\mathbf{i} + \cos t\,\mathbf{j}$, $\mathbf{r}'(t) = 2\cos 2t\,\mathbf{i} - \sin t\,\mathbf{j}$, $\mathbf{r}''(t) = -4\sin 2t\,\mathbf{i} - \cos t\,\mathbf{j}$,

and $\|\mathbf{r}'(t)\| = \sqrt{4\cos^2 2t + \sin^2 t}$. $a_{\mathsf{T}} = \dfrac{\mathbf{r}'(t) \cdot \mathbf{r}''(t)}{\|\mathbf{r}'(t)\|} = \dfrac{-8\cos 2t \sin 2t + \sin t \cos t}{(4\cos^2 2t + \sin^2 t)^{1/2}}$.

$$a_{\mathsf{N}} = \frac{\|\mathbf{r}'(t) \times \mathbf{r}''(t)\|}{\|\mathbf{r}'(t)\|} = \frac{\|-(2\cos 2t \cos t + 4\sin 2t \sin t)\,\mathbf{k}\|}{(4\cos^2 2t + \sin^2 t)^{1/2}} = \frac{2|\cos 2t \cos t + 2\sin 2t \sin t|}{(4\cos^2 2t + \sin^2 t)^{1/2}}.$$

Chapter 16: Partial Differentiation

Exercises 16.1

Note: In Exercises 1–6, let D denote the domain of f.

[1] f is defined for all ordered pairs $(x, y) \Rightarrow D = \mathbb{R}^2$. $f(x, y) = 2x - y^2 \Rightarrow$
$f(-2, 5) = 2(-2) - 5^2 = -29$, $f(5, -2) = 6$, and $f(0, -2) = -4$.

[5] f will be defined when $25 - x^2 - y^2 - z^2 \geq 0 \Rightarrow x^2 + y^2 + z^2 \leq 25$. Thus,
$D = \{(x, y, z): x^2 + y^2 + z^2 \leq 25\}$. $f(x, y, z) = \sqrt{25 - x^2 - y^2 - z^2} \Rightarrow$
$f(1, -2, 2) = \sqrt{25 - 1^2 - (-2)^2 - 2^2} = 4$ and $f(-3, 0, 2) = 2\sqrt{3}$.

Note: In Exercises 7–14, let $z = f(x, y)$.

[7] $z = \sqrt{1 - x^2 - y^2} \Rightarrow z^2 = 1 - x^2 - y^2 \Rightarrow x^2 + y^2 + z^2 = 1, z \geq 0$.
This is the top half of the sphere with radius 1 and center (0, 0, 0).

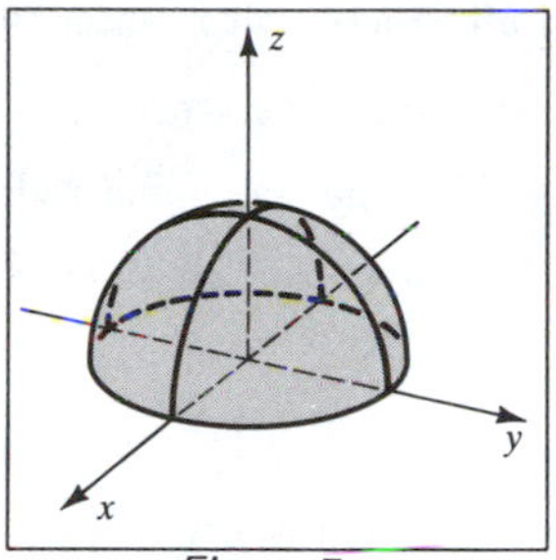

Figure 7

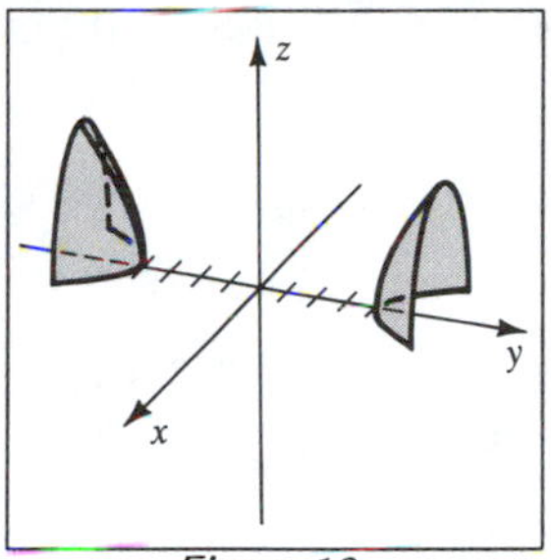

Figure 13

[13] $z = \sqrt{y^2 - 4x^2 - 16} \Rightarrow z^2 = y^2 - 4x^2 - 16 \Rightarrow y^2 - 4x^2 - z^2 = 16 \Rightarrow$
$\frac{y^2}{16} - \frac{x^2}{4} - \frac{z^2}{16} = 1, z \geq 0$. This is the upper half of a hyperboloid of two sheets with
axis along the y-axis. See the figure for Exercise 13, Section 14.6.

[15] $y^2 - x^2 = k$ are hyperbolas with asymptotes $y = \pm x$ and a vertical (horizontal)
transverse axis if $k > 0$ $(k < 0)$. If $k = 0$, we have $y = \pm x$.
See Example 4 and Figure 16.9.

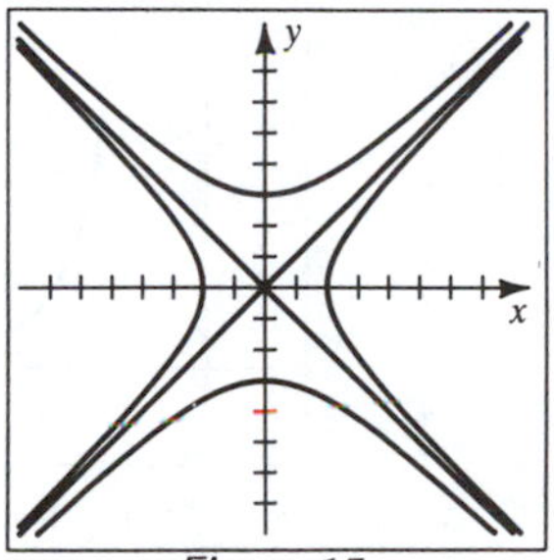

Figure 15

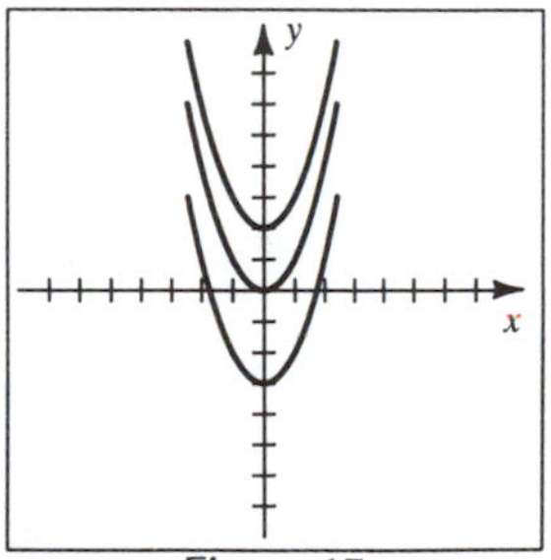

Figure 17

[17] $x^2 - y = k \Rightarrow y = x^2 - k$. These are parabolas with vertices $(0, -k)$.

[21] The level curves have equation $y \arctan x = k$.

For the level curve to pass through $P(1, 4)$, we must have $k = f(1, 4)$.

Since $f(1, 4) = 4 \arctan 1 = 4(\frac{\pi}{4}) = \pi$, an equation is $y \arctan x = \pi$.

[23] The level curves have equation $x^2 + 4y^2 - z^2 = k$.

For the level surface to pass through $P(2, 1, -3)$, we must have $k = f(2, 1, -3)$.

Since $f(2, -1, 3) = 4 + 4 - 9 = -1$, an equation is $x^2 + 4y^2 - z^2 = -1$.

[25] $\dfrac{1}{9x^2 + y^2} = k \Rightarrow 9x^2 + y^2 = \frac{1}{k}$. (a) $k > 0$ yields ellipses.

(b) & (c) $k = 0$ and $k < 0$ yield no curves since $9x^2 + y^2 \geq 0$ for all x and y.

[29] Let $u = xy$. Since $\lim\limits_{u \to 0} \dfrac{\sin u}{u} = 1$, we see that $z \to 1$ when $u \to 0$. u will be close to zero along both the x- and y-axes. The only graph satisfying this characteristic is (d).

[31] The only graph having circular level curves is (a).

[37] $x^2 + y^2 + z^2 = k$ include the origin ($k = 0$) and all spheres with center $(0, 0, 0)$ and radius $\sqrt{k}$. $k < 0$ yields no surface since $x^2 + y^2 + z^2 \geq 0$.

[41] $x^2 + y^2 = k$ are right circular cylinders with axis along the z-axis and radius $\sqrt{k}$.

$k = 0$ gives us the z-axis and $k < 0$ yields no surface since $x^2 + y^2 \geq 0$.

[43] The distance of a point (x, y) from the origin is $\sqrt{x^2 + y^2}$.

Since T is inversely proportional to this distance,

the temperature is given by $T = \dfrac{c}{\sqrt{x^2 + y^2}}$, where c is a constant.

(a) Isotherms are the level curves of T. $T = k \Rightarrow \dfrac{c}{\sqrt{x^2 + y^2}} = k \Rightarrow$

$x^2 + y^2 = (\frac{c}{k})^2$. These are circles with center $(0, 0)$ and radius $\frac{c}{k}$.

(b) $x = 4$, $y = 3$, $T = 40 \Rightarrow 40 = \dfrac{c}{\sqrt{4^2 + 3^2}} \Rightarrow c = 200$. At $T = 20$,

the equation in part (a) with $k = 20$ becomes $x^2 + y^2 = (\frac{200}{20})^2 = 100$.

[47] (a) $P = cAv^3$, where $c > 0$ is a constant.

(b) $k = cAv^3 \Rightarrow A = \dfrac{k}{cv^3}$, where $k \geq 0$ is a constant.

A typical level curve (see *Figure 47*) shows the combinations of areas and wind velocities that result in a fixed power $P = k$.

(c) $d = 10 \Rightarrow r = 5$.

$3000 = c(\pi 5^2)(20^3) \Rightarrow c = \frac{3}{200\pi}$.

Thus, $4000 = \frac{3}{200\pi} Av^3 \Rightarrow Av^3 = \frac{8}{3}\pi \times 10^5$.

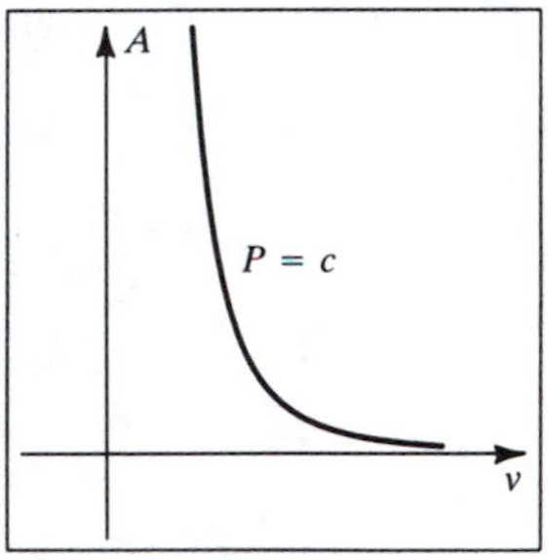

Figure 47

[49] Example: 5′11″ and 175 lb are approximately 180 cm and 80 kg.

From the graph, we have a surface area of approximately 2.0 m^2.

Using the formula, we obtain $S = 0.007184(80)^{0.425}(180)^{0.725} \approx 1.996 \text{ m}^2$.

Try a different example—perhaps your own height and weight.

Note: 1 inch $\approx$ 2.54 cm and 1 lb $\approx$ 0.454 kg.

[51] Without loss of generality, let the two cities be located at $F'(-c, 0)$ and $F(c, 0)$. If the plant is located at $P(x, y)$, then the sum of the distances from P to each city is equal to the constant M. By (12.5), this is the definition of an ellipse. See Figure 12.14.

Exercises 16.2

[1] Substituting 0 for x and 0 for y gives us $\lim_{(x,y)\to(0,0)} \dfrac{x^2 - 2}{3 + xy} = \dfrac{0 - 2}{3 + 0} = -\frac{2}{3}$.

[5] Substituting 0 for x and 0 for y yields a zero in the denominator.

Thus, we must try to change the form of the expression before substituting.

$$\lim_{(x,y)\to(0,0)} \frac{x^4 - y^4}{x^2 + y^2} = \lim_{(x,y)\to(0,0)} \frac{(x^2 + y^2)(x^2 - y^2)}{x^2 + y^2} = \lim_{(x,y)\to(0,0)} (x^2 - y^2) = 0.$$

[9] As in Exercise 5, we first change the form of the expression.

$$\lim_{(x,y,z)\to(2,3,1)} \frac{y^2 - 4y + 3}{x^2 z(y - 3)} = \lim_{(x,y,z)\to(2,3,1)} \frac{(y - 3)(y - 1)}{x^2 z(y - 3)} =$$

$$\lim_{(x,y,z)\to(2,3,1)} \frac{y - 1}{x^2 z} = \frac{3 - 1}{4(1)} = \frac{1}{2}$$

Note: In future exercises, L denotes the indicated limit.

[11] Let $f(x, y) = \dfrac{2x^2 - y^2}{x^2 + 2y^2}$. Along the path $x = 0$, $f(0, y) = \dfrac{-y^2}{2y^2} = -\frac{1}{2}$, and hence L $= -\frac{1}{2}$. Along the path $y = 0$, $f(x, 0) = \dfrac{2x^2}{x^2} = 2$, and hence L $= 2$.

Thus, by the two-path rule (16.4), L DNE.

[13] $\text{L} = \lim_{(x,y)\to(1,2)} \dfrac{xy - 2x - y + 2}{x^2 + y^2 - 2x - 4y + 5} = \lim_{(x,y)\to(1,2)} \dfrac{(x - 1)(y - 2)}{(x - 1)^2 + (y - 2)^2}$.

We will use the technique shown in Example 4 to show that the limit does not exist. In this case, we will consider all lines with slope m that pass through the point (1, 2). Hence, along the path $(y - 2) = m(x - 1)$,

$$\frac{(x - 1)(y - 2)}{(x - 1)^2 + (y - 2)^2} = \frac{m(x - 1)^2}{(x - 1)^2 + m^2(x - 1)^2} = \frac{m}{1 + m^2}.$$

Since different values for L can be obtained using different values of m, L DNE.

[17] Let $f(x, y, z) = \dfrac{xy + yz + xz}{x^2 + y^2 + z^2}$.

Along the z-axis ($x = y = 0$), $f(0, 0, z) = \dfrac{0}{z^2} = 0$, and hence L = 0.

Along the path $x = y = z$, $f(z, z, z) = \dfrac{3z^2}{3z^2} = 1$, and hence L = 1. Thus, L DNE.

[19] We will consider all lines that pass through (0, 0, 0) using a parametric equation representation. (See (14.34).) Along the path $x = at$, $y = bt$, and $z = ct$,

$$\mathrm{L} = \frac{a^3t^3 + b^3t^3 + c^3t^3}{abct^3} = \frac{(a^3 + b^3 + c^3)t^3}{abct^3} = \frac{a^3 + b^3 + c^3}{abc}.$$ Since different values for L can be obtained using different values of a, b, and c, L DNE.

Note: In Exercises 21–24 and 33–34, we use the fact that as $(x, y) \rightarrow (0, 0)$, $r \rightarrow 0$.

[21] Letting $x = r\cos\theta$ and $y = r\sin\theta$, $\dfrac{xy^2}{x^2 + y^2} = \dfrac{r^3\cos\theta\sin^2\theta}{r^2} = r\cos\theta\sin^2\theta$.

As $r \rightarrow 0$, $r\cos\theta\sin^2\theta \rightarrow 0$ for all θ. Thus, L = 0.

[23] $\dfrac{x^2 + y^2}{\sin(x^2 + y^2)} = \dfrac{r^2}{\sin r^2}$. As $r \rightarrow 0$, $r^2 \rightarrow 0$, and $\dfrac{r^2}{\sin r^2} \rightarrow 1$ for all θ. Thus, L = 1.

[25] f will be continuous wherever $\ln(x + y - 1)$ is defined.

$x + y - 1 > 0 \Rightarrow x + y > 1$. Thus, f is continuous on $\{(x, y): x + y > 1\}$.

[29] Since the denominator cannot equal 0, $x^2 + y^2 - z^2 \neq 0 \Rightarrow z^2 \neq x^2 + y^2$.

Thus, f is continuous on $\{(x, y, z): z^2 \neq x^2 + y^2\}$.

[33] Since $r^2 = x^2 + y^2$, $\displaystyle\lim_{(x,y)\rightarrow(0,0)} \frac{x^2 + y^2}{\ln(x^2 + y^2)} = \lim_{r\rightarrow 0} \frac{r^2}{\ln r^2} = 0$.

Recall that as $r \rightarrow 0$, $\ln r^2 \rightarrow -\infty$.

[35] $h(x, y) = g(f(x, y)) =$

$$\frac{[f(x, y)]^2 - 4}{f(x, y)} = \frac{(x^2 - y^2)^2 - 4}{x^2 - y^2} = \frac{x^4 - 2x^2y^2 + y^4 - 4}{x^2 - y^2}.$$

h is continuous on $\{(x, y): x^2 \neq y^2\}$.

[39] $g(f(x, y)) = e^{f(x,y)} = e^{x^2+2y}$.

$h(f(x, y)) = [f(x, y)]^2 - 3f(x, y) = (x^2 + 2y)^2 - 3(x^2 + 2y) =$

$(x^2 + 2y)(x^2 + 2y - 3)$.

$f(g(t), h(t)) = [g(t)]^2 + 2h(t) = (e^t)^2 + 2(t^2 - 3t) = e^{2t} + 2(t^2 - 3t)$.

[45] (a) Let $f(x, y) = x$ and $\epsilon > 0$ be given. Choose $\delta = \epsilon$.

If $0 < \sqrt{(x - a)^2 + (y - b)^2} < \delta$,

then $\sqrt{(x - a)^2} = |x - a| = |f(x, y) - a| < \delta = \epsilon$. Thus, $\displaystyle\lim_{(x,y)\rightarrow(a,b)} x = a$.

(b) Let $f(x, y) = y$ and $\epsilon > 0$ be given. Choose $\delta = \epsilon$.

If $0 < \sqrt{(x - a)^2 + (y - b)^2} < \delta$,

then $\sqrt{(y - b)^2} = |y - b| = |f(x, y) - b| < \delta = \epsilon$. Thus, $\displaystyle\lim_{(x,y)\rightarrow(a,b)} y = b$.

Exercises 16.3

Note: We sometimes denote $f_x(x, y)$ with f_x, $g_r(r, s)$ with g_r, etc.

1 $f(x, y) = 2x^4y^3 - xy^2 + 3y + 1 \Rightarrow$

$$f_x = \frac{\partial}{\partial x}(2x^4y^3) - \frac{\partial}{\partial x}(xy^2) + \frac{\partial}{\partial x}(3y) + \frac{\partial}{\partial x}(1)$$
$$= y^3 \cdot \frac{\partial}{\partial x}(2x^4) - y^2 \cdot \frac{\partial}{\partial x}(x) + 0 + 0 \text{ \{ treat } y \text{ as a constant \}}$$
$$= y^3(8x^3) - y^2(1) = 8x^3y^3 - y^2.$$
$$f_y = \frac{\partial}{\partial y}(2x^4y^3) - \frac{\partial}{\partial y}(xy^2) + \frac{\partial}{\partial y}(3y) + \frac{\partial}{\partial y}(1)$$
$$= 2x^4 \cdot \frac{\partial}{\partial y}(y^3) - x \cdot \frac{\partial}{\partial y}(y^2) + 3 + 0 \text{ \{ treat } x \text{ as a constant \}}$$
$$= 2x^4(3y^2) - x(2y) + 3 = 6x^4y^2 - 2xy + 3$$

5 $f(x, y) = xe^y + y \sin x \Rightarrow f_x = e^y + y \cos x$ and $f_y = xe^y + \sin x$.

7 In this exercise, it is easier to simplify the logarithm before differentiating.

$$f(t, v) = \ln \sqrt{\frac{t+v}{t-v}} = \ln(t+v)^{1/2} - \ln(t-v)^{1/2} = \tfrac{1}{2}\ln(t+v) - \tfrac{1}{2}\ln(t-v) \Rightarrow$$
$$f_t(t, v) = \frac{1}{2(t+v)} - \frac{1}{2(t-v)} = \frac{-2v}{2(t^2-v^2)} = -\frac{v}{t^2-v^2}.$$
$$f_v(t, v) = \frac{1}{2(t+v)} + \frac{1}{2(t-v)} = \frac{2t}{2(t^2-v^2)} = \frac{t}{t^2-v^2}.$$

9 $f(x, y) = x \cos \frac{x}{y} \Rightarrow f_x = \frac{\partial}{\partial x}(x) \cdot (\cos \frac{x}{y}) + (x) \cdot \frac{\partial}{\partial x}(\cos \frac{x}{y}) =$

$$(1)(\cos \tfrac{x}{y}) + x(-\sin \tfrac{x}{y})\frac{\partial}{\partial x}(\tfrac{x}{y}) = \cos \tfrac{x}{y} - x \sin \tfrac{x}{y}(\tfrac{1}{y}) = \cos \tfrac{x}{y} - \tfrac{x}{y}\sin \tfrac{x}{y}.$$
$$f_y = x \cdot \frac{\partial}{\partial y}(\cos \tfrac{x}{y}) = x(-\sin \tfrac{x}{y})\frac{\partial}{\partial y}(\tfrac{x}{y}) = (-x \sin \tfrac{x}{y})(x)\frac{\partial}{\partial y}(\tfrac{1}{y}) =$$
$$-x^2 \sin \tfrac{x}{y}(-1/y^2) = (\tfrac{x}{y})^2 \sin \tfrac{x}{y}.$$

11 $f(x, y, z) = 3x^2z + xy^2 \Rightarrow$

$$f_x(x, y, z) = 6xz + y^2,\ f_y(x, y, z) = 2xy, \text{ and } f_z(x, y, z) = 3x^2.$$

17 $f(q, v, w) = \sin^{-1}\sqrt{qv} + \sin vw \Rightarrow$

$$f_q(q, v, w) = \frac{1}{\sqrt{1-(\sqrt{qv})^2}} \cdot \tfrac{1}{2}(qv)^{-1/2} \cdot v = \frac{v}{2\sqrt{qv}\,\sqrt{1-qv}},$$
$$f_v(q, v, w) = \frac{q}{2\sqrt{qv}\,\sqrt{1-qv}} + w \cos vw, \text{ and } f_w(q, v, w) = v \cos vw.$$

19 $w = xy^4 - 2x^2y^3 + 4x^2 - 3y \Rightarrow w_x = y^4 - 4xy^3 + 8x \Rightarrow w_{xy} = 4y^3 - 12xy^2$ and

$w_y = 4xy^3 - 6x^2y^2 - 3 \Rightarrow w_{yx} = 4y^3 - 12xy^2$.

Thus, $w_{xy} = w_{yx}$ as guaranteed by (16.12).

25 $w = 3x^2y^3z + 2xy^4z^2 - yz \Rightarrow$

$$w_x = (3x^2y^3z + 2xy^4z^2 - yz)_x = 6xy^3z + 2y^4z^2 \Rightarrow$$
$$w_{xy} = (6xy^3z + 2y^4z^2)_y = 18xy^2z + 8y^3z^2 \Rightarrow$$
$$w_{xyz} = (18xy^2z + 8y^3z^2)_z = 18xy^2 + 16y^3z.$$

[29] The notation $\dfrac{\partial^3 w}{\partial z \partial y \partial x}$ is the same as w_{xyz}.

$$w = \sin xyz \Rightarrow w_x = (\cos xyz) \cdot (xyz)_x = yz \cos xyz \Rightarrow$$

$$\begin{aligned} w_{xy} &= (yz)_y \cdot \cos xyz + (yz)(\cos xyz)_y \\ &= z \cos xyz + (yz)(-\sin xyz)(xyz)_y = z \cos xyz - xyz^2 \sin xyz \Rightarrow \\ w_{xyz} &= \left[(z)_z \cos xyz + z(\cos xyz)_z\right] - \left[(xyz^2)_z \sin xyz + xyz^2(\sin xyz)_z\right] \\ &= \left[\cos xyz - z \sin xyz (xyz)_z\right] - \left[2xyz \sin xyz + xyz^2 \cos xyz (xyz)_z\right] \\ &= \cos xyz - xyz \sin xyz - 2xyz \sin xyz - x^2y^2z^2 \cos xyz \\ &= (1 - x^2y^2z^2) \cos xyz - 3xyz \sin xyz. \end{aligned}$$

[31] $w = r^4 s^3 t - 3s^2 e^{rt} \Rightarrow$

$w_r = 4r^3 s^3 t - 3s^2 t e^{rt} \Rightarrow w_{rr} = 12r^2 s^3 t - 3s^2 t^2 e^{rt} \Rightarrow w_{rrs} = 36r^2 s^2 t - 6st^2 e^{rt}$,

$w_{rs} = 12r^3 s^2 t - 6ste^{rt} \Rightarrow w_{rsr} = 36r^2 s^2 t - 6st^2 e^{rt}$, and

$w_s = 3r^4 s^2 t - 6se^{rt} \Rightarrow w_{sr} = 12r^3 s^2 t - 6ste^{rt} \Rightarrow w_{srr} = 36r^2 s^2 t - 6st^2 e^{rt}$.

Thus, $w_{rrs} = w_{rsr} = w_{srr}$.

[33] $w = f(x, y) = \ln \sqrt{x^2 + y^2} = \frac{1}{2} \ln (x^2 + y^2) \Rightarrow$

$$w_x = \frac{1}{2} \cdot \frac{1}{x^2 + y^2} \cdot 2x = \frac{x}{x^2 + y^2} \Rightarrow$$

$$w_{xx} = \frac{(x^2 + y^2)(x)_x - (x)(x^2 + y^2)_x}{(x^2 + y^2)^2} = \frac{x^2 + y^2 - 2x^2}{(x^2 + y^2)^2} = \frac{y^2 - x^2}{(x^2 + y^2)^2}.$$

$$w_y = \frac{1}{2} \cdot \frac{1}{x^2 + y^2} \cdot 2y = \frac{y}{x^2 + y^2} \Rightarrow$$

$$w_{yy} = \frac{(x^2 + y^2)(y)_y - (y)(x^2 + y^2)_y}{(x^2 + y^2)^2} = \frac{x^2 + y^2 - 2y^2}{(x^2 + y^2)^2} = \frac{x^2 - y^2}{(x^2 + y^2)^2}.$$

Thus, $w_{xx} + w_{yy} = \dfrac{\partial^2 f}{\partial x^2} + \dfrac{\partial^2 f}{\partial y^2} = \dfrac{y^2 - x^2}{(x^2 + y^2)^2} + \dfrac{x^2 - y^2}{(x^2 + y^2)^2} = 0.$

[35] $w = f(x, y) = \cos x \sinh y + \sin x \cosh y \Rightarrow$

$w_x = -\sin x \sinh y + \cos x \cosh y \Rightarrow w_{xx} = -\cos x \sinh y - \sin x \cosh y$, and

$w_y = \cos x \cosh y + \sin x \sinh y \Rightarrow w_{yy} = \cos x \sinh y + \sin x \cosh y$.

Thus, $w_{xx} + w_{yy} = 0$.

[39] $w = e^{-c^2 t} \sin cx \Rightarrow w_x = ce^{-c^2 t} \cos cx \Rightarrow$

$w_{xx} = -c^2 e^{-c^2 t} \sin cx$; $w_t = -c^2 e^{-c^2 t} \sin cx$. Thus, $w_{xx} = w_t$.

[41] $v = (\sin akt)(\sin kx) \Rightarrow v_x = k(\sin akt)(\cos kx) \Rightarrow v_{xx} = -k^2(\sin akt)(\sin kx)$;

$v_t = ak(\cos akt)(\sin kx) \Rightarrow v_{tt} = -a^2 k^2(\sin akt)(\sin kx)$. Thus, $v_{tt} = a^2 v_{xx}$.

[43] $u(x, y) = x^2 - y^2$; $v(x, y) = 2xy \Rightarrow$

$u_x = 2x$ and $u_y = -2y$; $v_x = 2y = -u_y$ and $v_y = 2x = u_x$.

[47] There are 9 second partial derivatives: w_{xx}, w_{xy}, w_{xz}, w_{yx}, w_{yy}, w_{yz}, w_{zx}, w_{zy}, w_{zz}

[49] (a) From the discussion on page 816, we see that the instantaneous rate of change of T in the direction of the x-axis is given by $T_x(x, y)$. $T = 10(x^2 + y^2)^2 \Rightarrow$

$$T_x(x, y) = 10 \cdot 2(x^2 + y^2)(2x) = 40x(x^2 + y^2) \Rightarrow T_x(1, 2) = 200 \text{ deg/cm}.$$

(b) Similarly, $T_y(x, y)$ gives the instantaneous rate of change of T in the direction of the y-axis. $T_y(x, y) = 40y(x^2 + y^2) \Rightarrow T_y(1, 2) = 400$ deg/cm.

[51] (a) $V = 100/(x^2 + y^2 + z^2) = 100(x^2 + y^2 + z^2)^{-1} \Rightarrow$

$$V_x(x, y, z) = -200x(x^2 + y^2 + z^2)^{-2} \Rightarrow V_x(2, -1, 1) = -\tfrac{100}{9} \text{ volts/in}$$

(b) $V_y(x, y, z) = -200y(x^2 + y^2 + z^2)^{-2} \Rightarrow V_y(2, -1, 1) = \frac{50}{9}$ volts/in

(c) $V_z(x, y, z) = -200z(x^2 + y^2 + z^2)^{-2} \Rightarrow V_z(2, -1, 1) = -\frac{50}{9}$ volts/in

[53] $C(x, y) = \dfrac{200}{x^2}\left[e^{-0.02(y-10)^2/x^2} + e^{-0.02(y+10)^2/x^2}\right] \Rightarrow$

$$\frac{\partial C}{\partial x} = \frac{\partial}{\partial x}\left(\frac{200}{x^2}\right)\left[e^{-0.02(y-10)^2/x^2} + e^{-0.02(y+10)^2/x^2}\right] +$$

$$\left(\frac{200}{x^2}\right)\frac{\partial}{\partial x}\left[e^{-0.02(y-10)^2/x^2} + e^{-0.02(y+10)^2/x^2}\right]$$

$$= -\frac{400}{x^3}\left[e^{-0.02(y-10)^2/x^2} + e^{-0.02(y+10)^2/x^2}\right] +$$

$$\frac{200}{x^2}\left[e^{-0.02(y-10)^2/x^2}\left(\frac{0.04(y-10)^2}{x^3}\right) + e^{-0.02(y+10)^2/x^2}\left(\frac{0.04(y+10)^2}{x^3}\right)\right]$$

Letting $x = 2$ and $y = 5$ in the last equation gives $\partial C/\partial x \approx -36.58$ $(\mu\text{g/m}^3)$/m, the rate at which the concentration changes in the horizontal direction at (2, 5).

Similarly,

$$\frac{\partial C}{\partial y} = \frac{200}{x^2}\left[e^{-0.02(y-10)^2/x^2}\left(\frac{-0.04(y-10)}{x^2}\right) + e^{-0.02(y+10)^2/x^2}\left(\frac{-0.04(y+10)}{x^2}\right)\right].$$

$\partial C/\partial y \approx -0.229$ $(\mu\text{g/m}^3)$/m

is the rate at which the concentration changes in the vertical direction at (2, 5).

[55] (a) $S(p, q) = \dfrac{p}{q + p(1 - q)} \Rightarrow S(10, 0.8) = \frac{25}{7} \approx 3.57,$

$S(100, 0.8) = \frac{125}{26} \approx 4.81$, and $S(1000, 0.8) = \frac{1250}{251} \approx 4.98$.

$$\lim_{p\to\infty} S(p, 0.8) = \lim_{p\to\infty}\frac{p}{0.8 + 0.2p} = \lim_{p\to\infty}\frac{5p}{p + 4} = 5$$

(b) $S(p, q) = p[q + p(1 - q)]^{-1} \Rightarrow \dfrac{\partial S}{\partial q} = (p)\dfrac{\partial}{\partial q}[q + p(1 - q)]^{-1} =$

$$(p)(-1)[q + p(1 - q)]^{-2}(1 - p) = -p\cdot\frac{1 - p}{[q + p(1 - q)]^2} = \frac{p(p - 1)}{[q + p(1 - q)]^2}.$$

(c) There is complete parallelism if $q = 1$. $q = 1 \Rightarrow \dfrac{\partial S}{\partial q} = p(p - 1) = p^2 - p$.

Since $\dfrac{\partial S}{\partial q} \geq 0$ for $p \geq 1$ when $q = 1$,

as the number of processors increases, the rate of change of the speedup increases.

[57] (a) $T = T_0 e^{-\lambda x} \sin(\omega t - \lambda x) \Rightarrow$

$$\frac{\partial T}{\partial t} = T_0 \omega e^{-\lambda x} \cos(\omega t - \lambda x)$$

is the rate of change of temperature w.r.t. time at a fixed depth x and

$$\frac{\partial T}{\partial x} = -T_0 \lambda e^{-\lambda x}\left[\cos(\omega t - \lambda x) + \sin(\omega t - \lambda x)\right]$$

is the rate of change of temperature w.r.t. the depth at a fixed time t.

(b) $\frac{\partial^2 T}{\partial x^2} = T_0 \lambda^2 e^{-\lambda x}\left[\cos(\omega t - \lambda x) + \sin(\omega t - \lambda x)\right] +$

$$T_0 \lambda^2 e^{-\lambda x}\left[-\sin(\omega t - \lambda x) + \cos(\omega t - \lambda x)\right] = 2\lambda^2 T_0 e^{-\lambda x} \cos(\omega t - \lambda x).$$

If $k = \frac{2\lambda^2}{\omega}$, then $\frac{\partial T}{\partial t} = k \frac{\partial^2 T}{\partial x^2}$.

[61] Show that $e_k = -a_{kk}$ for every k.

[63] Refer to Theorem (16.10)(i) and Figure 16.31.

If $a = 1$, $b = 2$, and $z = f(x, y)$, then

$z = 9 - x^2 - y^2 \Rightarrow \frac{\partial z}{\partial y} = -2y \Rightarrow$

slope of l at $P(1, 2, 4)$ is -4,

and an equation of l is $(z - 4) = -4(y - 2)$, or,

$z = -4y + 12$. Thus, parametric equations are

$x = 1$, $y = t$, $z = -4t + 12$.

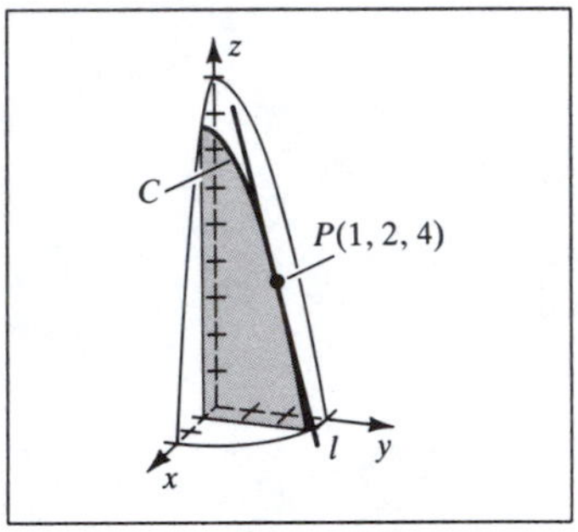

Figure 63

[65] $f_x(0.5, 0.2) \approx \frac{f(0.48, 0.2) - 4f(0.49, 0.2) + 3f(0.5, 0.2)}{2(0.01)} \approx 0.0079600438$

$f_y(0.5, 0.2) \approx \frac{f(0.5, 0.18) - 4f(0.5, 0.19) + 3f(0.5, 0.2)}{2(0.01)} \approx 0.0597349919$

$f(x, y) = y^2 \sin(xy) \Rightarrow f_x(x, y) = y^3 \cos(xy) \Rightarrow f_x(0.5, 0.2) \approx 0.0079600333.$

$f_y(x, y) = 2y\sin(xy) + xy^2 \cos(xy) \Rightarrow f_y(0.5, 0.2) \approx 0.0598334499.$

Exercises 16.4

[1] (a) Using (16.13) with $w = f(x, y) = 5y^2 - xy$,

$$\begin{aligned}\Delta w &= f(x + \Delta x, y + \Delta y) - f(x, y)\\ &= 5(y + \Delta y)^2 - (x + \Delta x)(y + \Delta y) - (5y^2 - xy)\\ &= 5y^2 + 10y\Delta y + 5(\Delta y)^2 - xy - x\Delta y - y\Delta x - \Delta x\Delta y - 5y^2 + xy\\ &= 10y\Delta y - x\Delta y - y\Delta x + 5(\Delta y)^2 - \Delta x\Delta y.\end{aligned}$$

(b) Using (16.15)(ii), $dw = f_x(x, y)\,dx + f_y(x, y)\,dy = -y\,dx + (10y - x)\,dy$.

(c) Since $dx = \Delta x$ and $dy = \Delta y$,

$$\begin{aligned}dw - \Delta w &= (-y\Delta x + 10y\Delta y - x\Delta y) -\\ &\qquad (10y\Delta y - x\Delta y - y\Delta x + 5(\Delta y)^2 - \Delta x\Delta y)\\ &= \Delta x\Delta y - 5(\Delta y)^2.\end{aligned}$$

Note: In Exercises 3–6, the expressions for ϵ_1 and ϵ_2 are not unique. See Example 2.

[3] $\Delta w = f(x + \Delta x,\ y + \Delta y) - f(x,\ y)$

$= \left[4(y + \Delta y)^2 - 3(x + \Delta x)(y + \Delta y) + 2(x + \Delta x)\right] - \left[4y^2 - 3xy + 2x\right]$

$= 8y\Delta y + 4(\Delta y)^2 - 3x\Delta y - 3y\Delta x - 3\Delta x\Delta y + 2\Delta x$

$= (-3y + 2)\Delta x + (8y - 3x)\Delta y - (3\Delta y)\Delta x + (4\Delta y)\Delta y.$

Since $f_x = -3y + 2$ and $f_y = 8y - 3x$, $\epsilon_1 = -3\Delta y$ and $\epsilon_2 = 4\Delta y$.

[5] $\Delta w = \left[(x + \Delta x)^3 + (y + \Delta y)^3\right] - \left[x^3 + y^3\right]$

$= 3x^2\Delta x + 3x(\Delta x)^2 + (\Delta x)^3 + 3y^2\Delta y + 3y(\Delta y)^2 + (\Delta y)^3$

$= (3x^2)\Delta x + (3y^2)\Delta y + \left[3x\Delta x + (\Delta x)^2\right]\Delta x + \left[3y\Delta y + (\Delta y)^2\right]\Delta y.$

Since $f_x = 3x^2$ and $f_y = 3y^2$, $\epsilon_1 = 3x\Delta x + (\Delta x)^2$ and $\epsilon_2 = 3y\Delta y + (\Delta y)^2$.

[7] Using (16.15)(ii) with $w = x^3 - x^2y + 3y^2$,

$$dw = w_x\,dx + w_y\,dy = (3x^2 - 2xy)\,dx + (-x^2 + 6y)\,dy.$$

[11] $w = x^2e^{xy} + (1/y^2) \Rightarrow$

$$dw = w_x\,dx + w_y\,dy = xe^{xy}(xy + 2)\,dx + (x^3e^{xy} - 2y^{-3})\,dy.$$

[13] Using (16.20)(ii) with $w = x^2\ln(y^2 + z^2)$, $dw = w_x\,dx + w_y\,dy + w_z\,dz =$

$$\left[2x\ln(y^2 + z^2)\right]dx + \left(\frac{2x^2y}{y^2 + z^2}\right)dy + \left(\frac{2x^2z}{y^2 + z^2}\right)dz.$$

[17] Using an extension to four variables of (16.20)(ii), $w = x^2z + 4yt^3 - xz^2t \Rightarrow$

$dw = w_x\,dx + w_y\,dy + w_z\,dz + w_t\,dt$

$= (2xz - z^2t)\,dx + (4t^3)\,dy + (x^2 - 2xzt)\,dz + (12yt^2 - xz^2)\,dt.$

[19] $f(x,\ y) = x^2 - 3x^3y^2 + 4x - 2y^3 + 6 \Rightarrow$

$$df = f_x\,dx + f_y\,dy = (2x - 9x^2y^2 + 4)\,dx + (-6x^3y - 6y^2)\,dy.$$

$x = -2$, $y = 3$, $dx = -2.02 - (-2) = -0.02$, and $dy = 3.01 - 3 = 0.01 \Rightarrow$

$$df = (-324)(-0.02) + (90)(0.01) = 7.38.$$

[21] $f(x,\ y,\ z) = x^2z^3 - 3yz^2 + x^{-3} + 2y^{1/2}z \Rightarrow$

$df = f_x\,dx + f_y\,dy + f_z\,dz$

$= (2xz^3 - 3x^{-4})\,dx + (-3z^2 + y^{-1/2}z)\,dy + (3x^2z^2 - 6yz + 2y^{1/2})\,dz.$

$x = 1$, $y = 4$, $z = 2$, $dx = 0.02$, $dy = -0.03$, and $dz = -0.04 \Rightarrow$

$$df = (13)(0.02) + (-11)(-0.03) + (-32)(-0.04) = 1.87.$$

[23] (a) Let x, y, and z denote the length, width, and height of the box.

The surface area S is given by $S(x,\ y,\ z) = 2xy + 2xz + 2yz$.

$dS = S_x\,dx + S_y\,dy + S_z\,dz$

$= (2y + 2z)\,dx + (2x + 2z)\,dy + (2x + 2y)\,dz.$

$x = 3$, $y = 4$, $z = 5$, and $dx = dy = dz = \pm\frac{1}{16}$ in. $= \pm\frac{1}{192}$ ft $\Rightarrow$

$$dS = (18 + 16 + 14)(\pm\tfrac{1}{192}) = \pm\tfrac{48}{192} = \pm\tfrac{1}{4}\ \text{ft}^2.$$

(b) The volume V of the box is given by $V(x, y, z) = xyz$.

$$dV = V_x\,dx + V_y\,dy + V_z\,dz$$
$$= (yz)\,dx + (xz)\,dy + (xy)\,dz \Rightarrow dV = (20 + 15 + 12)(\pm\tfrac{1}{192}) = \pm\tfrac{47}{192}\text{ ft}^3.$$

25 (a) $P = 15{,}700S^{5/2}RD = 15{,}700(0.54)^{5/2}(0.113/2)(2) \approx 380$ lb

(b) The maximum percentage error in the calculated value of P is dP/P.

$$dP = \frac{\partial P}{\partial S}\,dS + \frac{\partial P}{\partial R}\,dR + \frac{\partial P}{\partial D}\,dD$$
$$= 15{,}700\left[(\tfrac{5}{2})S^{3/2}RD\,dS + S^{5/2}D\,dR + S^{5/2}R\,dD\right] \Rightarrow$$

$$\frac{dP}{P} = \frac{15{,}700\left[(\tfrac{5}{2})S^{3/2}RD\,dS + S^{5/2}D\,dR + S^{5/2}R\,dD\right]}{15{,}700S^{5/2}RD}$$
$$= \frac{5}{2}\frac{dS}{S} + \frac{dR}{R} + \frac{dD}{D}$$
$$= \tfrac{5}{2}(\pm 3\%) + (\pm 2\%) + (\pm 2\%) = \pm 11.5\%.$$

31 Let k be a constant of proportionality.

Then $R = \dfrac{kL}{D^2}$ and $dR = R_L\,dL + R_D\,dD = \dfrac{k}{D^2}\,dL - \dfrac{2kL}{D^3}\,dD$.

Thus, $\dfrac{dR}{R} = \dfrac{k}{D^2R}\,dL - \dfrac{2kL}{D^3R}\,dD = \dfrac{dL}{L} - 2\dfrac{dD}{D}$ $\{D^2R = kL\}$.

$\dfrac{dL}{L} = \pm 1\%$ and $\dfrac{dD}{D} = \pm 3\%$. We choose $\dfrac{dL}{L}$ to be 1% and $\dfrac{dD}{D}$ to be -3%

to *maximize* $\dfrac{dR}{R}$, and then allow $\dfrac{dR}{R}$ to equal $\pm$ "this amount".

Hence, $\dfrac{dR}{R} = 1\% - 2(-3\%) = \pm 7\%$.

35 $h = \dfrac{x}{\cot\alpha - \cot\beta} \Rightarrow dh = h_\alpha\,d\alpha + h_\beta\,d\beta + h_x\,dx =$

$$-\frac{x\csc^2\alpha}{(\cot\alpha - \cot\beta)^2}\,d\alpha + \frac{x\csc^2\beta}{(\cot\alpha - \cot\beta)^2}\,d\beta + \frac{1}{(\cot\alpha - \cot\beta)^2}\,dx.$$

$d\alpha = d\beta = \pm 30'' = \pm\frac{30\pi}{60\cdot 60\cdot 180} = \pm\frac{\pi}{21{,}600} \approx \pm 0.000145.$

$$\alpha = 15^\circ,\ \beta = 20^\circ,\text{ and } x = 2000 \Rightarrow dh = -\frac{2000\csc^2 15^\circ}{(\cot 15^\circ - \cot 20^\circ)^2}\cdot\left(-\frac{\pi}{21{,}600}\right)$$
$$+\frac{2000\csc^2 20^\circ}{(\cot 15^\circ - \cot 20^\circ)^2}\cdot\frac{\pi}{21{,}600} + \frac{1}{(\cot 15^\circ - \cot 20^\circ)^2}\,dx.$$

In the first term of dh, $-\frac{\pi}{21{,}600}$ is chosen, and in the second term, $+\frac{\pi}{21{,}600}$.

This is done to *maximize* dh, and corresponds to the surveyor measuring α too small by $30''$ and β too large by $30''$. Approximating, $dh \approx 7.0448 + 1.0316\,dx$.

$|dh| \le 10 \Rightarrow |7.0448 + 1.0316\,dx| \le 10 \Rightarrow -16.527 \le dx \le 2.865.$

Thus, we choose $|dx| \le 2.865$ to assure us that $|dh| \le 10$, that is,

the maximum error in x must not exceed ± 2.9 ft.

[39] $f_x(x, y) = \dfrac{4xy^2}{(x^2 + y^2)^2}$ and $f_y(x, y) = \dfrac{-4yx^2}{(x^2 + y^2)^2}$ are continuous except at (0, 0) which is outside the domain of f. Thus, f is continuous on any rectangle R not containing the origin and by (16.17), f is differentiable at every point in its domain.

[41] (a) Using (16.8) with $(x, y) = (0, 0)$,

$$f_x(0, 0) = \lim_{h \to 0} \frac{f(0 + h, 0) - f(0, 0)}{h} = \lim_{h \to 0} \frac{0 - 0}{h} = 0$$

and

$$f_y(0, 0) = \lim_{h \to 0} \frac{f(0, 0 + h) - f(0, 0)}{h} = \lim_{h \to 0} \frac{0 - 0}{h} = 0.$$

Thus, $f_x(0, 0)$ and $f_y(0, 0)$ exist and equal 0.

(b) Consider $L = \displaystyle\lim_{(x, y) \to (0, 0)} \frac{xy}{x^2 + y^2}$. Along the path $y = 0$, $L = \dfrac{0}{x^2} = 0$, whereas along the path $y = x$, $L = \dfrac{x^2}{x^2 + x^2} = \frac{1}{2}$.

Thus, by the two-path rule, L DNE, and f cannot be continuous at (0, 0).

(c) Theorem (16.18) states "If a function f of two variables is differentiable at (x_0, y_0), then f is continuous at (x_0, y_0)." The contrapositive of (16.18) is "If a function f of two variables is *not* continuous at (x_0, y_0), then f is *not* differentiable at (x_0, y_0)." This is an equivalent statement. Thus, since f is not continuous at (0, 0), it cannot be differentiable at (0, 0).

[43] Let $f(x, y) = \frac{x^2}{4} + \frac{y^2}{9} - 1 = 0$ and $g(x, y) = \dfrac{(x - 1)^2}{10} + \dfrac{(y + 1)^2}{5} - 1 = 0$.

Then, $f_x(x, y) = \frac{1}{2}x$, $f_y(x, y) = \frac{2}{9}y$, $g_x(x, y) = \frac{1}{5}(x - 1)$, and $g_y(x, y) = \frac{2}{5}(y + 1)$.

For the first iteration with $(x_1, y_1) = (2, 1)$ we have:

$$\begin{bmatrix} f_x(2, 1) & f_y(2, 1) \\ g_x(2, 1) & g_y(2, 1) \end{bmatrix} \begin{bmatrix} \Delta x_1 \\ \Delta y_1 \end{bmatrix} \approx \begin{bmatrix} -f(2, 1) \\ -g(2, 1) \end{bmatrix} \Rightarrow$$

$$\begin{bmatrix} 1 & \frac{2}{9} \\ \frac{1}{5} & \frac{4}{5} \end{bmatrix} \begin{bmatrix} \Delta x_1 \\ \Delta y_1 \end{bmatrix} \approx \begin{bmatrix} -\frac{1}{9} \\ \frac{1}{10} \end{bmatrix} \Rightarrow \begin{bmatrix} \Delta x_1 \\ \Delta y_1 \end{bmatrix} \approx \begin{bmatrix} -0.1471 \\ 0.1618 \end{bmatrix}$$

Thus, $x_2 = x_1 + \Delta x_1 \approx 1.8529$ and $y_2 = y_1 + \Delta y_1 \approx 1.1618$.

In a similar manner, $(x_3, y_3) \approx (1.8460, 1.1546)$, and $(x_4, y_4) \approx (1.8460, 1.1546)$.

Exercises 16.5

[1] Using (16.21) with $w = u \sin v$, $u = x^2 + y^2$, and $v = xy$,

$$\frac{\partial w}{\partial x} = \frac{\partial w}{\partial u}\frac{\partial u}{\partial x} + \frac{\partial w}{\partial v}\frac{\partial v}{\partial x}$$
$$= (\sin v)(2x) + (u \cos v)(y) = 2x \sin(xy) + y(x^2 + y^2)\cos(xy).$$

$$\frac{\partial w}{\partial y} = \frac{\partial w}{\partial u}\frac{\partial u}{\partial y} + \frac{\partial w}{\partial v}\frac{\partial v}{\partial y}$$
$$= (\sin v)(2y) + (u \cos v)(x) = 2y \sin(xy) + x(x^2 + y^2)\cos(xy).$$

[3] Using (16.21) with $w = u^2 + 2uv$, $u = r \ln s$, and $v = 2r + s$,

$$\frac{\partial w}{\partial r} = \frac{\partial w}{\partial u}\frac{\partial u}{\partial r} + \frac{\partial w}{\partial v}\frac{\partial v}{\partial r}$$
$$= (2u + 2v)(\ln s) + (2u)(2)$$
$$= 2(r \ln s + 2r + s)(\ln s) + 4r \ln s = 2r(\ln s)^2 + 8r \ln s + 2s \ln s.$$

$$\frac{\partial w}{\partial s} = \frac{\partial w}{\partial u}\frac{\partial u}{\partial s} + \frac{\partial w}{\partial v}\frac{\partial v}{\partial s}$$
$$= (2u + 2v)(\tfrac{r}{s}) + (2u)(1)$$
$$= 2(r \ln s + 2r + s)(\tfrac{r}{s}) + 2r \ln s = \frac{2r^2 \ln s}{s} + \frac{4r^2}{s} + 2r + 2r \ln s.$$

[5] If $z = r^3 + s + v^2$, $r = xe^y$, $s = ye^x$, and $v = x^2 y$, then

$$\frac{\partial z}{\partial x} = \frac{\partial z}{\partial r}\frac{\partial r}{\partial x} + \frac{\partial z}{\partial s}\frac{\partial s}{\partial x} + \frac{\partial z}{\partial v}\frac{\partial v}{\partial x}$$
$$= (3r^2)(e^y) + (1)(ye^x) + (2v)(2xy) = 3x^2 e^{3y} + ye^x + 4x^3 y^2.$$

$$\frac{\partial z}{\partial y} = \frac{\partial z}{\partial r}\frac{\partial r}{\partial y} + \frac{\partial z}{\partial s}\frac{\partial s}{\partial y} + \frac{\partial z}{\partial v}\frac{\partial v}{\partial y}$$
$$= (3r^2)(xe^y) + (1)(e^x) + (2v)(x^2) = 3x^3 e^{3y} + e^x + 2x^4 y.$$

[7] If $r = x \ln y$, $x = 3u + vt$, and $y = uvt$, then

$$\frac{\partial r}{\partial u} = \frac{\partial r}{\partial x}\frac{\partial x}{\partial u} + \frac{\partial r}{\partial y}\frac{\partial y}{\partial u}$$
$$= (\ln y)(3) + (\tfrac{x}{y})(vt) = 3 \ln(uvt) + \frac{3u + vt}{u} = 3 \ln(uvt) + 3 + \tfrac{vt}{u}.$$

$$\frac{\partial r}{\partial v} = \frac{\partial r}{\partial x}\frac{\partial x}{\partial v} + \frac{\partial r}{\partial y}\frac{\partial y}{\partial v}$$
$$= (\ln y)(t) + (\tfrac{x}{y})(ut) = t \ln(uvt) + \frac{3u + vt}{v} = t \ln(uvt) + \tfrac{3u}{v} + t.$$

$$\frac{\partial r}{\partial t} = \frac{\partial r}{\partial x}\frac{\partial x}{\partial t} + \frac{\partial r}{\partial y}\frac{\partial y}{\partial t}$$
$$= (\ln y)(v) + (\tfrac{x}{y})(uv) = v \ln(uvt) + \frac{3u + vt}{t} = v \ln(uvt) + \tfrac{3u}{t} + v.$$

[9] If $p = u^2 + 3v^2 - 4w^2$, $u = x - 3y + 2r - s$,

$v = 2x + y - r + 2s$, and $w = -x + 2y + r + s$, then

$$\frac{\partial p}{\partial r} = \frac{\partial p}{\partial u}\frac{\partial u}{\partial r} + \frac{\partial p}{\partial v}\frac{\partial v}{\partial r} + \frac{\partial p}{\partial w}\frac{\partial w}{\partial r}$$
$$= (2u)(2) + (6v)(-1) + (-8w)(1)$$
$$= 4(x - 3y + 2r - s) - 6(2x + y - r + 2s) - 8(-x + 2y + r + s)$$
$$= (4x - 12y + 8r - 4s) + (-12x - 6y + 6r - 12s) + (8x - 16y - 8r - 8s)$$
$$= -34y + 6r - 24s.$$

[11] As in Example 3 with $w = x^3 - y^3$, $x = \dfrac{1}{t+1}$, and $y = \dfrac{t}{t+1}$,

$$\frac{dw}{dt} = \frac{\partial w}{\partial x}\frac{dx}{dt} + \frac{\partial w}{\partial y}\frac{dy}{dt} = (3x^2)\left[-\frac{1}{(t+1)^2}\right] + (-3y^2)\left[\frac{1}{(t+1)^2}\right] = -\frac{3(1+t^2)}{(t+1)^4}.$$

[13] If $w = r^2 - s\tan v$, $r = \sin^2 t$, $s = \cos t$, and $v = 4t$, then

$$\begin{aligned}\frac{dw}{dt} &= \frac{\partial w}{\partial r}\frac{dr}{dt} + \frac{\partial w}{\partial s}\frac{ds}{dt} + \frac{\partial w}{\partial v}\frac{dv}{dt}\\ &= (2r)(2\sin t\cos t) + (-\tan v)(-\sin t) + (-s\sec^2 v)(4)\\ &= 4\sin^3 t\cos t + \tan(4t)\sin t - 4\cos t\sec^2(4t).\end{aligned}$$

[15] Using (16.22) with $F(x, y) = 2x^3 + x^2y + y^3 - 1 = 0$,

$$\frac{dy}{dx} = -\frac{F_x(x, y)}{F_y(x, y)} = -\frac{6x^2 + 2xy}{x^2 + 3y^2}.$$

[19] Using (16.23) with $F(x, y, z) = 2xz^3 - 3yz^2 + x^2y^2 + 4z = 0$,

$$\frac{\partial z}{\partial x} = -\frac{F_x(x, y, z)}{F_z(x, y, z)} = -\frac{2z^3 + 2xy^2}{6xz^2 - 6yz + 4}$$

and

$$\frac{\partial z}{\partial y} = -\frac{F_y(x, y, z)}{F_z(x, y, z)} = -\frac{2x^2y - 3z^2}{6xz^2 - 6yz + 4}.$$

[23] (a) The rate at which the volume is increasing will be given by $\frac{dV}{dt}$.

We must determine a formula for V and differentiate it with respect to t using a chain rule. The volume of a right circular cylinder is $V = \pi r^2 h$.

Thus, $\frac{dV}{dt} = \frac{\partial V}{\partial r}\frac{dr}{dt} + \frac{\partial V}{\partial h}\frac{dh}{dt} = (2\pi rh)\frac{dr}{dt} + (\pi r^2)\frac{dh}{dt}$.

$r = 4$, $h = 7$, $\frac{dr}{dt} = 0.01$, and $\frac{dh}{dt} = 0.02 \Rightarrow$

$$\frac{dV}{dt} = (56\pi)(0.01) + (16\pi)(0.02) = 0.88\pi \approx 2.76 \text{ in.}^3/\text{min}.$$

(b) The curved surface area S does not include the top or bottom. $S = 2\pi rh \Rightarrow$

$$\begin{aligned}\frac{dS}{dt} &= \frac{\partial S}{\partial r}\frac{dr}{dt} + \frac{\partial S}{\partial h}\frac{dh}{dt}\\ &= (2\pi h)\frac{dr}{dt} + (2\pi r)\frac{dh}{dt} = (14\pi)(0.01) + (8\pi)(0.02) = 0.3\pi \approx 0.94 \text{ in.}^2/\text{min}.\end{aligned}$$

[27] We must find $\frac{dV}{dt}$ for the given conditions. Solving $PV = 8T$ for V, we have

$$V = \frac{8T}{P} \Rightarrow \frac{dV}{dt} = \frac{\partial V}{\partial T}\frac{dT}{dt} + \frac{\partial V}{\partial P}\frac{dP}{dt} = \left(\frac{8}{P}\right)\frac{dT}{dt} - \left(\frac{8T}{P^2}\right)\frac{dP}{dt}.$$

$$\frac{dT}{dt} = 2, \text{ and } \frac{dP}{dt} = \tfrac{1}{2} \Rightarrow \frac{dV}{dt} = (\tfrac{8}{10})(2) - (\tfrac{1600}{100})(\tfrac{1}{2}) = -6.4 \text{ in.}^3/\text{min}.$$

[31] $\frac{\partial}{\partial R_k}\left(\frac{1}{R}\right) = \frac{\partial}{\partial R_k}\left(\sum_{i=1}^{n}\frac{1}{R_i}\right) \Rightarrow$

$$\frac{\partial(1/R)}{\partial R}\frac{\partial R}{\partial R_k} = \frac{\partial(1/R_1)}{\partial R_k} + \frac{\partial(1/R_2)}{\partial R_k} + \cdots + \frac{\partial(1/R_k)}{\partial R_k} + \cdots + \frac{\partial(1/R_n)}{\partial R_k}.$$

On the right side of the last equation, only the kth term is not equal to zero.

$$\text{Thus, } -\frac{1}{R^2}\frac{\partial R}{\partial R_k} = -\frac{1}{R_k^2} \Rightarrow \frac{\partial R}{\partial R_k} = \frac{R^2}{R_k^2} = \left(\frac{R}{R_k}\right)^2.$$

[33] $f(x, y) = 2x^3 + 3x^2y + y^3 \Rightarrow$

$$f(tx, ty) = 2(tx)^3 + 3(tx)^2(ty) + (ty)^3$$
$$= t^3(2x^3 + 3x^2y + y^3) = t^3 f(x, y) \Rightarrow n = 3.$$
$$xf_x(x, y) + yf_y(x, y) = x(6x^2 + 6xy) + y(3x^2 + 3y^2)$$
$$= 3(2x^3 + 3x^2y + y^3) = 3f(x, y).$$

[37] Note that $w = f(x, y)$ does not specify any particular function,

it represents *any* function of x and y.

$$w_r = \frac{\partial w}{\partial x}\frac{\partial x}{\partial r} + \frac{\partial w}{\partial y}\frac{\partial y}{\partial r} = w_x x_r + w_y y_r = w_x(\cos\theta) + w_y(\sin\theta) \text{ and}$$
$$w_\theta = \frac{\partial w}{\partial x}\frac{\partial x}{\partial \theta} + \frac{\partial w}{\partial y}\frac{\partial y}{\partial \theta} = w_x x_\theta + w_y y_\theta = w_x(-r\sin\theta) + w_y(r\cos\theta) \Rightarrow$$
$$\left(\frac{\partial w}{\partial r}\right)^2 + \frac{1}{r^2}\left(\frac{\partial w}{\partial \theta}\right)^2 = w_r^2 + \frac{1}{r^2}w_\theta^2$$
$$= w_x^2\cos^2\theta + 2w_xw_y\cos\theta\sin\theta + w_y^2\sin^2\theta +$$
$$\frac{1}{r^2}\left[w_x^2 r^2\sin^2\theta - 2w_xw_y r^2\sin\theta\cos\theta + w_y^2 r^2\cos^2\theta\right]$$
$$= w_x^2(\sin^2\theta + \cos^2\theta) + w_y^2(\sin^2\theta + \cos^2\theta)$$
$$= w_x^2 + w_y^2 = \left(\frac{\partial w}{\partial x}\right)^2 + \left(\frac{\partial w}{\partial y}\right)^2.$$

[41] $w_x = -\sin(x + y) - \sin(x - y) \Rightarrow w_{xx} = -\cos(x + y) - \cos(x - y)$.

$w_y = -\sin(x + y) + \sin(x - y) \Rightarrow w_{yy} = -\cos(x + y) - \cos(x - y)$.

Thus, $w_{xx} - w_{yy} = 0$. Note that this is a special case of Exercise 40 with

$f(x) = \cos x$, $t = y$, and $a = 1$.

[43] $w_x = w_u u_x + w_v v_x \Rightarrow$

$$w_{xx} = (w_u u_x)_x + (w_v v_x)_x$$
$$= \left[(w_u)_x u_x + w_u(u_x)_x\right] + \left[(w_v)_x v_x + w_v(v_x)_x\right]$$
$$= \left[(w_{uu}u_x + w_{uv}v_x)u_x + w_u u_{xx}\right] + \left[(w_{vu}u_x + w_{vv}v_x)v_x + w_v v_{xx}\right]$$
$$= w_{uu}(u_x)^2 + (w_{uv} + w_{vu})u_x v_x + w_{vv}(v_x)^2 + w_u u_{xx} + w_v v_{xx}.$$

[45] Parametric equations for the line passing through $A(x_1, y_1)$ and $B(x_2, y_2)$ are $x(t) = x_1 + (x_2 - x_1)t$ and $y(t) = y_1 + (y_2 - y_1)t$ for $0 \le t \le 1$. On this line, f can be written as a function of the single variable t. Hence, the one-dimensional mean value theorem (4.12) applies and there exists $t^* \in (0, 1)$ such that if $x^* = x(t^*)$ and $y^* = y(t^*)$, then we have $f(x(1), y(1)) - f(x(0), y(0)) = \left[\frac{df}{dt}(x^*, y^*)\right](1 - 0) \Rightarrow$

$$f(x_2, y_2) - f(x_1, y_1) = \frac{\partial f}{\partial x}\frac{dx}{dt} + \frac{\partial f}{\partial y}\frac{dy}{dt} = f_x(x^*, y^*)(x_2 - x_1) + f_y(x^*, y^*)(y_2 - y_1).$$

Exercises 16.6

1 Using (16.26) with $f(x, y) = \sqrt{x^2 + y^2}$,

$$\nabla f(x, y) = f_x(x, y)\,\mathbf{i} + f_y(x, y)\,\mathbf{j} = \frac{x}{(x^2 + y^2)^{1/2}}\mathbf{i} + \frac{y}{(x^2 + y^2)^{1/2}}\mathbf{j}.$$

At $P(-4, 3)$, $\nabla f(-4, 3) = -\frac{4}{5}\mathbf{i} + \frac{3}{5}\mathbf{j}$.

5 Using (16.31) with $f(x, y, z) = yz^3 - 2x^2$,

$$\nabla f(x, y, z) = f_x(x, y, z)\,\mathbf{i} + f_y(x, y, z)\,\mathbf{j} + f_z(x, y, z)\,\mathbf{k} = -4x\,\mathbf{i} + z^3\,\mathbf{j} + 3yz^2\,\mathbf{k}.$$

At $P(2, -3, 1)$, $\nabla f(2, -3, 1) = -8\mathbf{i} + \mathbf{j} - 9\mathbf{k}$.

7 Using (16.25) with $f(x, y) = x^2 - 5xy + 3y^2$ and $\mathbf{u} = (\sqrt{2}/2)(\mathbf{i} + \mathbf{j})$,

$$D_{\mathbf{u}}f(x, y) = f_x(x, y)\,u_1 + f_y(x, y)\,u_2$$
$$= (2x - 5y)(\tfrac{\sqrt{2}}{2}) + (-5x + 6y)(\tfrac{\sqrt{2}}{2}).$$

At $P(3, -1)$, $D_{\mathbf{u}}f(3, -1) = (11)(\frac{\sqrt{2}}{2}) + (-21)(\frac{\sqrt{2}}{2}) = -5\sqrt{2} \approx -7.07$.

11 In order to find $D_{\mathbf{u}}f(x, y)$, we must first find a unit vector $\mathbf{u}$ in the direction of $\mathbf{a}$.

$\mathbf{a} = \mathbf{i} + 5\mathbf{j}$ and $\|\mathbf{a}\| = \sqrt{26} \Rightarrow \mathbf{u} = \frac{1}{\sqrt{26}}\mathbf{i} + \frac{5}{\sqrt{26}}\mathbf{j}$. $f(x, y) = \sqrt{9x^2 - 4y^2 - 1} \Rightarrow$

$$D_{\mathbf{u}}f(x, y) = \frac{9x}{\sqrt{9x^2 - 4y^2 - 1}}(\tfrac{1}{\sqrt{26}}) + \frac{-4y}{\sqrt{9x^2 - 4y^2 - 1}}(\tfrac{5}{\sqrt{26}}).$$

At $P(3, -2)$, $D_{\mathbf{u}}f(3, -2) = \frac{27}{8}(\frac{1}{\sqrt{26}}) + \frac{8}{8}(\frac{5}{\sqrt{26}}) = \frac{67}{8\sqrt{26}} \approx 1.64$.

15 $\mathbf{a} = \mathbf{i} + 2\mathbf{j} - 3\mathbf{k}$ and $\|\mathbf{a}\| = \sqrt{14} \Rightarrow \mathbf{u} = \frac{1}{\sqrt{14}}\mathbf{i} + \frac{2}{\sqrt{14}}\mathbf{j} - \frac{3}{\sqrt{14}}\mathbf{k}$.

Using (16.32) with $f(x, y, z) = xy^3z^2$,

$$D_{\mathbf{u}}f(x, y, z) = f_x(x, y, z)\,u_1 + f_y(x, y, z)\,u_2 + f_z(x, y, z)\,u_3$$
$$= (y^3z^2)(\tfrac{1}{\sqrt{14}}) + (3xy^2z^2)(\tfrac{2}{\sqrt{14}}) + (2xy^3z)(-\tfrac{3}{\sqrt{14}}).$$

At $P(2, -1, 4)$, $D_{\mathbf{u}}f(2, -1, 4) = (-16)(\frac{1}{\sqrt{14}}) + (96)(\frac{2}{\sqrt{14}}) + (-16)(-\frac{3}{\sqrt{14}}) =$

$$\frac{224}{\sqrt{14}} = 16\sqrt{14} \approx 59.87.$$

19 $\mathbf{a} = -3\mathbf{i} + \mathbf{k} \Rightarrow \mathbf{u} = -\frac{3}{\sqrt{10}}\mathbf{i} + \frac{1}{\sqrt{10}}\mathbf{k}$. $f(x, y, z) = (x + y)(y + z) \Rightarrow$

$$D_{\mathbf{u}}f(x, y, z) = f_x(x, y, z)\,u_1 + f_y(x, y, z)\,u_2 + f_z(x, y, z)\,u_3$$
$$= (y + z)(-\tfrac{3}{\sqrt{10}}) + f_y(x, y, z)\cdot 0 + (x + y)(\tfrac{1}{\sqrt{10}}).$$

At $P(5, 7, 1)$, $D_{\mathbf{u}}f(5, 7, 1) = (8)(-\frac{3}{\sqrt{10}}) + (12)(\frac{1}{\sqrt{10}}) = -\frac{12}{\sqrt{10}} \approx -3.79$.

Note: In Exercises 21–24, let v denote the unit vector in the direction of $\nabla f(x, y)$.

These exercises use (16.28) and (16.29) in parts (b) and (c), respectively.

[21] (a) $\overrightarrow{PQ} = (-3 - 2)\mathbf{i} + (1 - 0)\mathbf{j} = -5\mathbf{i} + \mathbf{j} \Rightarrow \mathbf{u} = -\frac{5}{\sqrt{26}}\mathbf{i} + \frac{1}{\sqrt{26}}\mathbf{j}$.

$f(x, y) = x^2 e^{-2y} \Rightarrow D_{\mathbf{u}}f(x, y) = (2xe^{-2y})(-\frac{5}{\sqrt{26}}) + (-2x^2 e^{-2y})(\frac{1}{\sqrt{26}})$.

At $P(2, 0)$, $D_{\mathbf{u}}f(2, 0) = (4)(-\frac{5}{\sqrt{26}}) + (-8)(\frac{1}{\sqrt{26}}) = -\frac{28}{\sqrt{26}}$,

which is the directional derivative of f at P in the direction from P to Q.

(b) $\nabla f(2, 0) = 4\mathbf{i} - 8\mathbf{j} \Rightarrow \mathbf{v} = \frac{4}{\sqrt{80}}\mathbf{i} - \frac{8}{\sqrt{80}}\mathbf{j} = \frac{1}{\sqrt{5}}\mathbf{i} - \frac{2}{\sqrt{5}}\mathbf{j}$,

which is a unit vector in the direction in which f increases most rapidly at P.

The rate of change of f in that direction is $\|\nabla f(2, 0)\| = \sqrt{80}$.

(c) $-\mathbf{v} = -\frac{1}{\sqrt{5}}\mathbf{i} + \frac{2}{\sqrt{5}}\mathbf{j}$ is a unit vector in the direction in which f decreases most rapidly at P. The rate of change of f in that direction is $-\|\nabla f(2, 0)\| = -\sqrt{80}$.

[23] (a) $\overrightarrow{PQ} = 2\mathbf{i} - 8\mathbf{j} + 3\mathbf{k} \Rightarrow \mathbf{u} = \frac{2}{\sqrt{77}}\mathbf{i} - \frac{8}{\sqrt{77}}\mathbf{j} + \frac{3}{\sqrt{77}}\mathbf{k}$.

$f(x, y, z) = \sqrt{x^2 + y^2 + z^2} \Rightarrow$

$$D_{\mathbf{u}}f(x, y, z) = \left[\frac{x}{(x^2 + y^2 + z^2)^{1/2}}\right](\tfrac{2}{\sqrt{77}}) + \left[\frac{y}{(x^2 + y^2 + z^2)^{1/2}}\right](-\tfrac{8}{\sqrt{77}}) + \left[\frac{z}{(x^2 + y^2 + z^2)^{1/2}}\right](\tfrac{3}{\sqrt{77}}).$$

At $P(-2, 3, 1)$, $D_{\mathbf{u}}f(-2, 3, 1) =$

$$(-\tfrac{2}{\sqrt{14}})(\tfrac{2}{\sqrt{77}}) + (\tfrac{3}{\sqrt{14}})(-\tfrac{8}{\sqrt{77}}) + (\tfrac{1}{\sqrt{14}})(\tfrac{3}{\sqrt{77}}) = \frac{-25}{\sqrt{14}\sqrt{77}} = -\frac{25}{7\sqrt{22}}.$$

(b) $\nabla f(-2, 3, 1) = -\frac{2}{\sqrt{14}}\mathbf{i} + \frac{3}{\sqrt{14}}\mathbf{j} + \frac{1}{\sqrt{14}}\mathbf{k} = \mathbf{v}$. $\|\nabla f(-2, 3, 1)\| = 1$.

(c) $-\mathbf{v} = \frac{2}{\sqrt{14}}\mathbf{i} - \frac{3}{\sqrt{14}}\mathbf{j} - \frac{1}{\sqrt{14}}\mathbf{k}$. $-\|\nabla f(-2, 3, 1)\| = -1$.

[25] (a) $T(x, y) = \dfrac{k}{(x^2 + y^2)^{1/2}}$ and $T(3, 4) = \dfrac{k}{\sqrt{3^2 + 4^2}} = 100 \Rightarrow k = 500$.

$\nabla T(x, y) = \dfrac{-500x}{(x^2 + y^2)^{3/2}}\mathbf{i} + \dfrac{-500y}{(x^2 + y^2)^{3/2}}\mathbf{j}$. $\mathbf{a} = \mathbf{i} + \mathbf{j} \Rightarrow \mathbf{u} = \frac{1}{\sqrt{2}}\mathbf{i} + \frac{1}{\sqrt{2}}\mathbf{j}$.

Thus, $D_{\mathbf{u}}T(3, 4) = (-12\mathbf{i} - 16\mathbf{j}) \cdot (\frac{1}{\sqrt{2}}\mathbf{i} + \frac{1}{\sqrt{2}}\mathbf{j}) = \frac{-28}{\sqrt{2}}$,

which is the rate of change of T at P in the direction of $\mathbf{i} + \mathbf{j}$.

(b) T increases most rapidly in the direction of $\nabla T(3, 4) = -12\mathbf{i} - 16\mathbf{j}$.

(c) T decreases most rapidly in the direction of $-\nabla T(3, 4) = 12\mathbf{i} + 16\mathbf{j}$.

(d) The rate of change will be 0 in the direction that is orthogonal to ∇T at (3, 4).

$\nabla T(3, 4) \cdot \mathbf{w} = 0 \Rightarrow (-12\mathbf{i} - 16\mathbf{j}) \cdot (w_1\mathbf{i} + w_2\mathbf{j}) = 0 \Rightarrow$

$-12w_1 - 16w_2 = 0 \Rightarrow w_1 = -\frac{4}{3}w_2$. If $w_2 = -3$, then $w_1 = 4$. Thus,

the rate of change is 0 in the direction of $\mathbf{w} = 4\mathbf{i} - 3\mathbf{j}$, or any multiple of $\mathbf{w}$.

27 (a) $V = x^2 + 4y^2 + 9z^2 \Rightarrow \nabla V(x, y, z) = 2x\mathbf{i} + 8y\mathbf{j} + 18z\mathbf{k}$.

$\mathbf{a} = \overrightarrow{PO} = -2\mathbf{i} + \mathbf{j} - 3\mathbf{k} \Rightarrow \mathbf{u} = -\frac{2}{\sqrt{14}}\mathbf{i} + \frac{1}{\sqrt{14}}\mathbf{j} - \frac{3}{\sqrt{14}}\mathbf{k}$.

$$\nabla V(2, -1, 3) \cdot \mathbf{u} = (4)(-\tfrac{2}{\sqrt{14}}) + (-8)(\tfrac{1}{\sqrt{14}}) + (54)(-\tfrac{3}{\sqrt{14}}) = -\tfrac{178}{\sqrt{14}}.$$

(b) The maximum rate of change is in the direction of

$$\nabla V(2, -1, 3) = 4\mathbf{i} - 8\mathbf{j} + 54\mathbf{k}.$$

(c) The maximum rate of change at P is

$$\|\nabla V(2, -1, 3)\| = \sqrt{4^2 + (-8)^2 + 54^2} = \sqrt{2996} \approx 54.7.$$

29 (a) See Example 5. Let the circle have radius r. Let $P(x, y)$ be a point on the upper boundary. Any radius of the semicircle can be described by the vector $\mathbf{a} = \overrightarrow{OP} = x\mathbf{i} + y\mathbf{j}$. Since a radius is always perpendicular to the tangent line, it follows that $\mathbf{a} = x\mathbf{i} + y\mathbf{j} = r\cos\theta\,\mathbf{i} + r\sin\theta\,\mathbf{j}$ is normal to the circle at the point (r, θ) and the unit normal vector is $\mathbf{n} = \cos\theta\,\mathbf{i} + \sin\theta\,\mathbf{j}$. The boundary is insulated iff $\nabla T(x, y) \cdot \mathbf{n} = 0 \Rightarrow \left(\frac{\partial T}{\partial x}\mathbf{i} + \frac{\partial T}{\partial y}\mathbf{j}\right) \cdot (\cos\theta\,\mathbf{i} + \sin\theta\,\mathbf{j}) = 0 \Rightarrow$ $\frac{\partial T}{\partial x}\cos\theta + \frac{\partial T}{\partial y}\sin\theta = 0$. Since $\frac{\partial T}{\partial r} = \frac{\partial T}{\partial x}\frac{\partial x}{\partial r} + \frac{\partial T}{\partial y}\frac{\partial y}{\partial r} = \frac{\partial T}{\partial x}\cos\theta + \frac{\partial T}{\partial y}\sin\theta$, the result follows.

(b) Since $\frac{\partial T}{\partial r} = \nabla T(x, y) \cdot \mathbf{n}$, $\frac{\partial T}{\partial r}$ is the rate of change of temperature in the direction normal to the circular boundary.

31 (a) $\displaystyle\lim_{h \to 0} \frac{f(x + h, y) - f(x - h, y)}{2h}$

$$= \lim_{h \to 0} \frac{f(x + h, y) - f(x, y) - f(x - h, y) + f(x, y)}{2h}$$

$$= \frac{1}{2}\lim_{h \to 0} \frac{f(x + h, y) - f(x, y)}{h} - \frac{1}{2}\lim_{h \to 0} \frac{f(x - h, y) - f(x, y)}{h}$$

$$= \frac{1}{2}f_x(x, y) + \frac{1}{2}\lim_{k \to 0} \frac{f(x + k, y) - f(x, y)}{k} \quad \{\text{where } k = -h\}$$

$= \frac{1}{2}f_x(x, y) + \frac{1}{2}f_x(x, y) = f_x(x, y)$. A similar solution holds for f_y.

(b) $f_x(1, 2) \approx \dfrac{f(1 + 0.01, 2) - f(1 - 0.01, 2)}{2(0.01)} \approx 1.00003333$ and

$f_y(1, 2) \approx \dfrac{f(1, 2 + 0.01) - f(1, 2 - 0.01)}{2(0.01)} \approx -0.11111235 \Rightarrow$

$$\nabla f(1, 2) \approx 1.00003333\mathbf{i} - 0.11111235\mathbf{j}.$$

$f_x(x, y) = \dfrac{3x^2}{1 + y} \Rightarrow f_x(1, 2) = 1$. $f_y(x, y) = -\dfrac{x^3}{(1 + y)^2} \Rightarrow f_y(1, 2) = -\frac{1}{9}$.

Thus, $\nabla f(1, 2) = \mathbf{i} - \frac{1}{9}\mathbf{j}$.

35 $\nabla(cu) = (cu)_x\mathbf{i} + (cu)_y\mathbf{j} = cu_x\mathbf{i} + cu_y\mathbf{j} = c(u_x\mathbf{i} + u_y\mathbf{j}) = c\nabla u$

[37] $\nabla(uv) = (uv)_x\,\mathbf{i} + (uv)_y\,\mathbf{j}$

$= (u_x v + u v_x)\,\mathbf{i} + (u_y v + u v_y)\,\mathbf{j}$

$= (u v_x\,\mathbf{i} + u v_y\,\mathbf{j}) + (u_x v\,\mathbf{i} + u_y v\,\mathbf{j})$

$= u(v_x\,\mathbf{i} + v_y\,\mathbf{j}) + v(u_x\,\mathbf{i} + u_y\,\mathbf{j}) = u\nabla v + v\nabla u$

[41] (a) The i-component of u is $\|\mathbf{u}\|\cos\theta$ and the j-component is $\|\mathbf{u}\|\sin\theta$.

Since $\|\mathbf{u}\| = 1$, $\mathbf{u} = \cos\theta\,\mathbf{i} + \sin\theta\,\mathbf{j} \Rightarrow D_{\mathbf{u}}f(x, y) = f_x(x, y)\cos\theta + f_y(x, y)\sin\theta$.

(b) $\theta = \frac{5\pi}{6} \Rightarrow \mathbf{u} = -\frac{1}{2}\sqrt{3}\,\mathbf{i} + \frac{1}{2}\mathbf{j}$. $D_{\mathbf{u}}f(x, y) = (2x + 2y)(-\frac{1}{2}\sqrt{3}) + (2x - 2y)(\frac{1}{2}) \Rightarrow$

$D_{\mathbf{u}}f(2, -3) = (-2)(-\frac{1}{2}\sqrt{3}) + (10)(\frac{1}{2}) = 5 + \sqrt{3}$.

Exercises 16.7

[1] $F(x, y, z) = 4x^2 - y^2 + 3z^2 - 10 \Rightarrow$

$\nabla F(x, y, z) = F_x(x, y, z)\mathbf{i} + F_y(x, y, z)\mathbf{j} + F_z(x, y, z)\mathbf{k} = 8x\mathbf{i} - 2y\mathbf{j} + 6z\mathbf{k} \Rightarrow$

$\nabla F(2, -3, 1) = F_x(2, -3, 1)\mathbf{i} + F_y(2, -3, 1)\mathbf{j} + F_z(2, -3, 1)\mathbf{k} = 16\mathbf{i} + 6\mathbf{j} + 6\mathbf{k}$.

Tangent plane: Using (16.34), $16(x - 2) + 6(y + 3) + 6(z - 1) = 0$.

Normal line: The normal line is parallel to $\nabla F(2, -3, 1) = 16\mathbf{i} + 6\mathbf{j} + 6\mathbf{k}$ and passes through $(2, -3, 1)$. $x = 2 + 16t$, $y = -3 + 6t$, $z = 1 + 6t$.

[5] $F(x, y, z) = xy + 2yz - xz^2 + 10 \Rightarrow \nabla F(x, y, z) =$

$(y - z^2)\mathbf{i} + (x + 2z)\mathbf{j} + (2y - 2xz)\mathbf{k} \Rightarrow \nabla F(-5, 5, 1) = 4\mathbf{i} - 3\mathbf{j} + 20\mathbf{k}$.

Tangent plane: $4(x + 5) - 3(y - 5) + 20(z - 1) = 0$.

Normal line: $x = -5 + 4t$, $y = 5 - 3t$, $z = 1 + 20t$.

[7] $F(x, y, z) = 2e^{-x}\cos y - z \Rightarrow \nabla F(x, y, z) =$

$(-2e^{-x}\cos y)\mathbf{i} + (-2e^{-x}\sin y)\mathbf{j} - \mathbf{k} \Rightarrow \nabla F(0, \frac{\pi}{3}, 1) = -\mathbf{i} - \sqrt{3}\mathbf{j} - \mathbf{k}$.

Tangent plane: $-1(x - 0) - \sqrt{3}(y - \frac{\pi}{3}) - 1(z - 1) = 0$, or, equivalently,

$x + \sqrt{3}(y - \frac{\pi}{3}) + (z - 1) = 0$.

Normal line: $x = t$, $y = \frac{\pi}{3} + \sqrt{3}t$, $z = 1 + t$. The normal line is parallel to $-\nabla F(0, \frac{\pi}{3}, 1) = \mathbf{i} + \sqrt{3}\mathbf{j} + \mathbf{k}$. (Note that $x = -t$, $y = \frac{\pi}{3} - \sqrt{3}t$, $z = 1 - t$ would be equivalent parametric equations for the normal line.)

[9] $F(x, y, z) = \ln\left(\frac{y}{2z}\right) - x \Rightarrow$

$\nabla F(x, y, z) = -\mathbf{i} + \frac{1}{y}\mathbf{j} - \frac{1}{z}\mathbf{k} \Rightarrow \nabla F(0, 2, 1) = -\mathbf{i} + \frac{1}{2}\mathbf{j} - \mathbf{k}$.

Tangent plane: $-x + \frac{1}{2}(y - 2) - (z - 1) = 0$.

Normal line: $x = -t$, $y = 2 + \frac{1}{2}t$, $z = 1 - t$.

[11] Level curves have the form $f(x, y) = y^2 - x^2 = k$.

$f(2, 1) = -3 = y^2 - x^2 \Rightarrow x^2 - y^2 = 3$. $\nabla f(x, y) = -2x\mathbf{i} + 2y\mathbf{j} \Rightarrow$

$\nabla f(2, 1) = -4\mathbf{i} + 2\mathbf{j}$. Graph the hyperbola $x^2 - y^2 = 3$.

$\nabla f(2, 1)$ is a vector normal to the hyperbola at the point $(2, 1)$. See *Figure 11.*

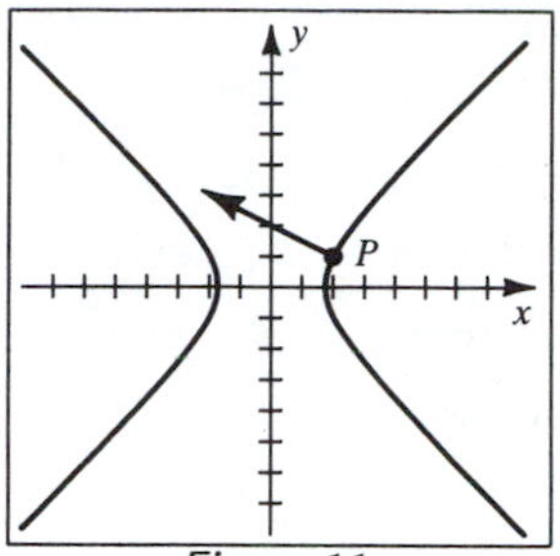

Figure 11

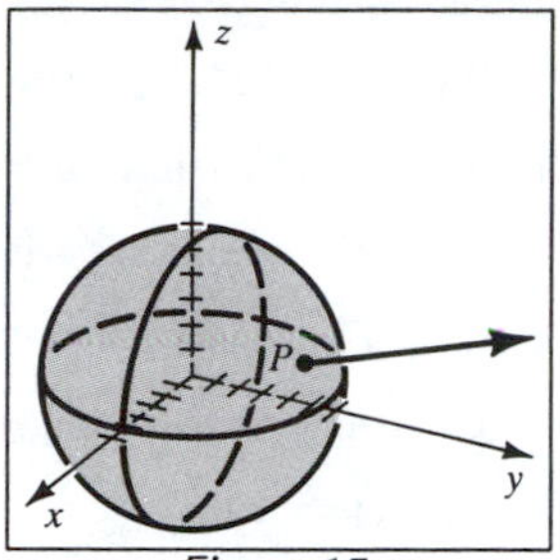

Figure 15

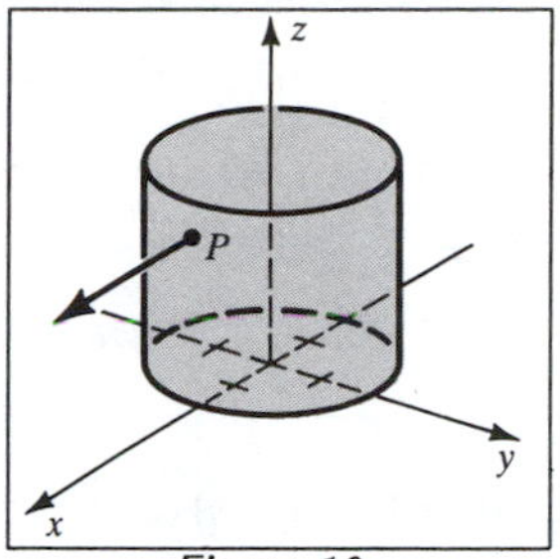

Figure 19

[15] Level surfaces have the form $F(x, y, z) = x^2 + y^2 + z^2 = k$.

$F(1, 5, 2) = 30 = x^2 + y^2 + z^2$, a sphere with radius $\sqrt{30}$ centered at O.

$\nabla F(x, y, z) = 2x\mathbf{i} + 2y\mathbf{j} + 2z\mathbf{k} \Rightarrow \nabla F(1, 5, 2) = 2\mathbf{i} + 10\mathbf{j} + 4\mathbf{k}$.

This vector is normal to the sphere at $(1, 5, 2)$.

[19] $F(2, 0, 3) = 4 = x^2 + y^2$, a circular cylinder along the z-axis.

$\nabla F(x, y, z) = 2x\mathbf{i} + 2y\mathbf{j} \Rightarrow \nabla F(2, 0, 3) = 4\mathbf{i}$.

[21] Let $F(x, y, z) = \dfrac{x^2}{a^2} + \dfrac{y^2}{b^2} + \dfrac{z^2}{c^2} - 1$.

$\nabla F(x_0, y_0, z_0) = \dfrac{2x_0}{a^2}\mathbf{i} + \dfrac{2y_0}{b^2}\mathbf{j} + \dfrac{2z_0}{c^2}\mathbf{k}$ gives the normal vector to the surface at

(x_0, y_0, z_0). By (16.34), the tangent plane is

$$\frac{2x_0}{a^2}(x - x_0) + \frac{2y_0}{b^2}(y - y_0) + \frac{2z_0}{c^2}(z - z_0) = 0 \Rightarrow$$

$$\frac{xx_0}{a^2} - \frac{x_0^2}{a^2} + \frac{yy_0}{b^2} - \frac{y_0^2}{b^2} + \frac{zz_0}{c^2} - \frac{z_0^2}{c^2} = 0 \Rightarrow$$

$$\frac{xx_0}{a^2} + \frac{yy_0}{b^2} + \frac{zz_0}{c^2} = \frac{x_0^2}{a^2} + \frac{y_0^2}{b^2} + \frac{z_0^2}{c^2},$$

which equals 1 because $P_0(x_0, y_0, z_0)$ satisfies the equation of the quadric surface.

[25] The normal to the hyperboloid is $2x\mathbf{i} - 4y\mathbf{j} - 8z\mathbf{k}$ and to the plane is $4\mathbf{i} - 2\mathbf{j} + 4\mathbf{k}$. The tangent plane of the hyperboloid is parallel to the given plane when these normal vectors are scalar multiples of each other, that is, when $2x = 4c$, $-4y = -2c$, and $-8z = 4c$ for some constant c. Substituting for x, y, and z in terms of c into the equation of the hyperboloid yields

$$(2c)^2 - 2(\tfrac{1}{2}c)^2 - 4(-\tfrac{1}{2}c)^2 = 16 \Rightarrow \tfrac{5}{2}c^2 = 16 \Rightarrow c = \pm\tfrac{4\sqrt{2}}{\sqrt{5}}.$$

There are two points: $\left(\frac{8\sqrt{2}}{\sqrt{5}}, \frac{2\sqrt{2}}{\sqrt{5}}, -\frac{2\sqrt{2}}{\sqrt{5}}\right)$ and $\left(-\frac{8\sqrt{2}}{\sqrt{5}}, -\frac{2\sqrt{2}}{\sqrt{5}}, \frac{2\sqrt{2}}{\sqrt{5}}\right)$.

[27] Without loss of generality, let the sphere have radius a, center at the origin, and

$F(x, y, z) = x^2 + y^2 + z^2 - a^2 = 0$. $\nabla F(x, y, z) = 2x\mathbf{i} + 2y\mathbf{j} + 2z\mathbf{k}$.

At (x_0, y_0, z_0) located on the sphere, the normal line is given by $x = x_0 + 2x_0t$,

$y = y_0 + 2y_0t$, $z = z_0 + 2z_0t$. At the origin, $x = 0 \Rightarrow 0 = x_0 + 2x_0t \Rightarrow t = -\frac{1}{2}$.

Letting $t = -\frac{1}{2}$ gives $y = z = 0$. Thus, this normal line always passes through the origin when $t = -\frac{1}{2}$ regardless of the choice of (x_0, y_0, z_0).

[31] Step 1: $f_x(x, y) = 2x \Rightarrow f_x(1.3, 1.1) = 2.6$.

$f_y(x, y) = -3y^2 \Rightarrow f_y(1.3, 1.1) = -3.63$.

$g_x(x, y) = 2x \Rightarrow g_x(1.3, 1.1) = 2.6$. $g_y(x, y) = 2y \Rightarrow g_y(1.3, 1.1) = 2.2$.

$f(1.3, 1.1) = 0.359$ and $g(1.3, 1.1) = -0.1$. Using (16.35) gives us

$z - 0.359 = 2.6(x - 1.3) - 3.63(y - 1.1)$ and

$z + 0.1 = 2.6(x - 1.3) + 2.2(y - 1.1)$.

Step 2: Letting $z = 0$ yields $2.6x - 3.63y = -0.972$ and

$2.6x + 2.2y = 5.9$ as the traces in the xy-plane.

Step 3: Solving yields $y \approx 1.1787$ and $x \approx 1.2718 \Rightarrow (x_2, y_2) = (1.2718, 1.1787)$.

Exercises 16.8

Note: As in (16.39), let $D(x, y) = f_{xx}(x, y)f_{yy}(x, y) - \left[f_{xy}(x, y)\right]^2$.

Also, let f_x, f_{xx}, f_y, and f_{yy} all be evaluated at (x, y).

[1] $f(x, y) = -x^2 - 4x - y^2 + 2y - 1 \Rightarrow f_x = -2x - 4$ and $f_y = -2y + 2$.

$f_x = 0$ and $f_y = 0 \Rightarrow x = -2$ and $y = 1$. $f_{xx} = -2$, $f_{xy} = 0$, and $f_{yy} = -2 \Rightarrow$

$D(-2, 1) = (-2)(-2) - 0^2 = 4 > 0$.

Since $f_{xx} < 0$, $f(-2, 1) = 4$ is a *LMAX* by 16.40(i).

[3] $f(x, y) = x^2 + 4y^2 - x + 2y \Rightarrow f_x = 2x - 1$ and $f_y = 8y + 2$.

$f_x = 0$ and $f_y = 0 \Rightarrow x = \frac{1}{2}$ and $y = -\frac{1}{4}$. $f_{xx} = 2$, $f_{xy} = 0$, and $f_{yy} = 8 \Rightarrow$

$D(\frac{1}{2}, -\frac{1}{4}) = 2(8) - 0^2 = 16 > 0$. Since $f_{xx} > 0$, $f(\frac{1}{2}, -\frac{1}{4}) = -\frac{1}{2}$ is a *LMIN*.

Note: We will use the notation SP for saddle point.

[7] $f(x, y) = x^3 + 3xy - y^3 \Rightarrow f_x = 3x^2 + 3y = 0$ and $f_y = 3x - 3y^2 = 0$.

From the first equation, $y = -x^2$. Substituting this into the second equation yields

$3x - 3(-x^2)^2 = 0 \Rightarrow x - x^4 \Rightarrow x(1 - x^3) = 0 \Rightarrow x = 0, 1$ and hence $y = 0, -1$.

$f_{xx} = 6x$, $f_{xy} = 3$, and $f_{yy} = -6y \Rightarrow$

$D(0, 0) = 0(0) - 3^2 = -9 < 0$ and $D(1, -1) = 6(6) - 3^2 = 27 > 0$.

Thus, $(0, 0, f(0, 0))$ is a SP and since $f_{xx}(1, -1) > 0$, $f(1, -1) = -1$ is a *LMIN*.

[9] $f(x, y) = \frac{1}{2}x^2 + 2xy - \frac{1}{2}y^2 + x - 8y \Rightarrow f_x = x + 2y + 1 = 0$ and

$f_y = 2x - y - 8 = 0 \Rightarrow x + 2y = -1$ and $4x - 2y = 16 \Rightarrow x = 3$ and $y = -2$.

$f_{xx} = 1$, $f_{xy} = 2$, and $f_{yy} = -1 \Rightarrow D(3, -2) = 1(-1) - 2^2 = -5 < 0 \Rightarrow$

$(3, -2, f(3, -2))$ is a SP.

[11] $f(x, y) = \frac{1}{3}x^3 - \frac{2}{3}y^3 + \frac{1}{2}x^2 - 6x + 32y + 4 \Rightarrow f_x = x^2 + x - 6 = 0$ and $f_y = -2y^2 + 32 = 0 \Rightarrow x = -3, 2$ and $y = \pm 4$. There are *four* critical points to test. $f_{xx} = 2x + 1$, $f_{xy} = 0$, and $f_{yy} = -4y \Rightarrow D(x, y) = -4y(2x + 1)$.

$D(2, 4) = -80 < 0 \Rightarrow (2, 4, f(2, 4))$ is a SP.

$D(-3, -4) = -80 < 0 \Rightarrow (-3, -4, f(-3, -4))$ is a SP.

$D(2, -4) = 80 > 0$ and $f_{xx}(2, -4) > 0 \Rightarrow f(2, -4) = -\frac{266}{3}$ is a *LMIN*.

$D(-3, 4) = 80 > 0$ and $f_{xx}(-3, 4) < 0 \Rightarrow f(-3, 4) = \frac{617}{6}$ is a *LMAX*.

[13] $f(x, y) = \frac{1}{2}x^4 - 2x^3 + 4xy + y^2 \Rightarrow f_x = 2x^3 - 6x^2 + 4y$ and $f_y = 4x + 2y$. $f_y = 0 \Rightarrow y = -2x$. $f_x = 0 \Rightarrow 2x^3 - 6x^2 = -4y \Rightarrow 6x^2 - 2x^3 + 8x = 0 \Rightarrow 2x(4 - x)(1 + x) = 0 \Rightarrow x = 0, 4, -1$ and hence the corresponding values for y are $y = 0, -8, 2$, respectively. There are three critical points to test. $f_{xx} = 6x^2 - 12x$, $f_{xy} = 4$, and $f_{yy} = 2 \Rightarrow D(x, y) = (6x^2 - 12x)(2) - 4^2 = 12x(x - 2) - 16$.

$D(0, 0) = -16 < 0 \Rightarrow (0, 0, f(0, 0))$ is a SP.

Note: f_{yy} may be used in place of f_{xx} in (16.40).

$D(4, -8) = 80 > 0$ and $f_{yy} > 0 \Rightarrow f(4, -8) = -64$ is a *LMIN*.

$D(-1, 2) = 20 > 0$ and $f_{yy} > 0 \Rightarrow f(-1, 2) = -\frac{3}{2}$ is a *LMIN*.

[17] $f_x = e^x \sin y = 0$ and $f_y = e^x \cos y = 0 \Rightarrow \sin y = \cos y = 0$ $\{e^x \neq 0\}$, which has no solution since $\sin^2 y + \cos^2 y = 1$ {$\sin y$ and $\cos y$ cannot both be zero at the same time}. Thus, f has no critical points and hence, no extrema or saddle points.

[21] Before starting to work this exercise, try to identify the five critical points on the graph. There are two maxima along the y-axis, two saddle points along the x-axis, and a minimum at the origin. $f(x, y) = (x^2 + 3y^2)e^{-(x^2+y^2)} \Rightarrow$

$f_x = (x^2 + 3y^2)e^{-(x^2+y^2)}(-2x) + 2xe^{-(x^2+y^2)} = 2xe^{-(x^2+y^2)}(1 - x^2 - 3y^2)$.

$f_y = (x^2 + 3y^2)e^{-(x^2+y^2)}(-2y) + 6ye^{-(x^2+y^2)} = 2ye^{-(x^2+y^2)}(3 - x^2 - 3y^2)$.

$f_x = 0$ and $f_y = 0 \Rightarrow x(1 - x^2 - 3y^2) = 0$ and $y(3 - x^2 - 3y^2) = 0$.

If $x = 0$, then $y = 0$ {from the factor y}, ± 1 {from the factor $3 - x^2 - 3y^2$}.

If $1 - x^2 - 3y^2 = 0$, then $x^2 = 1 - 3y^2$, $y = 0$, and hence $x = \pm 1$. $\{1 - x^2 - 3y^2 = 0 \Rightarrow 3 - x^2 - 3y^2 \neq 0$ and so we must have $y = 0.\}$

The five critical points are $(0, 0)$, $(0, 1)$, $(0, -1)$, $(1, 0)$, and $(-1, 0)$.

Since $f(x, y) \geq 0$ for all (x, y), $f(0, 0) = 0$ is a *LMIN*. Since $f(x, y) \to 0$ as $x^2 + y^2 \to \infty$, a maximum value will occur at one or more of the critical points.

$f(\pm 1, 0) = e^{-1}$ and $f(0, \pm 1) = 3e^{-1}$. Thus, $f(0, \pm 1) = 3e^{-1}$ are maximums.

$(\pm 1, 0, f(\pm 1, 0))$ are SP.

23 The minimum and maximum values of f either occur at local extrema or on the boundary determined by the ellipse. In Exercise 3, we determined all local extrema of f. We must now determine the maximum and minimum values of f on the boundary $x^2 + 4y^2 = 1$.

$x^2 + 4y^2 = 1 \Rightarrow -1 \le x \le 1$ and $-\frac{1}{2}\sqrt{1 - x^2} \le y \le \frac{1}{2}\sqrt{1 - x^2}$.

(1) On the upper boundary $y = \frac{1}{2}\sqrt{1 - x^2}$, $f(x, y) = x^2 + 4y^2 - x + 2y \Rightarrow$

$f(x, \frac{1}{2}\sqrt{1 - x^2}) = 1 - x + \sqrt{1 - x^2} = h(x)$. $h'(x) = -1 - x(1 - x^2)^{-1/2}$.

$h'(x) = 0 \Rightarrow x = -\sqrt{1 - x^2} \Rightarrow x = -\frac{1}{\sqrt{2}}$. $f(-\frac{1}{\sqrt{2}}, \frac{1}{2\sqrt{2}}) = 1 + \sqrt{2}$.

(2) On the lower boundary $y = -\frac{1}{2}\sqrt{1 - x^2}$,

$f(x, -\frac{1}{2}\sqrt{1 - x^2}) = 1 - x - \sqrt{1 - x^2} = h(x)$. $h'(x) = -1 + x(1 - x^2)^{-1/2}$.

$h'(x) = 0 \Rightarrow x = \sqrt{1 - x^2} \Rightarrow x = \frac{1}{\sqrt{2}}$. $f(\frac{1}{\sqrt{2}}, -\frac{1}{2\sqrt{2}}) = 1 - \sqrt{2}$.

At the intersection of the upper and lower boundaries, $f(-1, 0) = 2$ and $f(1, 0) = 0$.

Using Exercise 3, the *MIN* is $f(\frac{1}{2}, -\frac{1}{4}) = -\frac{1}{2}$ and the *MAX* is $f(-\frac{1}{\sqrt{2}}, \frac{1}{2\sqrt{2}}) = 1 + \sqrt{2}$.

27 In this exercise we must find the maximum and minimum of f on the edges and vertices of the triangle. These values are then compared to the local extrema found in Exercise 7. The boundaries of the triangle with vertices (1, 2), (1, −2), and (−1, −2) are $x = 1$, $y = -2$, and $y = 2x$. Thus, $-1 \le x \le 1$ and $-2 \le y \le 2x$.

(1) $f(x, y) = x^3 + 3xy - y^3$. On the line $x = 1$, $f(1, y) = 1 + 3y - y^3 = h(y)$.

$h'(y) = 3 - 3y^2 = 3(1 - y^2) = 0 \Rightarrow y = \pm 1$.

$f(1, 1) = 3$ and $f(1, -1) = -1$.

(2) On the line $y = -2$, $f(x, -2) = x^3 - 6x + 8 = h(x)$.

$h'(x) = 3x^2 - 6 = 3(x^2 - 2) = 0 \Rightarrow x = \pm\sqrt{2}$. These values are outside R.

(3) On the line $y = 2x$, $f(x, 2x) = -7x^3 + 6x^2 = h(x)$.

$h'(x) = -21x^2 + 12x = 3x(4 - 7x) = 0 \Rightarrow x = 0, \frac{4}{7}$.

$f(0, 0) = 0$ and $f(\frac{4}{7}, \frac{8}{7}) = \frac{32}{49}$.

At the vertices of R, $f(1, 2) = -1$, $f(1, -2) = 3$, and $f(-1, -2) = 13$. Using Exercise 7, the *MIN* are $f(1, 2) = f(1, -1) = -1$ and the *MAX* is $f(-1, -2) = 13$.

29 Let f be the square of the distance from the points $P(2, 1, -1)$ to the plane $4x - 3y + z = 5$, or, equivalently, $z = 5 - 4x + 3y$.

$$\begin{aligned} f(x, y) &= (x - 2)^2 + (y - 1)^2 + (z + 1)^2 \\ &= (x - 2)^2 + (y - 1)^2 + \left[(5 - 4x + 3y) + 1\right]^2 \\ &= (x - 2)^2 + (y - 1)^2 + (6 - 4x + 3y)^2. \end{aligned}$$

$f_x = 34x - 24y - 52 = 0$ and $f_y = -24x + 20y + 34 = 0$.

Five times the first equation and six times the second equation gives us

$170x - 120y = 260$ and $-144x + 120y = -204 \Rightarrow 26x = 56 \Rightarrow$

$x = \frac{28}{13}$ and $y = \frac{23}{26}$. $f(\frac{28}{13}, \frac{23}{26}) = \frac{1}{26} \Rightarrow d = 1/\sqrt{26}$.

[33] Since we are asked to find the relative dimensions, without loss of generality, let $V = 1$ unit and let z be the height of the box. Then, $xyz = 1$, or $z = \frac{1}{xy}$, and $A = xy + 2xz + 2yz = xy + \frac{2}{y} + \frac{2}{x}$. $A_x = y - \frac{2}{x^2} = 0$ and $A_y = x - \frac{2}{y^2} = 0 \Rightarrow$ $x^2y = 2 = xy^2 \Rightarrow xy(x - y) = 0 \Rightarrow x = 0,\ y = 0$, or $x = y$.

Since $x,\ y > 0$, we must have $x = y$. Thus, $z = \frac{1}{xy} = \frac{y}{2}$ $\{x = \frac{2}{y^2}\}$.

The box should have a square base with height $\frac{1}{2}$ the length of the side of the base.

[35] We could solve the equation $16x^2 + 4y^2 + 9z^2 = 144$ for z and substitute it into an equation for the volume V of the box. However, the algebra is more complicated than the method of solution shown. In order to differentiate the formula for V that appears later, we will need to find z_x and z_y, where z is assumed to be a function of x and y. We can do this by (16.23) with $F(x,\ y,\ z) = 16x^2 + 4y^2 + 9z^2 - 144 = 0$.

$$z_x = -\frac{F_x}{F_z} = -\frac{32x}{18z} \text{ and } z_y = -\frac{F_y}{F_z} = -\frac{8y}{18z}.$$

The volume V of the box is given by $V = (2x)(2y)(2z) = 8xyz$.

$$V_x = (8xy)_x\, z + 8xy(z)_x = 8yz + 8xy(z_x)$$

and

$$V_y = (8xy)_y\, z + 8xy(z)_y = 8xz + 8xy(z_y).$$

$V_x = 0$ and $V_y = 0 \Rightarrow 8yz + 8xy\left(-\frac{32x}{18z}\right) = 0$ and $8xz + 8xy\left(-\frac{8y}{18z}\right) = 0 \Rightarrow$

$18z^2 - 32x^2 = 0$ and $18z^2 - 8y^2 = 0 \Rightarrow x^2 = \frac{9}{16}z^2$ and $y^2 = \frac{9}{4}z^2$.

$16(\frac{9}{16}z^2) + 4(\frac{9}{4}z^2) + 9z^2 - 144 = 0 \Rightarrow z^2 = \frac{16}{3} \Rightarrow z = 4/\sqrt{3},\ x = 3/\sqrt{3}$, and $y = 6/\sqrt{3}$. Thus, the dimensions are $8/\sqrt{3}$, $6/\sqrt{3}$, and $12/\sqrt{3}$.

[39] Let x = base length, y = base width, and z = height. The cost will be equal to twice the area of the top and bottom plus the area of the sides.

$C = 2(2xy) + 2xz + 2yz$ and $8 = xyz \Rightarrow C = 4xy + \frac{16}{y} + \frac{16}{x}$ $\left\{z = \frac{8}{xy}\right\}$.

$C_x = 4y - \frac{16}{x^2} = 0$ and $C_y = 4x - \frac{16}{y^2} = 0 \Rightarrow$

$yx^2 = 4 = xy^2 \Rightarrow x = y = \sqrt[3]{4}$ ft, since $x,\ y > 0$. Also, $z = \frac{8}{xy} = 2\sqrt[3]{4}$ ft.

[41] Let ℓ, w, and h denote the length, width, and height of the box.

Then, the girth is $2w + 2h$ and $\ell + 2w + 2h = 108 \Rightarrow \ell = 108 - 2w - 2h$.

So, $V = \ell wh = wh(108 - 2w - 2h)$.

$V_w = 108h - 4hw - 2h^2 = 0$ and $V_h = 108w - 2w^2 - 4wh = 0 \Rightarrow$

$h(54 - 2w - h) = 0$ and $w(54 - 2h - w) = 0 \Rightarrow 54 - 2w - h = 0$ and

$54 - 2h - w = 0$ $\{h,\ w \neq 0\} \Rightarrow 2w + h = 2h + w \Rightarrow h = w$ and hence

$3h = 54$, or $h = 18$. Substituting gives $w = 18$ and $\ell = 36$.

[45] $f(m, b) = \sum_{k=1}^{n} (y_k - mx_k - b)^2 \Rightarrow$

$$f_m(m, b) = \sum_{k=1}^{n} 2(y_k - mx_k - b)(-x_k) = 0$$

{remember to differentiate w.r.t. m and b, *not* x_k} and

$$f_b(m, b) = \sum_{k=1}^{n} 2(y_k - mx_k - b)(-1) = 0 \Rightarrow$$

$\sum_{k=1}^{n} (mx_k^2 + bx_k) = \sum_{k=1}^{n} x_k y_k$ and $\sum_{k=1}^{n} (mx_k + b) = \sum_{k=1}^{n} y_k \Rightarrow$

$$\left(\sum_{k=1}^{n} x_k^2\right) m + \left(\sum_{k=1}^{n} x_k\right) b = \sum_{k=1}^{n} x_k y_k \text{ and } \left(\sum_{k=1}^{n} x_k\right) m + nb = \sum_{k=1}^{n} y_k$$

[47] $\sum_{k=1}^{3} x_k^2 = 1^2 + 4^2 + 7^2 = 66$, $\sum_{k=1}^{3} x_k = 1 + 4 + 7 = 12$,

$\sum_{k=1}^{3} x_k y_k = 1(3) + 4(5) + 7(6) = 65$, and $\sum_{k=1}^{3} y_k = 3 + 5 + 6 = 14 \Rightarrow$

$$66m + 12b = 65 \text{ and } 12m + 3b = 14 \Rightarrow m = \tfrac{1}{2} \text{ and } b = \tfrac{8}{3}. \text{ Thus, } y = \tfrac{1}{2}x + \tfrac{8}{3}.$$

[51] The least squares criterion is satisfied when the sum of the squares of the distances is minimal. The sum of the squares of the distances f from P to each city is $f(x, y) = \left[(x-2)^2 + (y-3)^2\right] + \left[(x-7)^2 + (y-2)^2\right] + \left[(x-5)^2 + (y-6)^2\right]$.

Then, $f_x(x, y) = 2(x-2) + 2(x-7) + 2(x-5) = 0$ and

$f_y(x, y) = 2(y-3) + 2(y-2) + 2(y-6) = 0 \Rightarrow 6x = 28$ and $6y = 22 \Rightarrow$

$$x = \tfrac{14}{3} \text{ and } y = \tfrac{11}{3}.$$

[55] $f_x(x, y) = 3x^2 \sin x + x^3 \cos x - y = 0 \Rightarrow$

$$y = 3x^2 \sin x + x^3 \cos x.$$

$f_y(x, y) = -x + 8y + 1 = 0 \Rightarrow y = \frac{1}{8}(x-1)$. Graph each equation on the same coordinate plane. A critical point will occur when $f_x = f_y = 0$ simultaneously. This corresponds to a point of intersection. Their point of intersection in R is approximately $(-0.35, -0.17)$, which is a critical point of f.

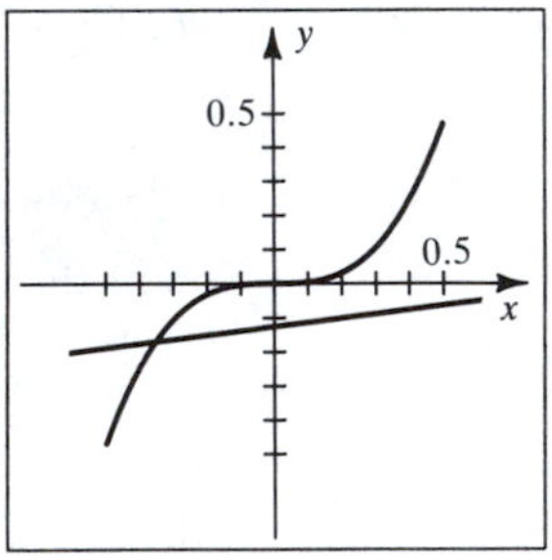

Figure 55

Exercises 16.9

Note: In Exercises 1–10, the equations listed first result from equating gradients and letting conditions equal 0. Let ⟦1⟧, ⟦2⟧, ⟦3⟧, etc., denote equation numbers.

1 By (16.43), $f_x(x, y) = \lambda g_x(x, y)$ ⟦1⟧, $f_y(x, y) = \lambda g_y(x, y)$ ⟦2⟧, and $g(x, y) = 0$ ⟦3⟧.

With $f(x, y) = y^2 - 4xy + 4x^2$ and $g(x, y) = x^2 + y^2 - 1$, these equations become $-4y + 8x = 2x\lambda$ ⟦1⟧, $2y - 4x = 2y\lambda$ ⟦2⟧, and $x^2 + y^2 - 1 = 0$ ⟦3⟧.

Adding ⟦1⟧ plus twice ⟦2⟧ $\Rightarrow 0 = 2x\lambda + 4y\lambda = 2\lambda(x + 2y) \Rightarrow \lambda = 0$ or $x = -2y$.

If $\lambda = 0$, then from ⟦1⟧, $y = 2x$ and using ⟦3⟧ we find that $x^2 + (2x)^2 - 1 = 0 \Rightarrow$

$5x^2 = 1 \Rightarrow x = \pm\frac{1}{\sqrt{5}} \Rightarrow (x, y) = (\frac{1}{\sqrt{5}}, \frac{2}{\sqrt{5}})$ or $(-\frac{1}{\sqrt{5}}, -\frac{2}{\sqrt{5}})$.

If $x = -2y$, substituting into ⟦3⟧ yields $(-2y)^2 + y^2 - 1 = 0 \Rightarrow$

$5y^2 = 1 \Rightarrow y = \pm\frac{1}{\sqrt{5}} \Rightarrow (x, y) = (\frac{2}{\sqrt{5}}, -\frac{1}{\sqrt{5}})$ or $(-\frac{2}{\sqrt{5}}, \frac{1}{\sqrt{5}})$.

$f(x, y) = 0$ at the first two points, which are *LMIN*,

and $f(x, y) = 5$ at the second two points, which are *LMAX*.

3 By (16.45), $f_x(x, y, z) = \lambda g_x(x, y, z)$ ⟦1⟧, $f_y(x, y, z) = \lambda g_y(x, y, z)$ ⟦2⟧, $f_z(x, y, z) = \lambda g_z(x, y, z)$ ⟦3⟧, and $g(x, y, z) = 0$ ⟦4⟧.

With $f(x, y, z) = x + y + z$ and $g(x, y, z) = x^2 + y^2 + z^2 - 25$, these equations become $1 = 2x\lambda$ ⟦1⟧, $1 = 2y\lambda$ ⟦2⟧, $1 = 2z\lambda$ ⟦3⟧, and $x^2 + y^2 + z^2 - 25 = 0$ ⟦4⟧.

From ⟦1⟧, ⟦2⟧, and ⟦3⟧, $\frac{1}{2\lambda} = x = y = z$, and using ⟦4⟧,

$x^2 + x^2 + x^2 = 25 \Rightarrow 3x^2 = 25 \Rightarrow x = y = z = \pm\frac{5}{\sqrt{3}}$.

$f(\frac{5}{\sqrt{3}}, \frac{5}{\sqrt{3}}, \frac{5}{\sqrt{3}}) = 5\sqrt{3}$ is a *LMAX* and $f(-\frac{5}{\sqrt{3}}, -\frac{5}{\sqrt{3}}, -\frac{5}{\sqrt{3}}) = -5\sqrt{3}$ is a *LMIN*.

7 See Example 4 for an example using 2 constraints. Let $g(x, y, z) = x - y = 1$ and $h(x, y, z) = y^2 - z^2 - 1$ with $f(x, y, z) = x^2 + y^2 + z^2$.

Then, $f_x = \lambda g_x + \mu h_x$ ⟦1⟧, $f_y = \lambda g_y + \mu h_y$ ⟦2⟧, $f_z = \lambda g_z + \mu h_z$ ⟦3⟧, $g = 0$ ⟦4⟧, and $h = 0$ ⟦5⟧ $\Rightarrow 2x = \lambda$ ⟦1⟧, $2y = -\lambda + 2y\mu$ ⟦2⟧, $2z = -2z\mu$ ⟦3⟧, $x - y - 1 = 0$ ⟦4⟧, and $y^2 - z^2 - 1 = 0$ ⟦5⟧. From ⟦3⟧, $z = 0$ or $\mu = -1$.

If $z = 0$, then from ⟦5⟧, $y = 1, -1$ and from ⟦4⟧, $x = 2, 0$.

This gives us $(2, 1, 0)$ and $(0, -1, 0)$.

If $\mu = -1$, then ⟦2⟧ gives $\lambda = -4y$. Using ⟦1⟧, $2x = -4y \Rightarrow x = -2y$.

Substituting into ⟦4⟧ yields $y = -\frac{1}{3}$. But substituting $y = -\frac{1}{3}$ into ⟦5⟧ gives $z^2 < 0$, which has no real solution, and hence $\mu \neq -1$.

$f(0, -1, 0) = 1$ and $f(2, 1, 0) = 5$ are both *LMIN*.

(Note that the intersection of $x - y = 1$ and $y^2 - z^2 = 1$ defines 2 separate curves when $y \geq 1$ or $y \leq -1$. The two points found give the square of the minimum distance between these curves and the origin. No maximum exists.)

9 Let $g(x, y, z) = x - z - 2$ and $h(x, y, z) = y^2 + t - 4$ with $f(x, y, z, t) = xyzt$.

Then, $f_x = \lambda g_x + \mu h_x$ [[1]], $f_y = \lambda g_y + \mu h_y$ [[2]], $f_z = \lambda g_z + \mu h_z$ [[3]],

$f_t = \lambda g_t + \mu h_t$ [[4]], $g = 0$ [[5]], and $h = 0$ [[6]] $\Rightarrow$ $yzt = \lambda$ [[1]], $xzt = 2y\mu$ [[2]],

$xyt = -\lambda$ [[3]], $xyz = \mu$ [[4]], $x - z - 2 = 0$ [[5]], and $y^2 + t - 4 = 0$ [[6]].

From [[1]] and [[3]], $yzt = -xyt \Rightarrow yt(x + z) = 0$.

If $yt = 0$, then $f(x, y, z, t) = 0$.

If $x = -z$, then from [[5]], $z = -1$ and $x = 1$.

Multiplying [[2]] by y and [[4]] by t yields

$$2y^2\mu = xyzt = \mu t \Rightarrow \mu(2y^2 - t) = 0 \Rightarrow \mu = 0 \text{ or } t = 2y^2.$$

If $\mu = 0$, then $xyz = 0$ {from [[4]]} and $f(x, y, z, t) = 0$.

If $t = 2y^2$, then from [[6]], $y = \pm\frac{2}{\sqrt{3}}$ and $t = \frac{8}{3}$.

$$f(1, \tfrac{2}{\sqrt{3}}, -1, \tfrac{8}{3}) = -\tfrac{16}{3\sqrt{3}} \text{ is a } LMIN \text{ and } f(1, -\tfrac{2}{\sqrt{3}}, -1, \tfrac{8}{3}) = \tfrac{16}{3\sqrt{3}} \text{ is a } LMAX.$$

11 If f is the square of the distance from $P(x, y, z)$ to $(2, 3, 4)$, then we have $f(x, y, z) =$ $(x - 2)^2 + (y - 3)^2 + (z - 4)^2$ and $g(x, y, z) = x^2 + y^2 + z^2 - 9 = 0$ [[1]].

$\nabla f = \lambda \nabla g \Rightarrow 2(x - 2) = 2x\lambda$ [[2]], $2(y - 3) = 2y\lambda$ [[3]], and $2(z - 4) = 2z\lambda$ [[4]].

From [[2]], [[3]], and [[4]], $\lambda = \frac{x - 2}{x} = \frac{y - 3}{y} = \frac{z - 4}{z}$. Using $\frac{x - 2}{x} = \frac{y - 3}{y}$ gives

us $xy - 2y = xy - 3x \Rightarrow y = \frac{3}{2}x$. Similarly, $\frac{x - 2}{x} = \frac{z - 4}{z}$ yields $z = 2x$.

Substituting into [[1]] yields $x^2 + (\frac{3}{2}x)^2 + (2x)^2 - 9 = 0 \Rightarrow \frac{29}{4}x^2 = 9 \Rightarrow$

$x = \pm\frac{6}{\sqrt{29}}$, and hence $y = \pm\frac{9}{\sqrt{29}}$, and $z = \pm\frac{12}{\sqrt{29}}$.

The positive values lead to a *minimum* distance since $(2, 3, 4)$ is also in the first octant, whereas the negative values give a *maximum* distance.

13 Let x and y be the base dimensions and z the height. Since the volume $V = xyz$ must be equal to 2, we have the constraint $g(x, y, z) = xyz - 2 = 0$ [[1]].

The cost C will be equal to the area of the 4 sides, plus 2 times the area of the bottom, plus 1.5 times the area of the top.

$$C(x, y, z) = (2yz + 2xz) + 2xy + \tfrac{3}{2}xy = 2yz + 2xz + \tfrac{7}{2}xy.$$

$\nabla C = \lambda \nabla g \Rightarrow 2z + \frac{7}{2}y = yz\lambda$ [[2]], $2z + \frac{7}{2}x = xz\lambda$ [[3]], and $2y + 2x = xy\lambda$ [[4]].

Multiplying [[2]] and [[3]] by x and y, respectively, and then subtracting yields $2z(x - y) = 0$, or $x = y$. Multiplying [[2]] and [[4]] by x and z, respectively, and then subtracting yields $\frac{7}{2}xy - 2yz = 0 \Rightarrow y(7x - 4z) = 0$, or $z = \frac{7}{4}x$.

Substituting into [[1]] gives $x = \frac{2}{\sqrt[3]{7}}$, $y = \frac{2}{\sqrt[3]{7}}$, and $z = \frac{7}{2\sqrt[3]{7}}$.

[19] In this exercise, we are maximizing the area A, subject to the constraint that the perimeter has a fixed value p. Without loss of generality, we will maximize A^2. Squaring both sides of Heron's formula in the hint gives us

$A^2 = f(x, y, z) = s(s - x)(s - y)(s - z)$ and $g(x, y, z) = x + y + z - p = 0$ ⟦1⟧.

$\nabla f = \lambda \nabla g \Rightarrow -s(s - y)(s - z) = \lambda$ ⟦2⟧, $-s(s - x)(s - z) = \lambda$ ⟦3⟧, and $-s(s - x)(s - y) = \lambda$ ⟦4⟧. Note that $f_x = s \cdot \dfrac{\partial}{\partial x}(s - x)(s - y)(s - z) = s(-1)(s - y)(s - z)$. f_y and f_z are found in a similar manner. From ⟦2⟧, ⟦3⟧, and ⟦4⟧, $s - x = s - y = s - z \Rightarrow x = y = z$. From ⟦1⟧, $x = y = z = \frac{1}{3}p$.

[21] Refer to *Figure 21*. The strength of the beam is given by $f(x, y) = k(2x)(2y)^2 = 8kxy^2$, where k is a constant. An equation of an ellipse with major axis 24 ($a = 12$) and minor axis 16 ($b = 8$) is $x^2/12^2 + y^2/8^2 = 1$. Hence, the constraint on the dimensions of the beam is given by $g(x, y) = x^2/12^2 + y^2/8^2 - 1 = 0$ ⟦1⟧.

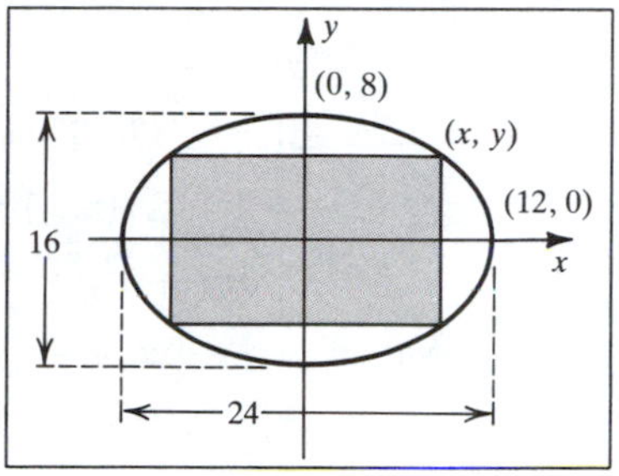

Figure 21

$\nabla f = \lambda \nabla g \Rightarrow 8ky^2 = \frac{1}{72}x\lambda$ ⟦2⟧ and $16kxy = \frac{1}{32}y\lambda$ ⟦3⟧.

Multiplying ⟦2⟧ by $2x$ and ⟦3⟧ by $y \Rightarrow \frac{1}{36}x^2\lambda = \frac{1}{32}y^2\lambda \Rightarrow$ $y^2 = \frac{8}{9}x^2$. Substituting into ⟦1⟧ yields $3x^2 = 144 \Rightarrow x = 4\sqrt{3}$ and $y = \frac{8}{3}\sqrt{6}$.

Thus, the width is $2x = 8\sqrt{3}$ in. and the depth is $2y = \frac{16}{3}\sqrt{6}$ in.

16.10 Review Exercises

[1] $f(x, y) = \sqrt{36 - 4x^2 + 9y^2}$ will be defined when $36 - 4x^2 + 9y^2 \geq 0$. Thus, the domain of f is $D = \{(x, y): 4x^2 - 9y^2 \leq 36\}$. $f(3, 4) = 12 \Rightarrow$ $36 - 4x^2 + 9y^2 = 12^2 = 144$. The level curve is the hyperbola $9y^2 - 4x^2 = 108$.

[3] $f(x, y, z) = (z^2 - x^2 - y^2)^{-3/2} = \dfrac{1}{(z^2 - x^2 - y^2)^{3/2}}$ will be defined when $z^2 - x^2 - y^2 > 0$. Thus, the domain of f is $D = \{(x, y, z): z^2 > x^2 + y^2\}$. $f(0, 0, 1) = 1$,

and the level surface is the hyperboloid of two sheets $z^2 - x^2 - y^2 = 1$.

[5] Substituting 0 for x and 0 for y gives us $\displaystyle\lim_{(x, y)\to(0, 0)} \frac{3xy + 5}{y^2 + 4} = \frac{3(0)(0) + 5}{0^2 + 4} = \frac{5}{4}$.

[7] Consider $\mathrm{L} = \displaystyle\lim_{(x, y)\to(0, 0)} \left(\frac{x^2 - y^2}{x^2 + y^2}\right)^2$. Along the path $y = x$,

$\mathrm{L} = \left(\dfrac{x^2 - x^2}{x^2 + x^2}\right)^2 = \left(\dfrac{0}{2x^2}\right)^2 = 0$. Along the path $y = 0$, $\mathrm{L} = \left(\dfrac{x^2 - 0}{x^2 + 0}\right)^2 = \dfrac{x^4}{x^4} = 1$.

Thus, by the two-path rule, the limit DNE.

[11] The first partial derivatives of f are f_x and f_y. $f(x, y) = x^3 \cos y - y^2 + 4x \Rightarrow$

$$f_x(x, y) = 3x^2 \cos y + 4 \text{ and } f_y(x, y) = -x^3 \sin y - 2y.$$

[15] $f(x, y, z, t) = x^2 z\sqrt{2y + t} \Rightarrow f_x(x, y, z, t) = 2xz\sqrt{2y + t}$.

$$f_y(x, y, z, t) = (x^2 z)\tfrac{1}{2}(2y + t)^{-1/2}(2) = \frac{x^2 z}{\sqrt{2y + t}}.$$

$$f_z(x, y, z, t) = x^2\sqrt{2y + t}.$$

$$f_t(x, y, z, t) = (x^2 z)\tfrac{1}{2}(2y + t)^{-1/2}(1) = \frac{x^2 z}{2\sqrt{2y + t}}.$$

[17] The second partial derivatives of f are f_{xx}, f_{xy}, f_{yx}, and f_{yy}.

$f(x, y) = x^3y^2 - 3xy^3 + x^4 - 3y + 2 \Rightarrow f_x = 3x^2y^2 - 3y^3 + 4x^3$ and

$f_y = 2x^3y - 9xy^2 - 3$. Thus, $f_{xx}(x, y) = 6xy^2 + 12x^2$,

$$f_{xy}(x, y) = f_{yx}(x, y) = 6x^2y - 9y^2, \text{ and } f_{yy}(x, y) = 2x^3 - 18xy.$$

[21] (a) $w = x^2 + 3xy - y^2 \Rightarrow$

$$\begin{aligned}\Delta w &= w(x + \Delta x, y + \Delta y) - w(x, y)\\ &= (x + \Delta x)^2 + 3(x + \Delta x)(y + \Delta y) - (y + \Delta y)^2 - (x^2 + 3xy - y^2)\\ &= 2x\Delta x + (\Delta x)^2 + 3x\Delta y + 3y\Delta x + 3\Delta x\Delta y - 2y\Delta y - (\Delta y)^2\\ &= (2x + 3y)\Delta x + (3x - 2y)\Delta y + (\Delta x)^2 + 3\Delta x\Delta y - (\Delta y)^2.\end{aligned}$$

$dw = w_x\,dx + w_y\,dy = (2x + 3y)\,dx + (3x - 2y)\,dy.$

(b) To find the exact change in w, calculate Δw: $x = -1$, $y = 2$,

$\Delta x = dx = -1.1 - (-1) = -0.1$, and $\Delta y = dy = 2.1 - 2 = 0.1 \Rightarrow$

$\Delta w = 4(-0.1) - 7(0.1) + (-0.1)^2 + 3(-0.1)(0.1) - (0.1)^2 = -1.13.$

To find an approximate change in w, calculate $dw = 4(-0.1) - 7(0.1) = -1.1$.

[23] We will use (16.17) to show that f is differentiable.

$$f(x, y) = \frac{xy}{y^2 - x^2} \Rightarrow f_x = \frac{y(x^2 + y^2)}{(y^2 - x^2)^2} \text{ and } f_y = \frac{-x(x^2 + y^2)}{(y^2 - x^2)^2}.$$

Since f_x and f_y are defined and continuous on the domain of f,

which is $\{(x, y): x^2 \neq y^2\}$, f is differentiable throughout its domain.

[25] If $s = uv + vw - uw$, $u = 2x + 3y$, $v = 4x - y$, and $w = -x + 2y$, then

$$\begin{aligned}s_x &= s_u u_x + s_v v_x + s_w w_x\\ &= (v - w)(2) + (u + w)(4) + (v - u)(-1) = v + 2w + 5u = 12x + 18y.\\ s_y &= s_u u_y + s_v v_y + s_w w_y\\ &= (v - w)(3) + (u + w)(-1) + (v - u)(2) = 5v - 4w - 3u = 18x - 22y.\end{aligned}$$

[29] By (16.23), $F(x, y, z) = x^2y + z\cos y - xz^3 = 0 \Rightarrow$

$$\frac{\partial z}{\partial x} = -\frac{F_x}{F_z} = \frac{z^3 - 2xy}{\cos y - 3xz^2} \text{ and } \frac{\partial z}{\partial y} = -\frac{F_y}{F_z} = \frac{z\sin y - x^2}{\cos y - 3xz^2}.$$

[31] (a) $P(-1, -3, 2)$ and $Q(-4, 1, -2) \Rightarrow \overrightarrow{PQ} = -3\mathbf{i} + 4\mathbf{j} - 4\mathbf{k} \Rightarrow$

$\mathbf{u} = -\frac{3}{\sqrt{41}}\mathbf{i} + \frac{4}{\sqrt{41}}\mathbf{j} - \frac{4}{\sqrt{41}}\mathbf{k}$. By (16.32), with $T(x, y, z) = 3x^2 + 2y^2 - 4z$,

$D_{\mathbf{u}}\, T(x, y, z) = (6x)(-\frac{3}{\sqrt{41}}) + (4y)(\frac{4}{\sqrt{41}}) + (-4)(-\frac{4}{\sqrt{41}})$.

At $P(-1, -3, 2)$, $D_{\mathbf{u}}\, T(-1, -3, 2) = -\frac{14}{\sqrt{41}}$.

(b) The maximum rate of change of T at P is $\|\nabla T(-1, -3, 2)\| =$

$\|-6\mathbf{i} - 12\mathbf{j} - 4\mathbf{k}\| = \sqrt{(-6)^2 + (-12)^2 + (-4)^2} = \sqrt{196} = 14$.

[33] $F(x, y, z) = 4x^2 - 2y^2 - 7z = 0 \Rightarrow$

$\nabla F(x, y, z) = 8x\mathbf{i} - 4y\mathbf{j} - 7\mathbf{k} \Rightarrow \nabla F(-2, -1, 2) = -16\mathbf{i} + 4\mathbf{j} - 7\mathbf{k}$.

Tangent plane: $-16(x + 2) + 4(y + 1) - 7(z - 2) = 0$ { (16.34) }

Normal line: $x = -2 - 16t$, $y = -1 + 4t$, $z = 2 - 7t$

[37] $F(x, y, z) = z + 4x^2 + 9y^2 \Rightarrow F(1, 0, 0) = 4$.

Thus, the level surface is $z = -4x^2 - 9y^2 + 4$,

a paraboloid.

$\nabla F(x, y, z) = 8x\mathbf{i} + 18y\mathbf{j} + \mathbf{k} \Rightarrow$

$\nabla F(1, 0, 0) = 8\mathbf{i} + \mathbf{k}$.

This vector is normal to the paraboloid at (1, 0, 0).

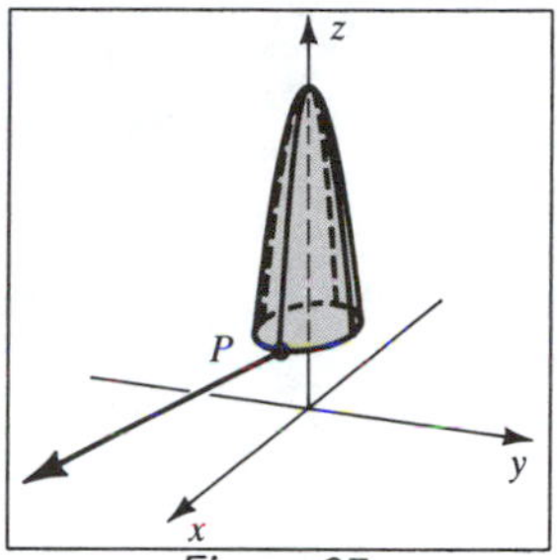

Figure 37

[39] $f(x, y) = x^2 + 3y - y^3 \Rightarrow f_x = 2x$ and $f_y = 3 - 3y^2$.

$f_x = 0$ and $f_y = 0 \Rightarrow x = 0$ and $y = \pm 1$.

$f_{xx} = 2$, $f_{xy} = 0$, and $f_{yy} = -6y \Rightarrow D(x, y) = f_{xx}f_{yy} - [f_{xy}]^2 = -12y$.

$D(0, 1) = -12 < 0 \Rightarrow (0, 1, f(0, 1))$ is a SP by (16.41).

$D(0, -1) = 12 > 0$ and $f_{xx} > 0 \Rightarrow f(0, -1) = -2$ is a *LMIN* by (16.40).

[41] $f(x, y, z) = xyz$ and $g(x, y, z) = x^2 + 4y^2 + 2z^2 - 8 = 0$ ⟦1⟧.

By (16.44) and (16.45), $\nabla f = \lambda \nabla g \Rightarrow yz = 2x\lambda$ ⟦2⟧, $xz = 8y\lambda$ ⟦3⟧, and

$xy = 4z\lambda$ ⟦4⟧. From ⟦2⟧, ⟦3⟧, and ⟦4⟧, $xyz = 2x^2\lambda = 8y^2\lambda = 4z^2\lambda \Rightarrow y^2 = \frac{1}{4}x^2$ and

$z^2 = \frac{1}{2}x^2$. Substituting into ⟦1⟧ yields $x^2 + 4(\frac{1}{4}x^2) + 2(\frac{1}{2}x^2) = 8 \Rightarrow 3x^2 = 8 \Rightarrow$

$x = \pm\sqrt{\frac{8}{3}}$, $y = \pm\sqrt{\frac{2}{3}}$, and $z = \pm\sqrt{\frac{4}{3}}$. These values represent eight points.

$f(-\sqrt{\frac{8}{3}}, -\sqrt{\frac{2}{3}}, -\sqrt{\frac{4}{3}}) = f(-\sqrt{\frac{8}{3}}, \sqrt{\frac{2}{3}}, \sqrt{\frac{4}{3}}) = f(\sqrt{\frac{8}{3}}, -\sqrt{\frac{2}{3}}, \sqrt{\frac{4}{3}}) =$

$f(\sqrt{\frac{8}{3}}, \sqrt{\frac{2}{3}}, -\sqrt{\frac{4}{3}}) = -\frac{8}{9}\sqrt{3}$ are *LMIN*.

$f(\sqrt{\frac{8}{3}}, \sqrt{\frac{2}{3}}, \sqrt{\frac{4}{3}}) = f(-\sqrt{\frac{8}{3}}, -\sqrt{\frac{2}{3}}, \sqrt{\frac{4}{3}}) =$

$f(-\sqrt{\frac{8}{3}}, \sqrt{\frac{2}{3}}, -\sqrt{\frac{4}{3}}) = f(\sqrt{\frac{8}{3}}, -\sqrt{\frac{2}{3}}, -\sqrt{\frac{4}{3}}) = \frac{8}{9}\sqrt{3}$ are *LMAX*.

[43] *Hint*: Consider a two-dimensional problem in the yz-plane and show that the point is approximately (0, 2.66, 1.76).

[45] $P = \dfrac{s}{(T + sk_0)(C + sk_1)} \Rightarrow$

$$\frac{\partial P}{\partial s} = \frac{(T + sk_0)(C + sk_1)(1) - (s)[(T + sk_0)k_1 + (C + sk_1)k_0]}{(T + sk_0)^2(C + sk_1)^2}.$$

$$\partial P/\partial s = 0 \Rightarrow TC + Tsk_1 + Csk_0 + s^2k_0k_1 = s(Tk_1 + sk_0k_1 + Ck_0 + sk_0k_1) \Rightarrow$$

$$TC = s^2k_0k_1 \Rightarrow s = \sqrt{\frac{TC}{k_0k_1}}.$$

Chapter 17: Multiple Integrals

Exercises 17.1

1. This is an R_x region. Recall that an R_x region lies between the graphs of two equations $y = f(x)$ and $y = g(x)$, with f and g continuous, and $f(x) \geq g(x)$ for every x in $[a, b]$, where a and b are the smallest and largest x-coordinates of the points (x, y) in the region. See §6.1 to review the descriptions of R_x and R_y regions.

3. This is an R_y region. Recall that an R_y **region** is a region that lies between the graphs of two equations of the form $x = f(y)$ and $x = g(y)$, with f and g continuous and $f(y) \geq g(y)$ for all y in $[c, d]$, where c and d are the smallest and largest y-coordinates of points in the region.

7. This region is symmetric with respect to both the x- and y-axes. Its boundary is determined by 4 line segments that could be written as either a function of x or as a function of y. This region could be considered as two R_x regions (one on the left of the y-axis and one on the right of the y-axis) or two R_y regions (one below the x-axis and one above the x-axis).

9. This region is neither an R_x region nor an R_y region because as we move from left to right (bottom to top), there is not one particular set of lower and upper (left and right) boundaries, that is, the boundaries change.

11. Note that $\Delta A_k = 1$ for all k since the area of every rectangle is 1.

(a) Since there are seven rectangles, the summation is from $k = 1$ to $k = 7$.

For each k, (u_k, v_k) are the coordinates of the lower left corner of square R_k.

$$\sum_{k=1}^{7} f(u_k, v_k)\Delta A_k = f(u_1, v_1) + f(u_2, v_2) + f(u_3, v_3) + f(u_4, v_4) + f(u_5, v_5) + f(u_6, v_6) + f(u_7, v_7)$$

$$= f(0, 2) + f(1, 2) + f(0, 1) + f(1, 1) + f(0, 0) + f(1, 0) + f(2, 0)$$

$$= 5 + 9 + 3 + 7 + 1 + 5 + 9 = 39.$$

(b) Similarly, with (u_k, v_k) in the upper right corner of R_k, $\sum_{k=1}^{7} f(u_k, v_k)\Delta A_k$

$$= f(1, 3) + f(2, 3) + f(1, 2) + f(2, 2) + f(1, 1) + f(2, 1) + f(3, 1)$$

$$= 11 + 15 + 9 + 13 + 7 + 11 + 15 = 81.$$

(c) Similarly, with (u_k, v_k) at the center of R_k, $\sum_{k=1}^{7} f(u_k, v_k)\Delta A_k$

$$= f(\tfrac{1}{2}, \tfrac{5}{2}) + f(\tfrac{3}{2}, \tfrac{5}{2}) + f(\tfrac{1}{2}, \tfrac{3}{2}) + f(\tfrac{3}{2}, \tfrac{3}{2}) + f(\tfrac{1}{2}, \tfrac{1}{2}) + f(\tfrac{3}{2}, \tfrac{1}{2}) + f(\tfrac{5}{2}, \tfrac{1}{2})$$

$$= 8 + 12 + 6 + 10 + 4 + 8 + 12 = 60.$$

13 $\mathrm{I} = \int_1^2 \int_{-1}^2 (12xy^2 - 8x^3)\,dy\,dx$ {given}

$= \int_1^2 \left[12x\frac{y^3}{3} - 8x^3 y \right]_{-1}^2 dx$ {treat x as a constant and integrate w.r.t. y}

$= \int_1^2 \left[4xy^3 - 8x^3 y \right]_{-1}^2 dx$ {simplify}

$= \int_1^2 \left[(32x - 16x^3) - (-4x + 8x^3) \right] dx$ {substitute 2 and then -1 for y}

$= \int_1^2 (36x - 24x^3)\,dx$ {simplify}

$= 6\left[3x^2 - x^4 \right]_1^2$ {integrate with respect to x, factor out 6}

$= 6\left[(12 - 16) - (3 - 1) \right] = 6(-6) = -36$

15 $\mathrm{I} = \int_1^2 \int_{1-x}^{\sqrt{x}} x^2 y\,dy\,dx$

$= \int_1^2 \left[\frac{1}{2}x^2 y^2 \right]_{1-x}^{\sqrt{x}} dx = \int_1^2 \left[\frac{1}{2}x^2(\sqrt{x})^2 - \frac{1}{2}x^2(1-x)^2 \right] dx$

$= \int_1^2 \frac{1}{2}(-x^4 + 3x^3 - x^2)\,dx = \frac{1}{2}\left[-\frac{1}{5}x^5 + \frac{3}{4}x^4 - \frac{1}{3}x^3 \right]_1^2$

$= \frac{1}{2}\left[(-\frac{32}{5} + 12 - \frac{8}{3}) - (-\frac{1}{5} + \frac{3}{4} - \frac{1}{3}) \right] = \frac{1}{2}(-\frac{31}{5} + \frac{45}{4} - \frac{7}{3}) = \frac{1}{2}(\frac{163}{60}) = \frac{163}{120}$.

As an alternative method in evaluating $\frac{1}{2}\left[-\frac{1}{5}x^5 + \frac{3}{4}x^4 - \frac{1}{3}x^3 \right]_1^2$, first factor the least common denominator—in this case, 60. The expression is then

$\frac{1}{2} \cdot \frac{1}{60}\left[-12x^5 + 45x^4 - 20x^3 \right]_1^2 = \frac{1}{120}\left[(-384 + 720 - 160) - (-12 + 45 - 20) \right]$

$= \frac{163}{120}$. The advantage of this evaluation technique is that you do not have to work as much with fractions.

19 If $u = \frac{y}{x}$, then $du = \frac{1}{x}dy$, or, equivalently, $x\,du = dy$. Thus, $\int_1^2 \int_{x^3}^{x} e^{y/x}\,dy\,dx =$

$\int_1^2 \left[xe^{y/x} \right]_{x^3}^{x} dx = \int_1^2 \left[xe^{x/x} - xe^{x^3/x} \right] dx = \int_1^2 x(e - e^{(x^2)})\,dx =$

$\left[\frac{1}{2}ex^2 - \frac{1}{2}e^{(x^2)} \right]_1^2 = \frac{1}{2}\left[(4e - e^4) - (e - e) \right] = \frac{1}{2}(4e - e^4) \approx -21.86$.

21 Refer to Examples 5 and 6 in the text.

(a) To apply Theorem (17.8)(i), we first examine the y-values since integration with respect to y is done before integration with respect to x. In this case, $0 \le y \le \sqrt{x}$. Next we must determine the limits of integration for x. From *Figure 21*, the smallest x-value in R is 0 and the largest x-value in R is 4. Thus, the integral can be expressed as the iterated integral $\int_0^4 \int_0^{\sqrt{x}} f(x, y)\,dy\,dx$.

(b) For Theorem (17.8)(ii), we integrate with respect to x first. Note that $y^2 \le x \le 4$ since the left boundary of R is determined by $x = y^2$ $\{\sqrt{x} = y\}$ and the right boundary is $x = 4$. The smallest y-value is 0 and the largest y-value is 2. The iterated integral is $\int_0^2 \int_{y^2}^4 f(x, y)\, dx\, dy$.

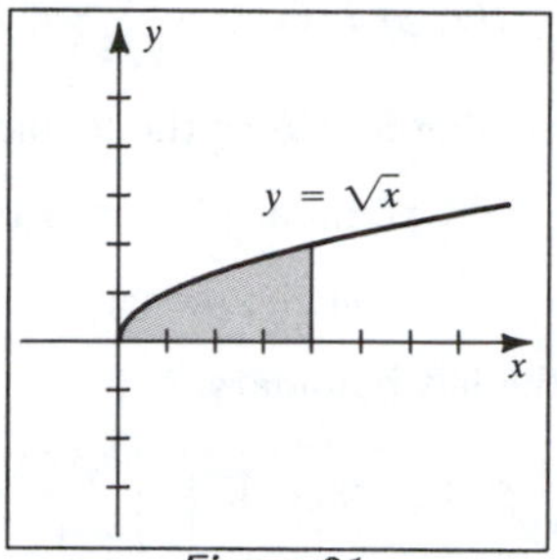

Figure 21

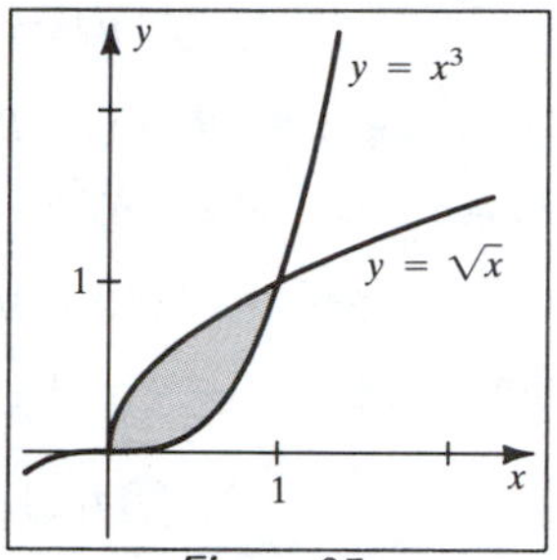

Figure 25

[25] (a) From *Figure 25* we see that y is bounded below by x^3 and above by $\sqrt{x}$. The values of x range from 0 to 1. Hence, $x^3 \le y \le \sqrt{x}$ and $0 \le x \le 1$. Integrating with respect to y first gives us $\iint_R f(x, y)\, dA = \int_0^1 \int_{x^3}^{\sqrt{x}} f(x, y)\, dy\, dx$.

(b) The left boundary of R is $x = y^2$ $\{\sqrt{x} = y\}$ and the right boundary is $x = y^{1/3}$ $\{x^3 = y\}$. Thus, $y^2 \le x \le y^{1/3}$ and $0 \le y \le 1$. Integrating with respect to x first gives us $\iint_R f(x, y)\, dA = \int_0^1 \int_{y^2}^{y^{1/3}} f(x, y)\, dx\, dy$.

[27] $-1 \le x \le 2$ and $-1 \le y \le 4 \Rightarrow \iint\limits_R (y + 2x)\, dA =$

$\int_{-1}^4 \int_{-1}^2 (y + 2x)\, dx\, dy = \int_{-1}^4 \Big[yx + x^2 \Big]_{-1}^2 dy = \int_{-1}^4 \Big[(2y + 4) - (-y + 1) \Big] dy =$

$\int_{-1}^4 (3y + 3)\, dy = 3\Big[\tfrac{1}{2}y^2 + y\Big]_{-1}^4 = 3\big[(8 + 4) - (\tfrac{1}{2} - 1)\big] = 3(\tfrac{25}{2}) = \tfrac{75}{2}$.

We could also use $\int_{-1}^2 \int_{-1}^4 (y + 2x)\, dy\, dx$.

[31] $0 \le y \le x^2$ and $0 \le x \le 2 \Rightarrow \iint\limits_R x^3 \cos xy\, dA =$

$\int_0^2 \int_0^{x^2} x^3 \cos xy\, dy\, dx$ $\{u = xy,\ du = x\, dy, \text{ or } 1/x\, du = dy\}$

$= \int_0^2 \Big[x^2 \sin xy \Big]_0^{x^2} dx = \int_0^2 x^2 \sin x^3\, dx = \Big[-\tfrac{1}{3}\cos x^3 \Big]_0^2 = -\tfrac{1}{3}(\cos 8 - \cos 0)$

$= \tfrac{1}{3}(1 - \cos 8) \approx 0.38$. Since $\sqrt{y} \le x \le 2$ and $0 \le y \le 4$,

the integral could also be expressed as $\int_0^4 \int_{\sqrt{y}}^2 x^3 \cos xy\, dx\, dy$.

However, it is *much* more difficult to integrate with respect to x first.

[33] (a) See *Figure 33.* We must express R as two R_x regions with the line $x = 2$ as the boundary between them. $y = x + 4$ is the upper boundary for both of these regions. On $1 \le x \le 2$, $y = 9 - 4x$ is the lower boundary.

On $2 \le x \le 4$, $y = \frac{1}{8}x^3$ is the lower boundary.

$$\iint_R f(x, y)\, dA = \int_1^2 \int_{9-4x}^{x+4} f(x, y)\, dy\, dx + \int_2^4 \int_{x^3/8}^{x+4} f(x, y)\, dy\, dx$$

(b) We must express R as two R_y regions with the line $y = 5$ as the boundary between them. $x = 2y^{1/3}$ $\{8y = x^3\}$ is the right boundary for both regions.

On $1 \le y \le 5$, $x = (9 - y)/4$ $\{y = 9 - 4x\}$ is the left boundary.

On $5 \le y \le 8$, $x = y - 4$ $\{y = x + 4\}$ is the left boundary.

$$\iint_R f(x, y)\, dA = \int_1^5 \int_{(9-y)/4}^{2y^{1/3}} f(x, y)\, dx\, dy + \int_5^8 \int_{y-4}^{2y^{1/3}} f(x, y)\, dx\, dy$$

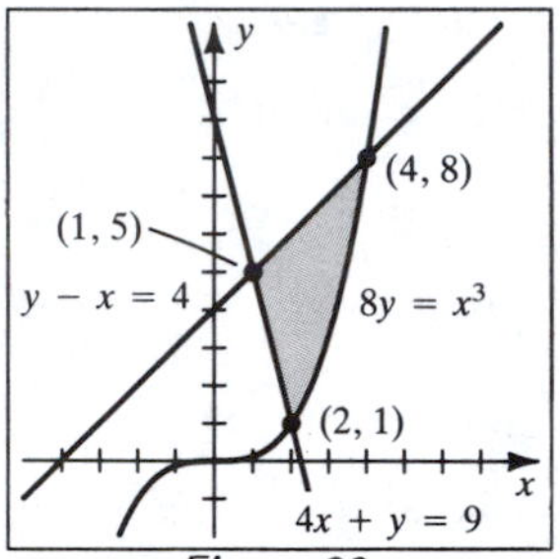

Figure 33

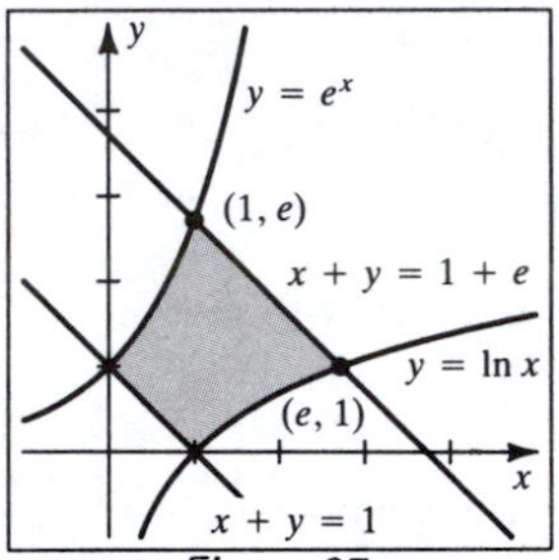

Figure 37

[37] (a) See *Figure 37.* To express R as two R_x regions, use $x = 1$ as a boundary.

In the left R_x region, the upper boundary is $y = e^x$ and the lower boundary is $y = 1 - x$. In the right R_x region, the upper boundary is $y = 1 + e - x$ and the lower boundary is $y = \ln x$.

Thus, $\iint_R f(x, y)\, dA = \int_0^1 \int_{1-x}^{e^x} f(x, y)\, dy\, dx + \int_1^e \int_{\ln x}^{1+e-x} f(x, y)\, dy\, dx.$

(b) To express R as two R_y regions, use $y = 1$ as a boundary. On the lower R_y region the right boundary is $x = e^y$ $\{y = \ln x\}$ and the left boundary is $x = 1 - y$. On the upper R_y region, the right boundary is $x = 1 + e - y$ and the left boundary is $x = \ln y$ $\{y = e^x\}$.

Thus, $\iint_R f(x, y)\, dA = \int_0^1 \int_{1-y}^{e^y} f(x, y)\, dx\, dy + \int_1^e \int_{\ln y}^{1+e-y} f(x, y)\, dx\, dy.$

[39] The upper boundary of the region is the parabola $y = 4 - x^2$ and the lower boundary is $y = -\sqrt{4 - x^2} \Leftrightarrow y^2 = 4 - x^2$, or the lower half of the circle $x^2 + y^2 = 4$ $\{y \le 0\}$ for $-1 \le x \le 2$.

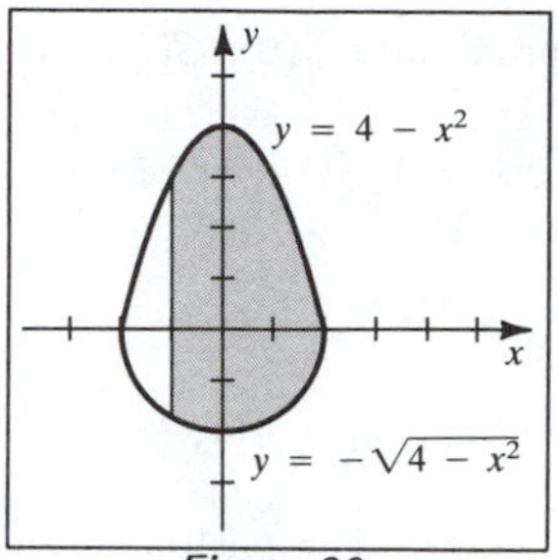

Figure 39

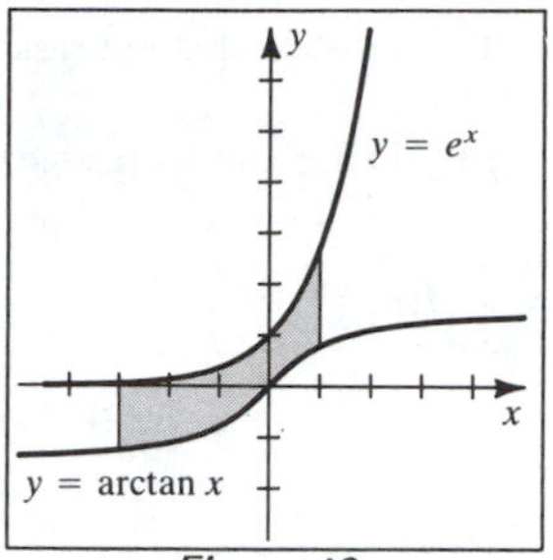

Figure 43

[43] The upper boundary of the region is $y = e^x$ and the lower boundary is $y = \arctan x$. Since $-3 \le x \le 1$, the left boundary is $x = -3$ and the right boundary is $x = 1$.

[45] The inequalities $2x \le y \le 2$ and $0 \le x \le 1$ determine R. Thus, R is the triangular region bounded by the lines $y = 2$, $x = 0$, and $y = 2x$. R can also be described by the inequalities $0 \le x \le \frac{1}{2}y$ and $0 \le y \le 2$. Thus,

$$\int_0^1 \int_{2x}^2 e^{(y^2)}\,dy\,dx = \int_0^2 \int_0^{y/2} e^{(y^2)}\,dx\,dy = \int_0^2 \Big[xe^{(y^2)}\Big]_0^{y/2} dy =$$

$$\tfrac{1}{2}\int_0^2 ye^{(y^2)}\,dy = \tfrac{1}{2}\Big[\tfrac{1}{2}e^{(y^2)}\Big]_0^2 = \tfrac{1}{4}(e^4 - e^0) = \tfrac{1}{4}(e^4 - 1) \approx 13.40.$$

[49] Since we are integrating with respect to x first, the inequalities $\sqrt[3]{y} \le x \le 2$ and $0 \le y \le 8$ determine R. Thus, R is the region bounded by $y = 0$, $x = 2$, and $x = \sqrt[3]{y}$ $\{$or $y = x^3\}$. R can also be described by the inequalities $0 \le y \le x^3$ and $0 \le x \le 2$.

$$\int_0^8 \int_{\sqrt[3]{y}}^2 \frac{y}{\sqrt{16 + x^7}}\,dx\,dy = \int_0^2 \int_0^{x^3} \frac{y}{(16 + x^7)^{1/2}}\,dy\,dx =$$

$$\int_0^2 \left[\frac{y^2}{2(16 + x^7)^{1/2}}\right]_0^{x^3} dx = \int_0^2 \frac{x^6}{2(16 + x^7)^{1/2}}\,dx = \Big[\tfrac{1}{7}(16 + x^7)^{1/2}\Big]_0^2 =$$

$$\tfrac{1}{7}(\sqrt{16 + 128} - \sqrt{16}) = \tfrac{1}{7}(12 - 4) = \tfrac{8}{7}.$$

51 Let $f(x,\,y) = e^{x^2 y^2}$, $G(y) = \int_c^d f(x,\,y)\,dx = \int_0^1 e^{x^2 y^2}\,dx$, and $I = \int_0^1 G(y)\,dy$. Using the trapezoidal rule to approximate I, $I \approx \frac{1-0}{2(3)}\Big[G(0) + 2G(\frac{1}{3}) + 2G(\frac{2}{3}) + G(1)\Big]$. In order to evaluate this expression, we must approximate $G(0)$, $G(\frac{1}{3})$, $G(\frac{2}{3})$, and $G(1)$ by using the trapezoidal rule again. $G(0) = \int_0^1 f(x,\,0)\,dx = \int_0^1 e^0\,dx = 1.$

$$G(\tfrac{1}{3}) = \int_0^1 f(x,\,\tfrac{1}{3})\,dx = \int_0^1 e^{x^2/9}\,dx$$
$$\approx \tfrac{1-0}{2(3)}\Big[f(0,\,\tfrac{1}{3}) + 2f(\tfrac{1}{3},\,\tfrac{1}{3}) + 2f(\tfrac{2}{3},\,\tfrac{1}{3}) + f(1,\,\tfrac{1}{3})\Big]$$
$$= \tfrac{1}{6}\Big[e^0 + 2e^{1/81} + 2e^{4/81} + e^{1/9}\Big] \approx 1.0406.$$

$$G(\tfrac{2}{3}) = \int_0^1 f(x,\,\tfrac{2}{3})\,dx = \int_0^1 e^{4x^2/9}\,dx$$
$$\approx \tfrac{1-0}{2(3)}\Big[f(0,\,\tfrac{2}{3}) + 2f(\tfrac{1}{3},\,\tfrac{2}{3}) + 2f(\tfrac{2}{3},\,\tfrac{2}{3}) + f(1,\,\tfrac{2}{3})\Big]$$
$$= \tfrac{1}{6}\Big[e^0 + 2e^{4/81} + 2e^{16/81} + e^{4/9}\Big] \approx 1.1829.$$

$$G(1) = \int_0^1 f(x,\,1)\,dx = \int_0^1 e^{(x^2)}\,dx$$
$$\approx \tfrac{1-0}{2(3)}\Big[f(0,\,1) + 2f(\tfrac{1}{3},\,1) + 2f(\tfrac{2}{3},\,1) + f(1,\,1)\Big]$$
$$= \tfrac{1}{6}\Big[e^0 + 2e^{1/9} + 2e^{4/9} + e^1\Big] \approx 1.5121.$$

Thus, $I \approx \frac{1-0}{2(3)}\Big[1 + 2(1.0406) + 2(1.1829) + 1.5121\Big] \approx 1.16.$

Exercises 17.2

1 This is an R_x region with upper boundary $y = 4x - x^2$ and lower boundary $y = -x$ on $0 \le x \le 5$. By (17.10), the area of the region is $A = \int_0^5 \int_{-x}^{4x-x^2} dy\,dx.$

3 This is an R_y region with right boundary $x = (9 - y)/2$ $\{y = -2x + 9\}$ and left boundary $x = -\sqrt{9 - y^2}$ on $-3 \le y \le 3$. $A = \int_{-3}^{3} \int_{-\sqrt{9-y^2}}^{(9-y)/2} dx\,dy.$

[7] See *Figure 7*. This is an R_y region with left boundary $x = -y^2$ and right boundary $x = y + 4$. The variable y varies between -1 and 2. Thus, $A = \int_{-1}^{2}\int_{-y^2}^{y+4} dx\,dy =$

$$\int_{-1}^{2}\Big[\,x\,\Big]_{-y^2}^{y+4} dy = \int_{-1}^{2} (y + 4 + y^2)\,dy = \Big[\tfrac{1}{3}y^3 + \tfrac{1}{2}y^2 + 4y\,\Big]_{-1}^{2} = \tfrac{33}{2}.$$

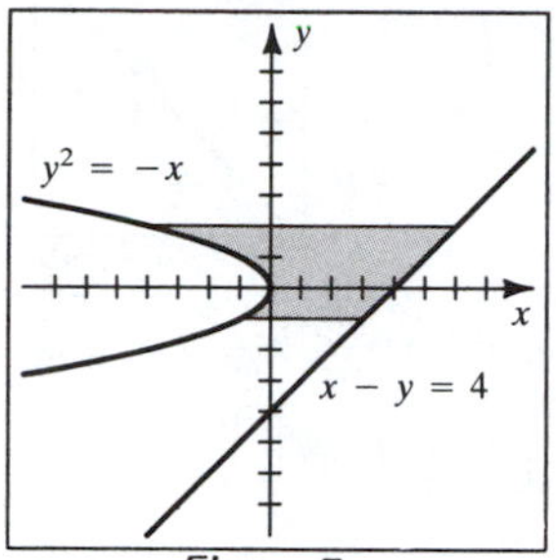

Figure 7

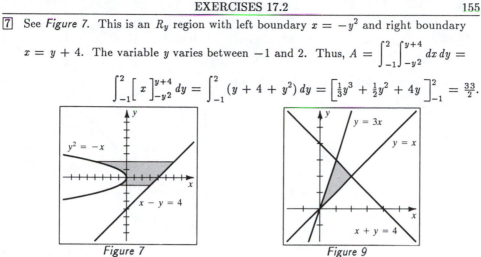

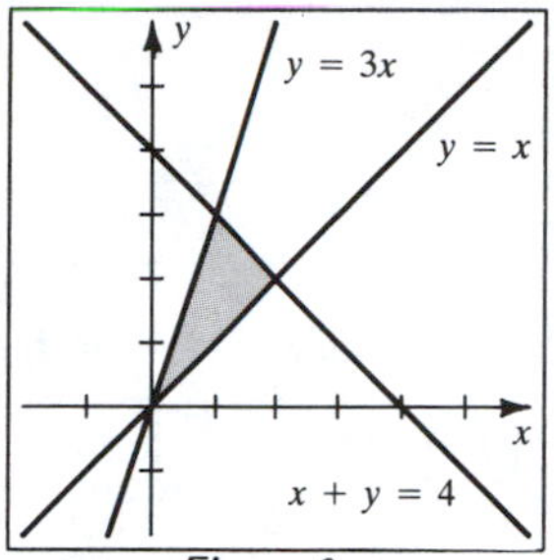

Figure 9

[9] See *Figure 9*. This could be treated as either two R_x regions or two R_y regions. We will treat it as two R_x regions with a boundary of $x = 1$ between them. The vertices of the triangular region are $A(0, 0)$, $B(1, 3)$, and $C(2, 2)$. On both regions the lower boundary is $y = x$. On the left R_x region, the upper boundary is $y = 3x$, whereas on the right R_x region, it is $y = 4 - x$. Thus,

$$A = \int_{0}^{1}\int_{x}^{3x} dy\,dx + \int_{1}^{2}\int_{x}^{4-x} dy\,dx = \int_{0}^{1}\Big[\,y\,\Big]_{x}^{3x} dx + \int_{1}^{2}\Big[\,y\,\Big]_{x}^{4-x} dx =$$

$$\int_{0}^{1} 2x\,dx + \int_{1}^{2} (4 - 2x)\,dx = 1 + 1 = 2.$$

[13] The plane has x-, y-, and z-intercepts at 3, 4, and 5, respectively.

An equation of the plane is $\frac{x}{3} + \frac{y}{4} + \frac{z}{5} = 1 \Leftrightarrow 20x + 15y + 12z = 60 \Leftrightarrow$ $z = f(x, y) = \frac{1}{12}(60 - 20x - 15y)$. The trace of the plane in the xy-plane $\{\, z = 0 \,\}$ is $20x + 15y = 60$, or $y = \frac{1}{3}(12 - 4x)$. R is the triangular region in the xy-plane bounded by the positive x- and y-axes and the line $y = \frac{1}{3}(12 - 4x)$. We will treat R as an R_x region.

By (17.9), $V = \iint_R f(x, y)\,dA = \int_{0}^{3}\int_{0}^{(12-4x)/3} \frac{1}{12}(60 - 20x - 15y)\,dy\,dx.$

[15] R is the region in the xy-plane with $0 \le x \le 4 - y^2$ and $0 \le y \le 2$. We will treat this as an R_y region which gives us $V = \iint_R f(x, y)\,dA = \int_{0}^{2}\int_{0}^{4-y^2} (6 - x)\,dx\,dy.$

[19] Since the integrand is $x^2 + y^2$, S is the paraboloid $z = x^2 + y^2$ over the region R in the xy-plane that has upper boundary $y = 1 - x^2$ and lower boundary $y = x - 1$ from $x = -2$ to $x = 1$.

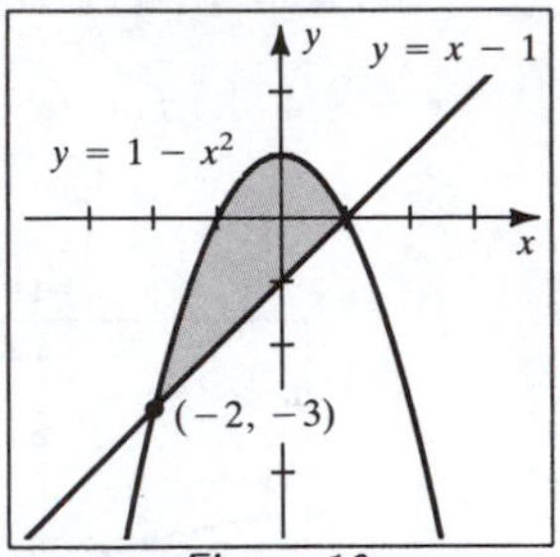

Figure 19

[21] R is the rectangular region in the xy-plane described by $0 \le x \le 2$ and $0 \le y \le 1$.

$$V = \iint_R f(x, y)\, dA = \int_0^2\int_0^1 (4x^2 + y^2)\, dy\, dx = \int_0^2 \Big[4x^2y + \tfrac{1}{3}y^3 \Big]_0^1 dx =$$

$$\int_0^2 (4x^2 + \tfrac{1}{3})\, dx = \Big[\tfrac{4}{3}x^3 + \tfrac{1}{3}x \Big]_0^2 = \tfrac{34}{3}.$$

[25] The plane $2x + y + z = 4$, or $z = f(x, y) = 4 - 2x - y$, has x-, y-, and z-intercepts of 2, 4, and 4, respectively. See *Figure 25*. Its trace in the xy-plane $\{ z = 0 \}$ is $2x + y = 4$, or $y = 4 - 2x$. R is a triangular region in the xy-plane with vertices (0, 0), (2, 0), and (0, 4). The volume underneath the plane and above the xy-plane in the first octant is given by

$$V = \int_0^2\int_0^{4-2x} (4 - 2x - y)\, dy\, dx = \int_0^2 \Big[(4 - 2x)y - \tfrac{1}{2}y^2 \Big]_0^{4-2x} dx$$

$$= \int_0^2 \Big[(4 - 2x)^2 - \tfrac{1}{2}(4 - 2x)^2 \Big] dx = \tfrac{1}{2}\int_0^2 (4 - 2x)^2\, dx = 2\int_0^2 (2 - x)^2\, dx$$

$$= 2\int_0^2 (x^2 - 4x + 4)\, dx = 2\Big[\tfrac{1}{3}x^3 - 2x^2 + 4x \Big]_0^2 = 2(\tfrac{8}{3}) = \tfrac{16}{3}.$$

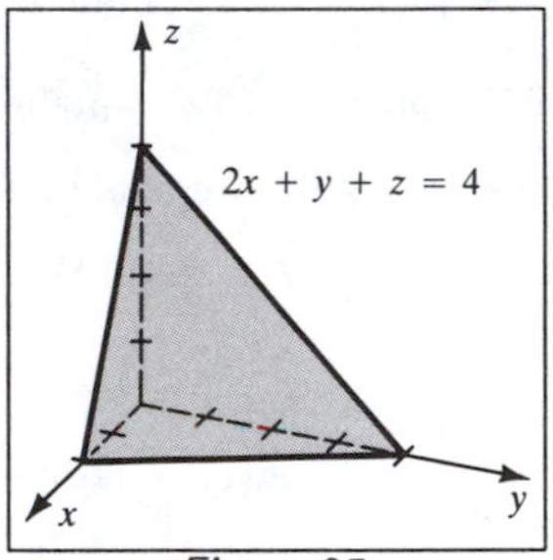

Figure 25

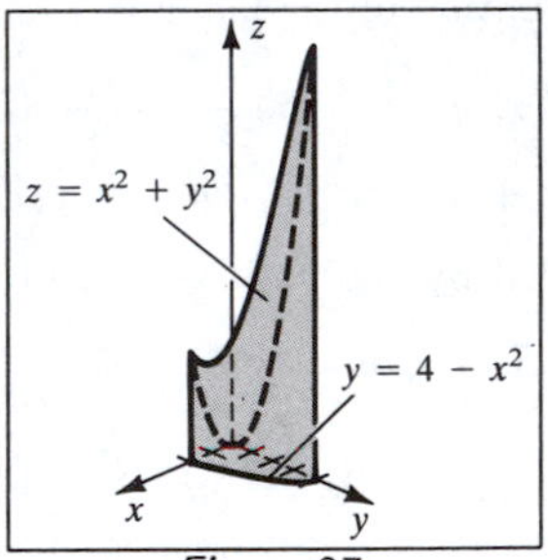

Figure 27

[27] The volume described is bounded by the paraboloid $z = f(x, y) = x^2 + y^2$ above and the xy-plane below. R is the region in the xy-plane bounded by the parabola $y = 4 - x^2$ and the lines $x = 0$ and $y = 0$. We will treat R as an R_x region.

$$V = \int_0^2\int_0^{4-x^2} (x^2 + y^2)\, dy\, dx = \int_0^2 \Big[x^2y + \tfrac{1}{3}y^3 \Big]_0^{4-x^2} dx =$$

$$\int_0^2 \Big[x^2(4 - x^2) + \tfrac{1}{3}(4 - x^2)^3 \Big] dx = \int_0^2 (\tfrac{64}{3} - 12x^2 + 3x^4 - \tfrac{1}{3}x^6)\, dx = \tfrac{832}{35}.$$

[31] The volume described is bounded above by the cylinder $z = f(x, y) = x^2 + 4$ and bounded below by the xy-plane $\{z = 0\}$. In the xy-plane, $y = 4 - x^2$ intersects $x + y = 2$ at $(-1, 3)$ and $(2, 0)$.

$$V = \int_{-1}^{2}\int_{2-x}^{4-x^2} (x^2 + 4)\, dy\, dx = \int_{-1}^{2} (8 + 4x - 2x^2 + x^3 - x^4)\, dx = \tfrac{423}{20}.$$

[33] Note that $\Delta y_j\, \Delta x_k = \frac{1}{4}(\frac{1}{4}) = \frac{1}{16}$ for all j and k and that the (u_k, v_j) are midpoints of the rectangles in the grid. $(u_1, v_1) = (\frac{1}{4}(1) - \frac{1}{8}, \frac{1}{4}(1) - \frac{1}{8}) = (\frac{1}{8}, \frac{1}{8})$.
$(u_1, v_2) = (\frac{1}{4}(1) - \frac{1}{8}, \frac{1}{4}(2) - \frac{1}{8}) = (\frac{1}{8}, \frac{3}{8})$. $(u_2, v_1) = (\frac{1}{4}(2) - \frac{1}{8}, \frac{1}{4}(1) - \frac{1}{8}) = (\frac{3}{8}, \frac{1}{8})$.
The coordinates for the other five points are found in a similar manner.

$$V = \int_0^1\int_0^{1/2} f(x, y)\, dy\, dx = \int_0^1\int_0^{1/2} \sin\big[\cos(xy)\big]\, dy\, dx \approx \sum_{k=1}^{4}\sum_{j=1}^{2} f(u_k, v_j)\,\Delta y_j\,\Delta x_k =$$

$$(\tfrac{1}{16})\sum_{k=1}^{4}\sum_{j=1}^{2} f(u_k, v_j) = \tfrac{1}{16}[f(\tfrac{1}{8}, \tfrac{1}{8}) + f(\tfrac{1}{8}, \tfrac{3}{8}) + f(\tfrac{3}{8}, \tfrac{1}{8}) + f(\tfrac{3}{8}, \tfrac{3}{8}) +$$

$$f(\tfrac{5}{8}, \tfrac{1}{8}) + f(\tfrac{5}{8}, \tfrac{3}{8}) + f(\tfrac{7}{8}, \tfrac{1}{8}) + f(\tfrac{7}{8}, \tfrac{3}{8})] \approx (\tfrac{1}{16})6.6751 \approx 0.4172.$$

Exercises 17.3

[1] As θ varies from 0 to $\frac{\pi}{2}$, $0 \le r \le 4\sin\theta$, and this is one-half of the area.

Using symmetry and (17.13), $A = 2\int_0^{\pi/2}\int_0^{4\sin\theta} r\, dr\, d\theta$.

[3] One-half of the area lies above the polar axis and one-half below. When $\theta = 0$, $r = 3$. We want to know the value of θ when $r = 0$—this range on θ will represent the area above the polar axis. $r = 0 \Rightarrow 1 + 2\cos\theta = 0 \Rightarrow \cos\theta = -\frac{1}{2} \Rightarrow$
$\theta = \frac{2\pi}{3}, \frac{4\pi}{3}$ for $0 \le \theta < 2\pi$. As θ varies from 0 to $\frac{2\pi}{3}$, $0 \le r \le 1 + 2\cos\theta$, and this is one-half of the area. $A = 2\int_0^{2\pi/3}\int_0^{1 + 2\cos\theta} r\, dr\, d\theta$. (If we let θ vary between 0 and π, the resulting area would include the area of the inner loop twice.)

[7] The graph of $r = 4\sin 3\theta$ is a three-leaved rose. One loop is formed as θ varies from 0 to $\frac{\pi}{3}$ and $0 \le r \le 4\sin 3\theta$. $A = \int_0^{\pi/3}\int_0^{4\sin 3\theta} r\, dr\, d\theta = \int_0^{\pi/3}\Big[\frac{1}{2}r^2\Big]_0^{4\sin 3\theta} d\theta =$
$\int_0^{\pi/3} 8\sin^2 3\theta\, d\theta = 8\int_0^{\pi/3} \frac{1}{2}(1 - \cos 6\theta)\, d\theta$ {half-angle identity} $=$

$$4\Big[\theta - \tfrac{1}{6}\sin 6\theta\Big]_0^{\pi/3} = \tfrac{4\pi}{3}.$$

[9] From *Figure 9*, we see that $2 - 2\cos\theta = 3 \Rightarrow \cos\theta = -\frac{1}{2} \Rightarrow \theta = \frac{2\pi}{3}, \frac{4\pi}{3}$. We will double the area produced as θ varies between $\frac{2\pi}{3}$ and π. r is bounded by 3 and $2 - 2\cos\theta$.

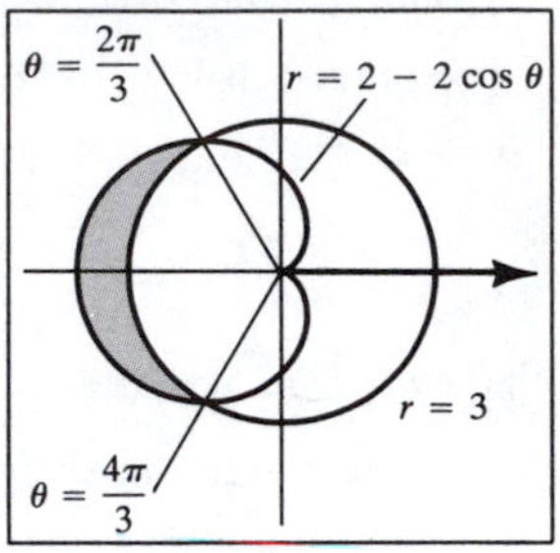

Figure 9

$$A = 2\int_{2\pi/3}^{\pi}\int_{3}^{2-2\cos\theta} r\,dr\,d\theta$$
$$= 2\int_{2\pi/3}^{\pi}\Big[\tfrac{1}{2}r^2\Big]_3^{2-2\cos\theta}\,d\theta$$
$$= \int_{2\pi/3}^{\pi}\Big[(2 - 2\cos\theta)^2 - 3^2\Big]\,d\theta$$
$$= \int_{2\pi/3}^{\pi}\Big[4\cos^2\theta - 8\cos\theta - 5\Big]\,d\theta = \int_{2\pi/3}^{\pi}\Big[4\Big(\frac{1-\cos 2\theta}{2}\Big) - 8\cos\theta - 5\Big]\,d\theta$$
$$= \int_{2\pi/3}^{\pi}\Big[2\cos 2\theta - 8\cos\theta - 3\Big]\,d\theta = \Big[\sin 2\theta - 8\sin\theta - 3\theta\Big]_{2\pi/3}^{\pi} =$$
$$\tfrac{9}{2}\sqrt{3} - \pi \approx 4.65.$$

[11] The graph of $r^2 = 9\cos 2\theta$ is a lemniscate {figure-8} with its loops vertically centered on the polar axis. One-half of a loop is traced out as θ varies between 0 and $\frac{\pi}{4}$ while r is bounded by 0 and $3\sqrt{\cos 2\theta}$. $A = 2\int_0^{\pi/4}\int_0^{3\sqrt{\cos 2\theta}} r\,dr\,d\theta =$

$$2\int_0^{\pi/4}\Big[\tfrac{1}{2}r^2\Big]_0^{3\sqrt{\cos 2\theta}}\,d\theta = \int_0^{\pi/4} 9\cos 2\theta\,d\theta = 9\Big[\tfrac{1}{2}\sin 2\theta\Big]_0^{\pi/4} = \tfrac{9}{2}.$$

[13] In polar coordinates, R can be described by $0 \le \theta \le 2\pi$ and $0 \le r \le 2$. Since $x^2 + y^2 = r^2$ and $dA = r\,dr\,d\theta$, $\iint_R (x^2 + y^2)^{3/2}\,dA =$

$$\int_0^{2\pi}\int_0^2 (r^3)\,r\,dr\,d\theta = \int_0^{2\pi}\Big[\tfrac{1}{5}r^5\Big]_0^2\,d\theta = \int_0^{2\pi}\tfrac{32}{5}\,d\theta = \tfrac{64\pi}{5}.$$

[17] The triangle is bounded by the lines $y = 0$, $y = x$, and $x = 3$. In polar coordinates, these are $\theta = 0$, $\theta = \frac{\pi}{4}$, and $r\cos\theta = 3$ or $r = 3\sec\theta$. $\iint_R \sqrt{x^2 + y^2}\,dA$

$$= \int_0^{\pi/4}\int_0^{3\sec\theta}(r)\,r\,dr\,d\theta = \int_0^{\pi/4}\Big[\tfrac{1}{3}r^3\Big]_0^{3\sec\theta}\,d\theta = \tfrac{1}{3}\int_0^{\pi/4} 27\sec^3\theta\,d\theta$$
$$= \tfrac{9}{2}\Big[\sec\theta\tan\theta + \ln|\sec\theta + \tan\theta|\Big]_0^{\pi/4}$$ {Formula 71 or Example 6, §9.1}
$$= \tfrac{9}{2}\Big[\sqrt{2} + \ln(\sqrt{2} + 1)\Big] \approx 10.33.$$

Note: In Exercises 19–24, let R denote the region in the xy-plane.

[19] $-a \le x \le a$, $0 \le y \le \sqrt{a^2 - x^2}$ describes the region R, which is bounded by the upper half of the circle $x^2 + y^2 = a^2$. In polar coordinates, this region can be described by $0 \le \theta \le \pi$, $0 \le r \le a$. Thus,

$$\int_{-a}^{a}\int_{0}^{\sqrt{a^2-x^2}} e^{-(x^2+y^2)}\,dy\,dx = \int_{0}^{\pi}\int_{0}^{a} e^{-r^2}\,r\,dr\,d\theta = \int_{0}^{\pi} -\tfrac{1}{2}e^{-a^2}\,d\theta = \tfrac{\pi}{2}(1 - e^{-a^2}).$$

[21] R is the region under the line $y = x$ and above the x-axis from $x = 1$ to 2.
In polar coordinates, $x = 1 \Leftrightarrow r\cos\theta = 1$, or $r = \sec\theta$, and $x = 2 \Leftrightarrow r = 2\sec\theta$.
Thus, θ varies between 0 and $\frac{\pi}{4}$ $\{y = x\}$ while $\sec\theta \le r \le 2\sec\theta$.

$$\int_{1}^{2}\int_{0}^{x} \frac{1}{\sqrt{x^2 + y^2}}\,dy\,dx = \int_{0}^{\pi/4}\int_{\sec\theta}^{2\sec\theta} (\tfrac{1}{r})\,r\,dr\,d\theta =$$

$$\int_{0}^{\pi/4}\Big[\, r \,\Big]_{\sec\theta}^{2\sec\theta} d\theta = \int_{0}^{\pi/4} \sec\theta\,d\theta = \Big[\ln|\sec\theta + \tan\theta|\,\Big]_{0}^{\pi/4} =$$

$$\ln(\sqrt{2} + 1) - \ln 1 = \ln(\sqrt{2} + 1) \approx 0.88.$$

[25] R is the region in the xy-plane between the circle $x^2 + y^2 = 9$ and the trace of the sphere in the xy-plane $\{z = 0\}$, which is $x^2 + y^2 = 25$. The height of the solid in the *first* octant is $z = \sqrt{25 - x^2 - y^2} = \sqrt{25 - r^2}$ and the projection of the solid onto the xy-plane is the region between the circles of radii 3 and 5 centered at the origin. Thus, by symmetry in each octant, $V =$

$$8\int_{0}^{\pi/2}\int_{3}^{5} (25 - r^2)^{1/2}\,r\,dr\,d\theta = 8\int_{0}^{\pi/2}\Big[-\tfrac{1}{3}(25 - r^2)^{3/2}\Big]_{3}^{5} d\theta = 8\int_{0}^{\pi/2} \tfrac{64}{3}\,d\theta = \tfrac{256\pi}{3}.$$

[27] R is determined by $x^2 + y^2 = 2x$, which is a circle of radius 1 and center (1, 0). In polar coordinates, this equation is $r^2 = 2r\cos\theta$, or $r = 2\cos\theta$. The upper half of the solid is the cone $z = r$ $\{z \ge 0\}$. Thus, by symmetry in 4 octants,

$$V = 4\int_{0}^{\pi/2}\int_{0}^{2\cos\theta} (r)\,r\,dr\,d\theta = \tfrac{32}{3}\int_{0}^{\pi/2} \cos^3\theta\,d\theta = \tfrac{32}{3}\int_{0}^{\pi/2} (1 - \sin^2\theta)\cos\theta\,d\theta =$$

$$\tfrac{32}{3}\int_{0}^{1} (1 - u^2)\,du\ \{u = \sin\theta,\ du = \cos\theta\,d\theta\} = \tfrac{32}{3}(\tfrac{2}{3}) = \tfrac{64}{9}.$$

[31] In polar coordinates, R_a can be described by $0 \le \theta \le 2\pi$ and $0 \le r \le a$.

$$\lim_{a\to\infty} \iint_{R_a} e^{-(x^2+y^2)}\,dA = \lim_{a\to\infty} \int_{0}^{2\pi}\int_{0}^{a} e^{-r^2}\,r\,dr\,d\theta =$$

$$\lim_{a\to\infty} \int_{0}^{2\pi}\Big[-\tfrac{1}{2}e^{-r^2}\Big]_{0}^{a} d\theta = -\tfrac{1}{2}\lim_{a\to\infty}(e^{-a^2} - 1)\int_{0}^{2\pi} d\theta = -\tfrac{1}{2}(0 - 1)(2\pi) = \pi.$$

Exercises 17.4

Note: Let $\mathbb{S}$ denote surface area. Also, let $z = f(x, y)$.

[1] Since $z \geq 0$, $x^2 + y^2 + z^2 = 4 \Rightarrow z = \sqrt{4 - x^2 - y^2} \Rightarrow f_x = \dfrac{-x}{\sqrt{4 - x^2 - y^2}}$ and

$f_y = \dfrac{-y}{\sqrt{4 - x^2 - y^2}}$. R can be described by $-1 \leq x \leq 1$ and $-1 \leq y \leq 1$.

Using the first octant portion and symmetry about the x- and y-axes,

$$\mathbb{S} = \iint_R \sqrt{[f_x(x, y)]^2 + [f_y(x, y)]^2 + 1}\, dA =$$

$$4\int_0^1 \int_0^1 \sqrt{\left(\frac{-x}{\sqrt{4 - x^2 - y^2}}\right)^2 + \left(\frac{-y}{\sqrt{4 - x^2 - y^2}}\right)^2 + 1}\, dy\, dx.$$

[5] R can be described by $0 \leq x \leq 1$ and $0 \leq y \leq 1$. $z = y + \frac{1}{2}x^2 \Rightarrow f_x = x$ and

$f_y = 1$. $\mathbb{S} = \iint_R \sqrt{[f_x(x, y)]^2 + [f_y(x, y)]^2 + 1}\, dA = \int_0^1 \int_0^1 \sqrt{x^2 + 1^2 + 1}\, dy\, dx =$

$\int_0^1 \sqrt{x^2 + 2}\, dx = \frac{1}{2}\left[x\sqrt{x^2 + 2} + 2\ln(x + \sqrt{x^2 + 2})\right]_0^1$ {using Formula 21 or trig

substitution with $x = \sqrt{2}\tan\theta$} $= \frac{1}{2}\left[\sqrt{3} + 2\ln(1 + \sqrt{3}) - \ln 2\right] \approx 1.52$.

[7] R is bounded by the circle $x^2 + y^2 = k^2$, and can be described by $0 \leq \theta \leq 2\pi$ and

$0 \leq r \leq k$. $(x/a) + (y/b) + (z/c) = 1 \Rightarrow z = 1 - \frac{c}{a}x - \frac{c}{b}y \Rightarrow$

$f_x = -\frac{c}{a}$ and $f_y = -\frac{c}{b} \Rightarrow \mathbb{S} = \iint_R \sqrt{(-\frac{c}{a})^2 + (-\frac{c}{b})^2 + 1}\, dA =$

$\iint_R \sqrt{c^2\left[(\frac{1}{a})^2 + (\frac{1}{b})^2 + (\frac{1}{c})^2\right]}\, dA = c\sqrt{(\frac{1}{a})^2 + (\frac{1}{b})^2 + (\frac{1}{c})^2} \int_0^{2\pi} \int_0^k r\, dr\, d\theta =$

$$c\sqrt{(\tfrac{1}{a})^2 + (\tfrac{1}{b})^2 + (\tfrac{1}{c})^2} \int_0^{2\pi} \left[\tfrac{1}{2}r^2\right]_0^k d\theta = \pi c k^2 \sqrt{(\tfrac{1}{a})^2 + (\tfrac{1}{b})^2 + (\tfrac{1}{c})^2}.$$

[9] The plane $z = 1$ intersects $z = x^2 + y^2$ when $x^2 + y^2 = 1$. Inside this circle is the region R, which can be described by $0 \leq \theta \leq 2\pi$ and $0 \leq r \leq 1$.

$z = f(x, y) = x^2 + y^2 \Rightarrow f_x = 2x$ and $f_y = 2y \Rightarrow \mathbb{S} = \iint_R \sqrt{(2x)^2 + (2y)^2 + 1}\, dA$

$$= \int_0^{2\pi} \int_0^1 (4r^2 + 1)^{1/2}\, r\, dr\, d\theta = \tfrac{1}{8}\int_0^{2\pi} \tfrac{2}{3}(5^{3/2} - 1)\, d\theta = \tfrac{\pi}{6}(5^{3/2} - 1) \approx 5.33.$$

[13] $z = f(x, y) = 7 - \frac{7}{25}(x^2 + y^2) \Rightarrow f_x = -\frac{14}{25}x$ and $f_y = -\frac{14}{25}y \Rightarrow$

$\mathbb{S}_{\text{roof}} = \iint_R \sqrt{(-\frac{14}{25}x)^2 + (-\frac{14}{25}y)^2 + 1}\, dA = \iint_R \sqrt{1 + \frac{196}{625}(x^2 + y^2)}\, dA$

$$= \int_0^{2\pi} \int_0^5 \sqrt{1 + \tfrac{196}{625}r^2}\, r\, dr\, d\theta = 2\pi\left[\tfrac{2}{3}\left(\tfrac{625}{2 \cdot 196}\right)(1 + \tfrac{196}{625}r^2)^{3/2}\right]_0^5$$

$$= \tfrac{625\pi}{294}\left[(8.84)^{3/2} - 1\right] \approx 168.9 \text{ ft}^2.$$

The area of the circular floor is $\pi r^2 = \pi(5)^2 = 25\pi \approx 78.5$ ft^2.

The total amount of material is approximately $168.9 + 78.5 = 247.4$ ft^2.

Exercises 17.5

1 $\text{I} = \int_0^3\int_{-1}^0\int_1^2 (x + 2y + 4z)\,dx\,dy\,dz$ {given}

$= \int_0^3\int_{-1}^0 \Big[\tfrac{1}{2}x^2 + 2yx + 4zx\Big]_1^2 dy\,dz$ {treat y and z as constants}

$= \int_0^3\int_{-1}^0 \Big[(2 + 4y + 8z) - (\tfrac{1}{2} + 2y + 4z)\Big] dy\,dz$

{substitute 2 for x and then 1 for x}

$= \int_0^3\int_{-1}^0 (\tfrac{3}{2} + 2y + 4z)\,dy\,dz$ {simplify}

$= \int_0^3 \Big[\tfrac{3}{2}y + y^2 + 4zy\Big]_{-1}^0 dz$ {treat z as a constant}

$= \int_0^3 \Big[0 - (-\tfrac{3}{2} + 1 - 4z)\Big] dz$ {substitute 0 for y and then -1 for y}

$= \int_0^3 (\tfrac{1}{2} + 4z)\,dz$ {simplify}

$= \Big[\tfrac{1}{2}z + 2z^2\Big]_0^3 = (\tfrac{3}{2} + 18) - 0 = \tfrac{39}{2}$

5 $\text{I} = \int_{-1}^2\int_1^{x^2}\int_0^{x+y} 2x^2y\,dz\,dy\,dx$ {given}

$= \int_{-1}^2\int_1^{x^2}\Big[2x^2yz\Big]_0^{x+y} dy\,dx$ {treat y and x as constants}

$= \int_{-1}^2\int_1^{x^2} (2x^3y + 2x^2y^2)\,dy\,dx$ {substitute $x + y$ for z and simplify}

$= \int_{-1}^2 \Big[x^3y^2 + \tfrac{2}{3}x^2y^3\Big]_1^{x^2} dx$ {treat x as a constant}

$= \int_{-1}^2 \Big[(x^7 + \tfrac{2}{3}x^8) - (x^3 + \tfrac{2}{3}x^2)\Big] dx$ {substitute x^2 for y and then 1 for y}

$= \int_{-1}^2 (x^7 + \tfrac{2}{3}x^8 - x^3 - \tfrac{2}{3}x^2)\,dx$ {simplify}

$= \Big[\tfrac{1}{8}x^8 + \tfrac{2}{27}x^9 - \tfrac{1}{4}x^4 - \tfrac{2}{9}x^3\Big]_{-1}^2$ {antiderivative}

$= \Big[(32 + \tfrac{1024}{27} - 4 - \tfrac{16}{9}) - (\tfrac{1}{8} - \tfrac{2}{27} - \tfrac{1}{4} + \tfrac{2}{9})\Big]$

$= \Big[(32 - 4) + (\tfrac{1024}{27} + \tfrac{2}{27}) - (\tfrac{16}{9} + \tfrac{2}{9}) + (\tfrac{1}{4} - \tfrac{1}{8})\Big]$ {group terms}

$= \Big[28 + 38 - 2 + \tfrac{1}{8}\Big] = 64 + \tfrac{1}{8} = \tfrac{513}{8} = 64.125$

7 (1) z integrated first The region R is in the xy-plane.

The trace of $x + 2y + 3z = 6$ in the xy-plane is $x + 2y = 6$ { $z = 0$ }.

Thus, R is the triangular region in the xy-plane with vertices (0, 0), (6, 0), and (0, 3). z varies between 0 and $\tfrac{1}{3}(6 - x - 2y)$.

$\int_0^6\int_0^{(6-x)/2}\int_0^{(6-x-2y)/3} f(x, y, z)\,dz\,dy\,dx$; $\int_0^3\int_0^{6-2y}\int_0^{(6-x-2y)/3} f(x, y, z)\,dz\,dx\,dy$

(2) x integrated first The region R is in the yz-plane.

The trace of $x + 2y + 3z = 6$ in the yz-plane is $2y + 3z = 6$ $\{x = 0\}$.

Thus, R is the triangular region in the yz-plane with vertices (0, 0), (3, 0), and (0, 2). x varies between 0 and $6 - 2y - 3z$.

$$\int_0^3 \int_0^{(6-2y)/3} \int_0^{6-2y-3z} f(x, y, z)\,dx\,dz\,dy; \quad \int_0^2 \int_0^{(6-3z)/2} \int_0^{6-2y-3z} f(x, y, z)\,dx\,dy\,dz$$

(3) y integrated first The region R is in the xz-plane.

The trace of $x + 2y + 3z = 6$ in the xz-plane is $x + 3z = 6$ $\{y = 0\}$.

Thus, R is the triangular region in the xz-plane with vertices (0, 0), (6, 0), and (0, 2). y varies between 0 and $\frac{1}{2}(6 - x - 3z)$.

$$\int_0^6 \int_0^{(6-x)/3} \int_0^{(6-x-3z)/2} f(x, y, z)\,dy\,dz\,dx; \quad \int_0^2 \int_0^{6-3z} \int_0^{(6-x-3z)/2} f(x, y, z)\,dy\,dx\,dz$$

[11] See *Figure 11*. Q is bounded below by the xy-plane $\{z = 0\}$ and on the left by the xz-plane $\{y = 0\}$. The shape of Q is determined by the cylinder $z = 4 - x^2$ with rulings parallel to the y-axis and the plane $z = 4 - y$ that is orthogonal to the yz-plane. We will let R be the triangular region in the yz-plane with vertices (0, 0), (4, 0), and (0, 4). By the symmetry of the figure, we can double the volume in the first octant. In this case x will vary between 0 and $x = \sqrt{4 - z}$ $\{z + x^2 = 4\}$.

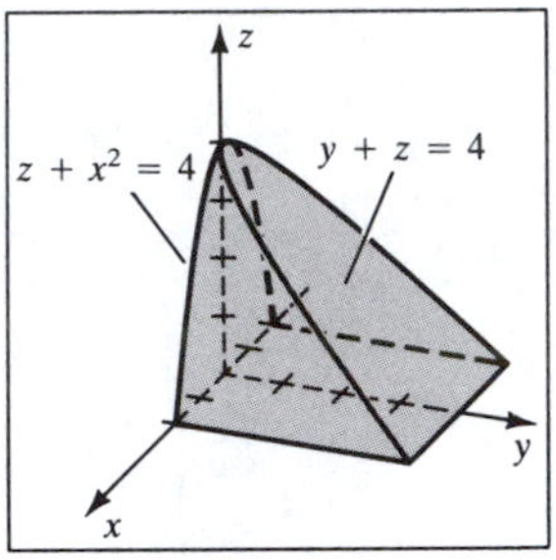

Figure 11

$$V = 2\int_0^4 \int_0^{4-y} \int_0^{\sqrt{4-z}} dx\,dz\,dy = 2\int_0^4 \int_0^{4-y} (4 - z)^{1/2}\,dz\,dy =$$

$$2\int_0^4 \left[-\tfrac{2}{3}(4 - z)^{3/2}\right]_0^{4-y} dy = -\tfrac{4}{3}\int_0^4 (y^{3/2} - 8)\,dy = -\tfrac{4}{3}(\tfrac{64}{5} - 32) = \tfrac{128}{5}.$$

Note: Many students have the most difficulty with one of the first steps—how to determine the order of integration. As a visual aid, think of the region as a three-dimensional object that is collapsible in all 3 directions. Also, think of the surfaces each having a different color—similar to many of the figures in the text.

For this particular case, let the parabolic cylinder $z = 4 - x^2$ be red, and the plane $z = 4 - y$ be green. To help determine the order of integration, we want to view the object from 3 directions and determine which of these directions gives us a "one-color" view.

1) *Viewing from the positive x-axis*: We would see only red, and the two-dimensional collapsed shape would be a triangle in the yz-plane. Hence, integrating with respect to x first would be a good choice.

2) *Viewing from the positive y-axis*: We would see only green, and the two-dimensional collapsed shape would be a parabolic arch in the xz-plane. Hence, integrating with respect to y first is also reasonable.

3) *Viewing from the positive z-axis*: We would see both red and green, and the two-dimensional collapsed shape would be a square in the xy-plane. In order to compute the volume by integrating with respect to z first, we would have to represent the volume as a sum of two integrals $\{$one with $0 \le z \le 4 - y$ and one with $0 \le z \le 4 - x^2\}$. Since this is usually more difficult than evaluating one integral, this is not a preferred method of solution.

In summary, the best choices for dV would be $dx\,dz\,dy$ or $dy\,dz\,dx$. We chose the former. In future exercises, you should try a similar analysis after sketching the region. Hopefully, this will lead you to the easiest solution.

15 Q is bounded "on the back" by the yz-plane $\{x = 0\}$. Its shape is determined by the cylinder $y^2 + z^2 = 1$ with rulings parallel to the x-axis and the plane $x + y + z = 2$, which has x-, y-, and z-intercepts at 2. We will let R be the region in the yz-plane determined by the circle $y^2 + z^2 = 1$. x varies between 0 and $2 - y - z$.

$V = \int_{-1}^{1}\int_{-\sqrt{1-y^2}}^{\sqrt{1-y^2}}\int_{0}^{2-y-z} dx\,dz\,dy = \int_{-1}^{1}\int_{-\sqrt{1-y^2}}^{\sqrt{1-y^2}} (2 - y - z)\,dz\,dy$

$= \int_{-1}^{1}\Big[\,2z - yz - \tfrac{1}{2}z^2\,\Big]_{-\sqrt{1-y^2}}^{\sqrt{1-y^2}}\,dy$

$= \int_{-1}^{1}\Big[\,(2\sqrt{1-y^2} - y\sqrt{1-y^2}) - (-2\sqrt{1-y^2} + y\sqrt{1-y^2})\,\Big]\,dy$

$= 4\int_{-1}^{1}\sqrt{1-y^2}\,dy - 2\int_{-1}^{1} y\sqrt{1-y^2}\,dy.$ The integrand of the second integral is odd, so that integral is equal to zero. The first integral is 4 times the area of a semicircle with radius 1. Thus, $V = 4(\frac{1}{2}\pi \cdot 1^2) = 2\pi$.

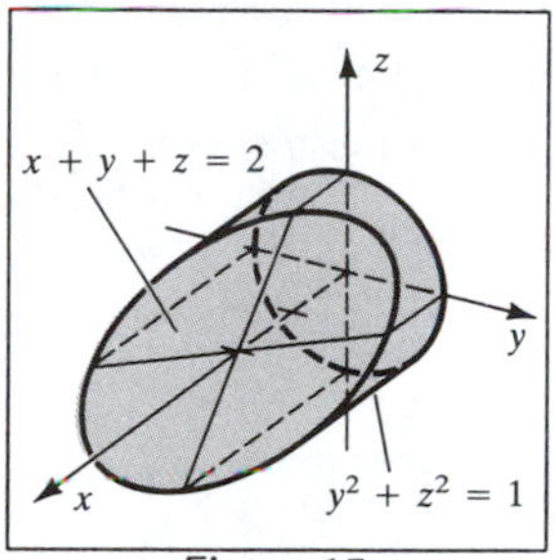

Figure 15

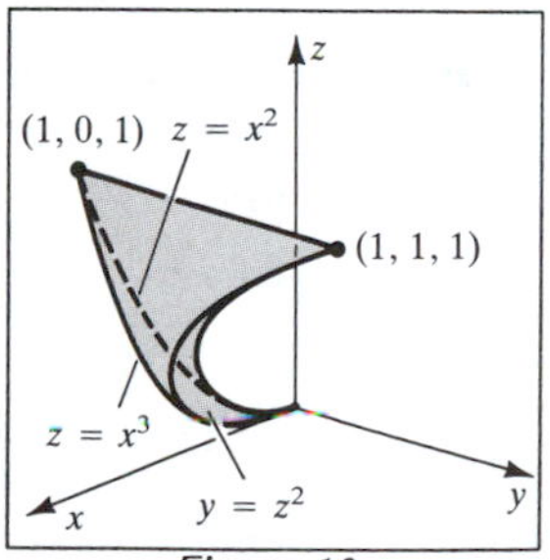

Figure 19

19 The solid lies between the two parallel cylinders $z = x^2$ and $z = x^3$ from $y = 0$ to $y = z^2$. R is the region in the xz-plane determined by $x^3 \le z \le x^2$ and $0 \le x \le 1$.

Thus, $V = \int_{0}^{1}\int_{x^3}^{x^2}\int_{0}^{z^2} dy\,dz\,dx = \int_{0}^{1}\int_{x^3}^{x^2} z^2\,dz\,dx = \frac{1}{3}\int_{0}^{1}(x^6 - x^9)\,dx = \frac{1}{70}$.

[21] The graph of $\frac{x}{a} + \frac{y}{b} + \frac{z}{c} = 1$ is a plane with x-, y-, and z-intercepts of a, b, and c, respectively. Without loss of generality, we assume that a, b, and c are positive. This makes it easier to visualize the tetrahedron. We will let R be the triangular region in the first quadrant of the xy-plane determined by the vertices (0, 0), (a, 0), and (0, b). The trace of the plane in the xy-plane is $\frac{x}{a} + \frac{y}{b} = 1 \Leftrightarrow y = b(1 - x/a)$ $\Leftrightarrow y = \frac{b}{a}(a - x)$. $\frac{x}{a} + \frac{y}{b} + \frac{z}{c} = 1 \Rightarrow bcx + acy + abz = abc \Rightarrow$

$z = \dfrac{c(ab - bx - ay)}{ab}$. z varies between 0 and $\dfrac{c(ab - bx - ay)}{ab}$.

See the figure in Exercise 7 for a similar tetrahedron.

$$V = \int_0^a \int_0^{b(a-x)/a} \int_0^{c(ab-bx-ay)/(ab)} dz\,dy\,dx = \frac{c}{ab}\int_0^a \int_0^{b(a-x)/a} (ab - bx - ay)\,dy\,dx$$

$$= \frac{c}{ab}\int_0^a \left[aby - bxy - \tfrac{1}{2}ay^2 \right]_0^{b(a-x)/a} dx$$

$$= \frac{c}{ab}\int_0^a \left[b^2(a - x) - \frac{b^2}{a}x(a - x) - \frac{b^2}{2a}(a^2 - 2ax + x^2) \right] dx$$

$$= \frac{c}{2a^2b}\int_0^a \left[2ab^2(a - x) - 2b^2x(a - x) - b^2(a^2 - 2ax + x^2) \right] dx$$

$$= \frac{c}{2a^2b}\left[a^2b^2x - ab^2x^2 + \tfrac{1}{3}b^2x^3 \right]_0^a = \frac{c}{2a^2b}\cdot\frac{a^3b^2}{3} = \frac{abc}{6}$$

[23] Q is the region bounded by the planes $z = 0$, $z = 1$, $x = 2$, $x = 3$ and the cylinders $y = \sqrt{1 - z}$ and $y = \sqrt{4 - z}$.

[27] $y = \pm\sqrt{z - x^2} \Rightarrow y^2 = z - x^2 \Rightarrow z = x^2 + y^2$. In the xz-plane, the trace of $z = x^2 + y^2$ is $z = x^2$, or $x = \pm\sqrt{z}$. Thus, Q is the region bounded by the paraboloid $z = x^2 + y^2$ and the planes $z = 1$ and $z = 2$.

[29] The region in the xy-plane is bounded by $0 \le y \le e^{-x}$ and $0 \le x \le 1$.

$$m = \iint_R \delta(x, y)\,dA = \int_0^1 \int_0^{e^{-x}} y^2\,dy\,dx.$$

[31] The solid described is a tetrahedron similar to the one shown in the figure for Exercise 7. $m = \iiint_Q \delta(x, y, z)\,dV = \int_0^2 \int_0^{4-2y} \int_0^{4-x-2y} (x^2 + y^2)\,dz\,dx\,dy$

[33] $m = \iiint_Q \delta(x, y, z)\, dV$

$$= \int_0^{1000}\int_0^{1000}\int_0^{1000} (1.225 - 0.000113z)\, dz\, dy\, dx$$

$$= \int_0^{1000}\int_0^{1000} \left[1.225z - \frac{0.000113}{2} z^2 \right]_0^{1000} dy\, dx$$

$$= \int_0^{1000}\int_0^{1000} (1.1685 \times 10^3)\, dy\, dx$$

$$= \int_0^{1000} (1.1685 \times 10^6)\, dx = 1.1685 \times 10^9 \text{ kg.}$$

Alternatively, we could change the density from kg/m^3 to kg/km^3 by multiplying by 10^9, and then the limits of integration would each be 0 to 1.

Exercises 17.6

Note: Let k be a constant of proportionality.

[1] By (17.24), $m = \iint_R \delta(x, y)\, dA = \int_0^9\int_0^{\sqrt{x}} (x + y)\, dy\, dx = \int_0^9 (x^{3/2} + \frac{1}{2}x)\, dx = \frac{2349}{20}$.

$M_y = \iint_R x\,\delta(x, y)\, dA = \int_0^9\int_0^{\sqrt{x}} x(x + y)\, dy\, dx = \int_0^9 (x^{5/2} + \frac{1}{2}x^2)\, dx = \frac{31,347}{42}$.

$M_x = \iint_R y\,\delta(x, y)\, dA = \int_0^9\int_0^{\sqrt{x}} y(x + y)\, dy\, dx = \int_0^9 (\frac{1}{2}x^2 + \frac{1}{3}x^{3/2})\, dx = \frac{1539}{10}$.

Thus, $\bar{x} = \frac{M_y}{m} = \frac{1290}{203}$ and $\bar{y} = \frac{M_x}{m} = \frac{38}{29}$.

[3] The distance between $P(x, y)$ and the y-axis is $|x|$.

Thus, $\delta(x, y) = k|x|$, where k is a constant of proportionality. By symmetry, we can double the mass in the first quadrant to find m and let $\delta(x, y) = kx$, $x \geq 0$.

$m = 2\int_0^2\int_{x^2}^4 kx\, dy\, dx = 2k\int_0^2 (4x - x^3)\, dx = 8k$.

$M_y = \int_{-2}^2\int_{x^2}^4 x\,k|x|\, dy\, dx = \int_{-2}^2 kx|x|(4 - x^2)\, dx = 0$ {see following note}

Note: The integrand is odd since x is odd, $|x|$ is even, $4 - x^2$ is even, and an odd function times an even function times an even function is an odd function.

$$M_x = \int_{-2}^2\int_{x^2}^4 y\,k|x|\, dy\, dx$$

$$= \int_{-2}^2 \tfrac{1}{2}(16 - x^4)\,k|x|\, dx \text{ \{the integrand is even, apply (5.34)(i)\}}$$

$$= k\int_0^2 (16 - x^4)\,x\, dx = \tfrac{64}{3}k.$$

Thus, $\bar{x} = \frac{M_y}{m} = 0$ and $\bar{y} = \frac{M_x}{m} = \frac{8}{3}$.

7 $m = \int_{-\pi/4}^{\pi/4}\int_{1/2}^{\sec x} 4\,dy\,dx = \int_{-\pi/4}^{\pi/4}(4\sec x - 2)\,dx$

$= 2\cdot 2\int_0^{\pi/4}(2\sec x - 1)\,dx$ {factor out 2, even integrand}

$= 4\Big[\,2\ln|\sec x + \tan x| - x\,\Big]_0^{\pi/4} = 8\ln(\sqrt{2}+1) - \pi \approx 3.91.$

$M_x = \int_{-\pi/4}^{\pi/4}\int_{1/2}^{\sec x} 4y\,dy\,dx = \int_{-\pi/4}^{\pi/4}(2\sec^2 x - \tfrac{1}{2})\,dx$

$= \int_0^{\pi/4}(4\sec^2 x - 1)\,dx$ {even integrand, multiply integrand by 2}

$= \Big[\,4\tan x - x\,\Big]_0^{\pi/4} = 4 - \tfrac{\pi}{4}.$

By symmetry, $\bar{x} = 0$ and $\bar{y} = \dfrac{M_x}{m} = \dfrac{4 - (\pi/4)}{8\ln(\sqrt{2}+1) - \pi} \approx 0.82.$

9 By (17.25), $I_x = \iint_R y^2\,\delta(x, y)\,dA =$

$\int_0^9\int_0^{\sqrt{x}} y^2(x + y)\,dy\,dx = \int_0^9 (\tfrac{1}{3}x^{5/2} + \tfrac{1}{4}x^2)\,dx = 3^5(\tfrac{31}{28}) = \tfrac{7533}{28}.$

$I_y = \iint_R x^2\,\delta(x, y)\,dA =$

$\int_0^9\int_0^{\sqrt{x}} x^2(x + y)\,dy\,dx = \int_0^9 (x^{7/2} + \tfrac{1}{2}x^3)\,dx = 3^7(\tfrac{19}{8}) = \tfrac{41{,}553}{8}.$

$I_O = I_x + I_y = 3^5(\tfrac{1259}{56}) = \tfrac{305{,}937}{56}.$

13 (a) Position the square so that its vertices are (0, 0), (0, a), (a, 0), and (a, a).

$I_x = I_y$ are the moments of inertia with respect to a side and by (17.25),

$$I_x = \delta\int_0^a\int_0^a y^2\,dy\,dx = \tfrac{1}{3}a^3\delta\int_0^a dx = \tfrac{1}{3}a^4\delta.$$

(b) Position the square as shown in *Figure 13*.

The diagonals correspond to the x- and y-axes.

We will calculate I_y from $x = 0$ to $(a/\sqrt{2})$ and $y = 0$ to $(a/\sqrt{2}) - x$, and then quadruple it.

$I_y = 4\delta\int_0^{(a/\sqrt{2})}\int_0^{(a/\sqrt{2})-x} x^2\,dy\,dx$

$= 4\delta\int_0^{(a/\sqrt{2})}\left(\dfrac{a}{\sqrt{2}}x^2 - x^3\right)dx = \tfrac{1}{12}a^4\delta.$

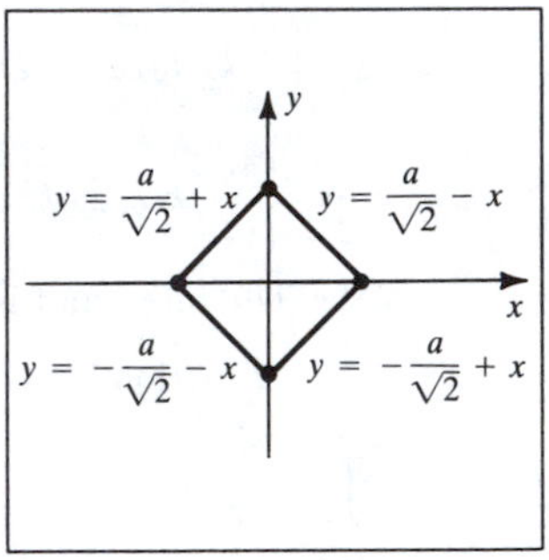

Figure 13

(c) If we position the square as in *Figure 13*, then the center of mass is at the origin and the moment of inertia with respect to the center of mass corresponds to I_O.

$I_O = I_x + I_y = 2I_y$ {since $I_x = I_y$} $= \tfrac{1}{6}a^4\delta.$

15 The area of the square is a^2, and so the mass is $m = a^2\delta$.

Thus, $I = md^2 \Rightarrow d = \left(\dfrac{I_x}{m}\right)^{1/2} = \left(\dfrac{\frac{1}{3}a^4\delta}{a^2\delta}\right)^{1/2} = \dfrac{a}{\sqrt{3}}.$

[17] Let the cube have vertices (0, 0, 0), (a, 0, 0), (a, a, 0), (0, a, 0), (0, 0, a), (a, 0, a), (a, a, a), and (0, a, a) with $0 \le x, y, z \le a$. Position the fixed corner at the origin. The square of the distance from the origin is $x^2 + y^2 + z^2$ and $\delta(x, y, z) = k(x^2 + y^2 + z^2)$, where k is a constant of proportionality. By symmetry, $\bar{x} = \bar{y} = \bar{z}$. Using (17.26),

$$m = \iiint_Q \delta(x, y, z)\, dV$$

$$= \int_0^a\int_0^a\int_0^a k(x^2 + y^2 + z^2)\, dx\, dy\, dz$$

$$= \int_0^a\int_0^a k(\tfrac{1}{3}a^3 + ay^2 + az^2)\, dy\, dz = \int_0^a k(\tfrac{1}{3}a^4 + \tfrac{1}{3}a^4 + a^2z^2)\, dz = ka^5.$$

$$M_{yz} = \iiint_Q x\delta(x, y, z)\, dV$$

$$= \int_0^a\int_0^a\int_0^a k(x^3 + xy^2 + xz^2)\, dx\, dy\, dz$$

$$= \int_0^a\int_0^a k(\tfrac{1}{4}a^4 + \tfrac{1}{2}a^2y^2 + \tfrac{1}{2}a^2z^2)\, dy\, dz$$

$$= \int_0^a k(\tfrac{1}{4}a^5 + \tfrac{1}{6}a^5 + \tfrac{1}{2}a^3z^2)\, dz = \tfrac{7}{12}ka^6.$$

Thus, $\bar{x} = \bar{y} = \bar{z} = \dfrac{M_{yz}}{m} = \tfrac{7}{12}a.$

[19] See Figure 14.70. Let R be the region in the xz-plane described by $x = 4z^2$ and $x = 4$. $m = \int_0^4\int_{-\sqrt{x}/2}^{\sqrt{x}/2}\int_{-\sqrt{x-4z^2}}^{\sqrt{x-4z^2}} (x^2 + z^2)\, dy\, dz\, dx$ since $y = \pm\sqrt{x - 4z^2}$ and when $y = 0$, $z = \pm\frac{1}{2}\sqrt{x}$. The integrals for M_{yz}, M_{xz}, and M_{xy} have the same limits, but the integrands are $x(x^2 + z^2)$, $y(x^2 + z^2)$, and $z(x^2 + z^2)$, respectively.

Finally, use (17.26)(iii) to find $\bar{x}$, $\bar{y}$, and $\bar{z}$.

Note: In Exer. 21-24, without loss of generality let $\delta(x, y, z) = 1$.

[21] The boundary of the solid shown is the hemisphere $z = \sqrt{a^2 - x^2 - y^2}$. Its trace in the xy-plane is $x^2 + y^2 = a^2$, which is a circle of radius a. By (17.26) with $\delta = 1$, $m = \int_{-a}^a\int_{-\sqrt{a^2-x^2}}^{\sqrt{a^2-x^2}}\int_0^{\sqrt{a^2-x^2-y^2}} dz\, dy\, dx$. The integral for M_{xy} has the same limits, but the integrand is z. $\bar{z} = M_{xy}/m$ and by symmetry, $\bar{x} = \bar{y} = 0$.

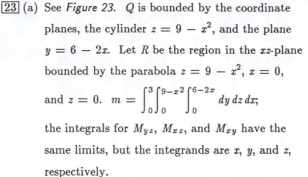

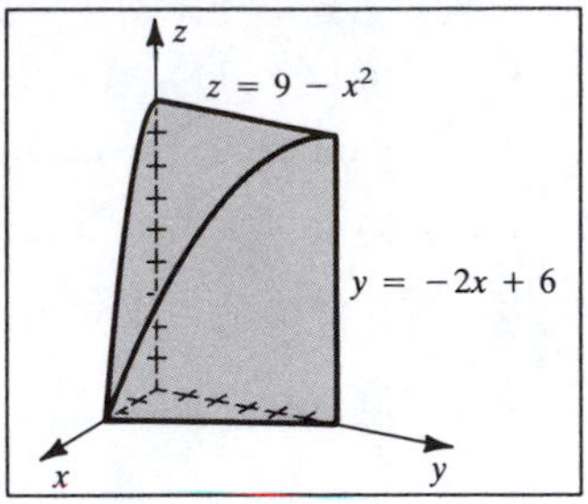

Figure 23

[23] (a) See *Figure 23*. Q is bounded by the coordinate planes, the cylinder $z = 9 - x^2$, and the plane $y = 6 - 2x$. Let R be the region in the xz-plane bounded by the parabola $z = 9 - x^2$, $x = 0$, and $z = 0$. $m = \int_0^3 \int_0^{9-x^2} \int_0^{6-2x} dy\,dz\,dx$; the integrals for M_{yz}, M_{xz}, and M_{xy} have the same limits, but the integrands are x, y, and z, respectively.

(b) $$m = \int_0^3 \int_0^{9-x^2} \int_0^{6-2x} dy\,dz\,dx$$
$$= \int_0^3 \int_0^{9-x^2} (6 - 2x)\,dz\,dx$$
$$= \int_0^3 (2x^3 - 6x^2 - 18x + 54)\,dx = \tfrac{135}{2}.$$

$$M_{xy} = \int_0^3 \int_0^{9-x^2} \int_0^{6-2x} z\,dy\,dz\,dx = \int_0^3 \int_0^{9-x^2} z(6 - 2x)\,dz\,dx =$$
$$\int_0^3 (-x^5 + 3x^4 + 18x^3 - 54x^2 - 81x + 243)\,dx = \tfrac{2673}{10}.$$

$$M_{yz} = \int_0^3 \int_0^{9-x^2} \int_0^{6-2x} x\,dy\,dz\,dx = \int_0^3 \int_0^{9-x^2} x(6 - 2x)\,dz\,dx =$$
$$\int_0^3 (2x^4 - 6x^3 - 18x^2 + 54x)\,dx = \tfrac{567}{10}.$$

$$M_{xz} = \int_0^3 \int_0^{9-x^2} \int_0^{6-2x} y\,dy\,dz\,dx = \int_0^3 \int_0^{9-x^2} (18 - 12x + 2x^2)\,dz\,dx =$$
$$2\int_0^3 (-x^4 + 6x^3 - 54x + 81)\,dx = \tfrac{729}{5}.$$

Thus, $\bar{x} = \dfrac{M_{yz}}{m} = \tfrac{21}{25}$, $\bar{y} = \dfrac{M_{xz}}{m} = \tfrac{54}{25}$, and $\bar{z} = \dfrac{M_{xy}}{m} = \tfrac{99}{25}$.

[25] The sphere has the equation $x^2 + y^2 + z^2 = a^2$ and its trace in the xy-plane is $x^2 + y^2 = a^2$. Thus, $-\sqrt{a^2 - x^2 - y^2} \le z \le \sqrt{a^2 - x^2 - y^2}$, $-\sqrt{a^2 - x^2} \le y \le \sqrt{a^2 - x^2}$, $-a \le x \le a$, and $I_z = \iiint_Q (x^2 + y^2)\,\delta(x, y, z)\,dV =$
$$\int_{-a}^{a} \int_{-\sqrt{a^2-x^2}}^{\sqrt{a^2-x^2}} \int_{-\sqrt{a^2-x^2-y^2}}^{\sqrt{a^2-x^2-y^2}} (x^2 + y^2)(x^2 + y^2 + z^2)\,dz\,dy\,dx.$$

Exercises 17.7

1 (a) Each ordered triple is of the form $(4,\ \theta,\ z)$. The values of θ and z are arbitrary.

Hence, $r = 4$ is the right circular cylinder of radius 4 with axis along the z-axis.

(b) Each ordered triple is of the form $(r,\ -\frac{\pi}{2},\ z)$.

The values of r and z are arbitrary. Hence, $\theta = -\frac{\pi}{2}$ is the yz-plane.

In general, this plane can be described by $\theta = \frac{\pi}{2} + \pi n$.

(c) Each ordered triple is of the form $(r,\ \theta,\ 1)$. The values of r and θ are arbitrary.

Hence, $z = 1$ is the plane parallel to the xy-plane with z-intercept 1.

3 $r = -3\sec\theta \Rightarrow \dfrac{r}{\sec\theta} = -3 \Rightarrow r\cos\theta = -3 \Rightarrow x = -3$ {from (17.29)},

the plane parallel to the yz-plane with x-intercept -3.

7 Multiplying both sides of $r = 6\sin\theta$ by r in order to obtain r^2 gives us

$r^2 = 6r\sin\theta \Rightarrow x^2 + y^2 = 6y \Rightarrow x^2 + y^2 - 6y + 9 = 9 \Rightarrow x^2 + (y-3)^2 = 9.$

This is the right circular cylinder with trace $x^2 + (y-3)^2 = 9$ in the xy-plane.

9 Squaring both sides of $z = 2r$ yields $z^2 = 4r^2$.

Substituting $x^2 + y^2$ for r^2 gives us the cone $z^2 = 4x^2 + 4y^2$.

13 $r = 2\csc\theta\cot\theta \Rightarrow r = 2\cdot\dfrac{1}{\sin\theta}\cdot\dfrac{\cos\theta}{\sin\theta} \Rightarrow r\sin^2\theta = 2\cos\theta \Rightarrow$

$r^2\sin^2\theta = 2r\cos\theta$ {multiplying both sides by r} $\Rightarrow (r\sin\theta)^2 = 2(r\cos\theta)$. This is

the cylinder with trace $y^2 = 2x$ in the xy-plane and rulings parallel to the z-axis.

15 Using (17.29), $x^2 + y^2 + z^2 = 4 \Rightarrow (x^2 + y^2) + z^2 = 4 \Rightarrow r^2 + z^2 = 4$.

17 $3x + y - 4z = 12 \Rightarrow 3r\cos\theta + r\sin\theta - 4z = 12$

23 $y^2 + z^2 = 9 \Rightarrow (r\sin\theta)^2 + z^2 = 9 \Rightarrow r^2\sin^2\theta + z^2 = 9$

25 The trace of $z = 16 - r^2$ in the xy-plane $\{z = 0\}$ is $r = 4$. z varies between 0 and $16 - r^2$ while the bounds on r and θ are $0 \le r \le 4$ and $0 \le \theta \le 2\pi$.

By (17.30), $\iiint_Q f(r,\ \theta,\ z)\,dV = \displaystyle\int_0^{2\pi}\int_0^4\int_0^{16-r^2} f(r,\ \theta,\ z)\,r\,dz\,dr\,d\theta.$

27 In order to set up this integral, we must divide Q into two separate regions—call them Q_1 and Q_2. Q_1 is the region below the blue disk with equation $z = 4$. Q_2 is the region below the red curved surface. The boundary between Q_1 and Q_2 occurs when $4 = z = \sqrt{25 - r^2} \Rightarrow 16 = 25 - r^2 \Rightarrow r^2 = 9 \Rightarrow r = 3$. Thus, on Q_1, $0 \le r \le 3$ and $0 \le z \le 4$, and on Q_2, $3 \le r \le 4$ and $0 \le z \le \sqrt{25 - r^2}$.

$\iiint_Q f(r,\ \theta,\ z)\,dV = \iiint_{Q_1} f(r,\ \theta,\ z)\,dV + \iiint_{Q_2} f(r,\ \theta,\ z)\,dV$

$$= \int_0^{2\pi}\int_0^3\int_0^4 f(r,\ \theta,\ z)\,r\,dz\,dr\,d\theta + \int_0^{2\pi}\int_3^4\int_0^{\sqrt{25-r^2}} f(r,\ \theta,\ z)\,r\,dz\,dr\,d\theta.$$

Note: In Exercises 29–30, without loss of generality, let $\delta(x, y, z) = 1$.

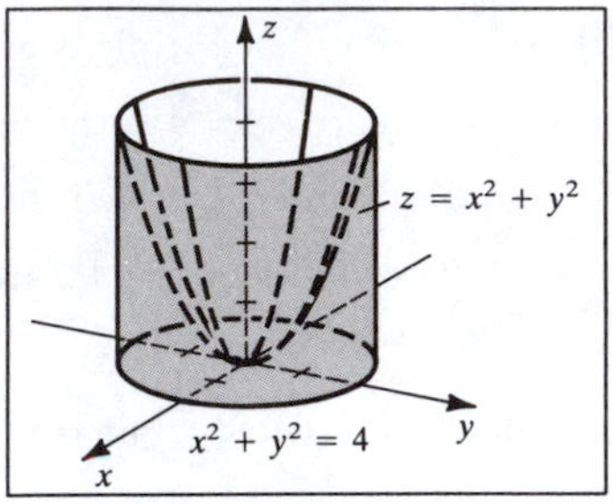

Figure 29

[29] (a) In cylindrical coordinates:

$z = x^2 + y^2 \Leftrightarrow z = r^2$,

$x^2 + y^2 = 4 \Leftrightarrow r^2 = 4 \Rightarrow r = 2$,

and the xy-plane $\Leftrightarrow z = 0$.

The solid is the region between the xy-plane and the paraboloid shown in *Figure 29.*

$$V = \int_0^{2\pi}\int_0^2\int_0^{r^2} r\,dz\,dr\,d\theta = \int_0^{2\pi}\int_0^2 r^3\,dr\,d\theta = 4\int_0^{2\pi} d\theta = 8\pi.$$

(b) $M_{xy} = \iiint\limits_Q z\,\delta(x, y, z)\,dV = \int_0^{2\pi}\int_0^2\int_0^{r^2} zr\,dz\,dr\,d\theta = \int_0^{2\pi}\int_0^2 \frac{1}{2}r^5\,dr\,d\theta =$

$\frac{16}{3}\int_0^{2\pi} d\theta = \frac{32\pi}{3}$. By symmetry, $\bar{x} = \bar{y} = 0$. Since $\delta(x, y, z) = 1$ can be assumed, $m = V \Rightarrow \bar{z} = \dfrac{M_{xy}}{V} = \frac{4}{3}$, and the centroid is $(0, 0, \frac{4}{3})$.

[31] Let the cylinder have equation $r = a$ for $0 \leq z \leq h$.

(a) $I_z = \iiint\limits_Q (x^2 + y^2)\,\delta\,dV = \delta\int_0^{2\pi}\int_0^a\int_0^h (r^2)\,r\,dz\,dr\,d\theta = h\delta\int_0^{2\pi}\int_0^a r^3\,dr\,d\theta =$

$\frac{1}{4}ha^4\delta\int_0^{2\pi} d\theta = \frac{1}{2}\pi ha^4\delta.$

(b) Let the diameter of the base coincide with the x-axis.

Thus, its moment of inertia with respect to a diameter of the base will be I_x.

$$\begin{aligned} I_x &= \iiint\limits_Q (y^2 + z^2)\,\delta\,dV \\ &= \delta\int_0^{2\pi}\int_0^a\int_0^h (r^2\sin^2\theta + z^2)\,r\,dz\,dr\,d\theta \\ &= h\delta\int_0^{2\pi}\int_0^a (r^3\sin^2\theta + \tfrac{1}{3}h^2 r)\,dr\,d\theta \\ &= ha^2\delta\int_0^{2\pi} (\tfrac{1}{4}a^2\sin^2\theta + \tfrac{1}{6}h^2)\,d\theta = \pi ha^2(\tfrac{1}{4}a^2 + \tfrac{1}{3}h^2)\delta. \end{aligned}$$

Note: Let k be a constant of proportionality.

[37] The volume of air is contained in a cylindrical region of height 10,000 m and a radius of 3 m. Thus,

$$\begin{aligned} m &= \iiint\limits_Q \delta\,dV = \int_0^{2\pi}\int_0^3\int_0^{10{,}000} \delta\,r\,dz\,dr\,d\theta \\ &= \int_0^{2\pi}\int_0^3 \Big[1.2z - \tfrac{1}{2}(1.05 \times 10^{-4})z^2 + \tfrac{1}{3}(2.6 \times 10^{-9})z^3\Big]_0^{10{,}000} r\,dr\,d\theta \\ &= \frac{22{,}850}{3}\int_0^{2\pi}\int_0^3 r\,dr\,d\theta = \frac{22{,}850}{3}\cdot\frac{9}{2}\int_0^{2\pi} d\theta = 68{,}550\pi \approx 215{,}356.2, \text{ or } 215{,}360 \text{ kg}. \end{aligned}$$

[39] The region in the xy-plane is $x = \sqrt{1 - y^2}$ from $y = 0$ to 1, which is the quarter circle $r = 1$ for $\theta = 0$ to $\frac{\pi}{2}$. $\int_0^1 \int_0^{\sqrt{1-y^2}} \int_0^{\sqrt{4-x^2-y^2}} z\,dz\,dx\,dy =$

$$\int_0^{\pi/2} \int_0^1 \int_0^{\sqrt{4-r^2}} z\,r\,dz\,dr\,d\theta = \tfrac{1}{2}\int_0^{\pi/2} \int_0^1 (4r - r^3)\,dr\,d\theta = \tfrac{1}{2} \cdot \tfrac{7}{4} \int_0^{\pi/2} d\theta = \tfrac{7\pi}{16}.$$

Exercises 17.8

[1] (a) Using (17.31) with $\rho = 4$, $\phi = \frac{\pi}{6}$, and $\theta = \frac{\pi}{2}$:

$x = \rho \sin\phi \cos\theta = 4\sin\frac{\pi}{6} \cos\frac{\pi}{2} = 4(\frac{1}{2})(0) = 0,$

$y = \rho \sin\phi \sin\theta = 4\sin\frac{\pi}{6} \sin\frac{\pi}{2} = 4(\frac{1}{2})(1) = 2,$ and

$z = \rho \cos\phi = 4\cos\frac{\pi}{6} = 4(\frac{1}{2}\sqrt{3}) = 2\sqrt{3}.$

The rectangular coordinates are $(0, 2, 2\sqrt{3})$.

(b) Using (17.29) with $x = 0$ and $y = 2 \Rightarrow r = \sqrt{x^2 + y^2} = \sqrt{0^2 + 2^2} = 2$, $\theta = \frac{\pi}{2}$ from the given spherical coordinates, and $z = 2\sqrt{3}$ from part (a).

The cylindrical coordinates are $(2, \frac{\pi}{2}, 2\sqrt{3})$.

[3] (a) Using (17.31) with $x = 1$, $y = 1$, and $z = -2\sqrt{2}$:

$\rho^2 = x^2 + y^2 + z^2 = 1^2 + 1^2 + (-2\sqrt{2})^2 = 10 \Rightarrow \rho = \sqrt{10},$

$\cos\phi = \frac{z}{\rho} = \frac{-2\sqrt{2}}{\sqrt{10}} = \frac{-2}{\sqrt{5}} \Rightarrow \phi = \cos^{-1}(-\frac{2}{\sqrt{5}})$, and

$\tan\theta = \frac{y}{x} = 1 \Rightarrow \theta = \frac{\pi}{4}$ since x and y are both positive.

The spherical coordinates are $(\sqrt{10}, \cos^{-1}(-\frac{2}{\sqrt{5}}), \frac{\pi}{4})$.

(b) $r = \sqrt{x^2 + y^2} = \sqrt{1^2 + 1^2} = \sqrt{2}$, $\theta = \frac{\pi}{4}$ from part (a), and $z = -2\sqrt{2}$ from the given rectangular coordinates. The cylindrical coordinates are $(\sqrt{2}, \frac{\pi}{4}, -2\sqrt{2})$.

[5] (a) Each ordered triple is of the form $(3, \phi, \theta)$. The values of ϕ and θ are arbitrary.

Hence, $\rho = 3$ is the sphere of radius 3 and center O. See Figure 17.72(i).

(b) Each ordered triple is of the form $(\rho, \frac{\pi}{6}, \theta)$. The values of ρ and θ are arbitrary. Hence, $\phi = \frac{\pi}{6}$ is a half-cone with vertex O and vertex angle $2 \cdot \frac{\pi}{6} = \frac{\pi}{3}$. See Figure 17.72(ii).

(c) Each ordered triple is of the form $(\rho, \phi, \frac{\pi}{3})$. The values of ρ and ϕ are arbitrary. Hence, $\theta = \frac{\pi}{3}$ is a half-plane with edge on the z-axis and making an angle of $\frac{\pi}{3}$ with the xz-plane. See Figure 17.72(iii).

[7] Multiplying both sides of the equation by ρ to obtain ρ^2 gives us $\rho = 4\cos\phi \Rightarrow \rho^2 = 4\rho \cos\phi \Rightarrow x^2 + y^2 + z^2 = 4z \Rightarrow x^2 + y^2 + (z - 2)^2 = 4.$

This is the sphere of radius 2 and center $(0, 0, 2)$.

[11] Multiplying both sides of the equation by ρ gives us $\rho = 6\sin\phi\cos\theta \Rightarrow$

$\rho^2 = 6\rho\sin\phi\cos\theta \Rightarrow x^2 + y^2 + z^2 = 6x \Rightarrow (x-3)^2 + y^2 + z^2 = 9.$

This is the sphere of radius 3 and center (3, 0, 0).

[15]

$$\begin{aligned}
\rho &= 5\csc\phi && \{\text{given equation}\}\\
\rho\sin\phi &= 5 && \{\text{multiply both sides by } \sin\theta\}\\
\rho^2\sin^2\phi &= 25 && \{\text{square both sides}\}\\
\rho^2\sin^2\phi(\cos^2\theta + \sin^2\theta) &= 25 && \{\text{associate } \rho\sin\phi \text{ with } \cos\theta \text{ for } x\}\\
\rho^2\sin^2\phi\cos^2\theta + \rho^2\sin^2\phi\sin^2\theta &= 25 && \{\text{distribute}\}\\
x^2 + y^2 &= 25 && \{x = \rho\sin\phi\cos\theta,\ y = \rho\sin\phi\sin\theta\}
\end{aligned}$$

This is the right circular cylinder of radius 5 with axis along the z-axis.

[17]

$$\begin{aligned}
\tan\phi &= 2 && \{\text{given equation}\}\\
\sin\phi &= 2\cos\phi && \{\text{multiply both sides by } \cos\phi\}\\
\rho\sin\phi &= 2\rho\cos\phi && \{\text{associate } \cos\phi \text{ with } \rho \text{ for } z\}\\
\rho^2\sin^2\phi &= 4\rho^2\cos^2\phi && \{\text{need } \rho^2 \text{ for (17.31)(ii)}\}\\
\rho^2\sin^2\phi(\cos^2\theta + \sin^2\theta) &= 4z^2 && \{\text{associate } \rho\sin\phi \text{ with } \cos\theta \text{ for } x\}\\
\rho^2\sin^2\phi\cos^2\theta + \rho^2\sin^2\phi\sin^2\theta &= 4z^2 && \{\text{distribute}\}
\end{aligned}$$

This is the cone $x^2 + y^2 = 4z^2$.

[19] *Note*: Use $\csc\phi$, not $\csc\theta$. $\rho = 6\cot\phi\csc\phi \Rightarrow \rho = 6\cdot\dfrac{\cos\phi}{\sin\phi}\cdot\dfrac{1}{\sin\phi} \Rightarrow$

$\rho\sin^2\phi = 6\cos\phi \Rightarrow \rho^2\sin^2\phi(\cos^2\theta + \sin^2\theta) = 6\rho\cos\phi \Rightarrow$

$\rho^2\sin^2\phi\cos^2\theta + \rho^2\sin^2\phi\sin^2\theta = 6z.$ This is the paraboloid $6z = x^2 + y^2$.

[21] $x^2 + y^2 + z^2 = 4 \Rightarrow \rho^2 = 4 \Rightarrow \rho = 2\ \{\rho \geq 0, \text{ so } \rho \neq -2\}$.

[25] $x^2 = 4 - y^2 \Rightarrow x^2 + y^2 = 4 \Rightarrow \rho^2\sin^2\phi\cos^2\theta + \rho^2\sin^2\phi\sin^2\theta = 4 \Rightarrow$

$\rho^2\sin^2\phi(\cos^2\theta + \sin^2\theta) = 4 \Rightarrow \rho^2\sin^2\phi = 4 \Rightarrow \rho^2 = 4\csc^2\phi \Rightarrow$

$\rho = 2\csc\phi\ \{\rho > 0,\ 0 < \phi < \pi\}$.

[31] Let the hemisphere have the equation $x^2 + y^2 + z^2 = a^2$ for $z \geq 0$.

The distance from the point $P(x, y, z)$ to the origin is $(x^2 + y^2 + z^2)^{1/2}$.

Thus, $\delta(x, y, z) = k(x^2 + y^2 + z^2)^{1/2} = k\rho$, where k is a constant of proportionality.

By (17.32), $m = \iiint_Q \delta(\rho, \phi, \theta)\,dV$

$$= \int_0^{\pi/2}\int_0^a\int_0^{2\pi} (k\rho)(\rho^2\sin\phi)\,d\theta\,d\rho\,d\phi$$

$$= 2k\pi\int_0^{\pi/2}\int_0^a \rho^3\sin\phi\,d\rho\,d\phi = \tfrac{1}{2}k\pi a^4\int_0^{\pi/2}\sin\phi\,d\phi = \tfrac{1}{2}k\pi a^4. \text{ (cont.)}$$

Since $z = \rho\cos\phi$,

$$M_{xy} = \int_0^{\pi/2}\int_0^a\int_0^{2\pi} (\rho\cos\phi)(k\rho)(\rho^2\sin\phi)\,d\theta\,d\rho\,d\phi$$

$$= 2k\pi\int_0^{\pi/2}\int_0^a \rho^4\sin\phi\cos\phi\,d\rho\,d\phi$$

$$= \tfrac{2}{5}k\pi a^5\int_0^{\pi/2}\sin\phi\cos\phi\,d\phi = \tfrac{1}{5}k\pi a^5\int_0^{\pi/2}\sin 2\phi\,d\phi = \tfrac{1}{5}k\pi a^5.$$

By symmetry, $\bar{x} = \bar{y} = 0$. $\bar{z} = \dfrac{M_{xy}}{m} = \tfrac{2}{5}a$.

[35] The cone and the sphere intersect when $4z - z^2 = x^2 + y^2 = z^2 \Rightarrow z = 0, 2$. Thus, "above the cone" means above the plane $z = 2$. See *Figure 35*. As ϕ varies between 0 and $\frac{\pi}{4}$, ρ varies between the plane $z = 2$ ($\rho\cos\phi = 2$ or $\rho = 2\sec\phi$) and the surface of the sphere ($\rho^2 = 4\rho\cos\phi$ or $\rho = 4\cos\phi$). Hence,

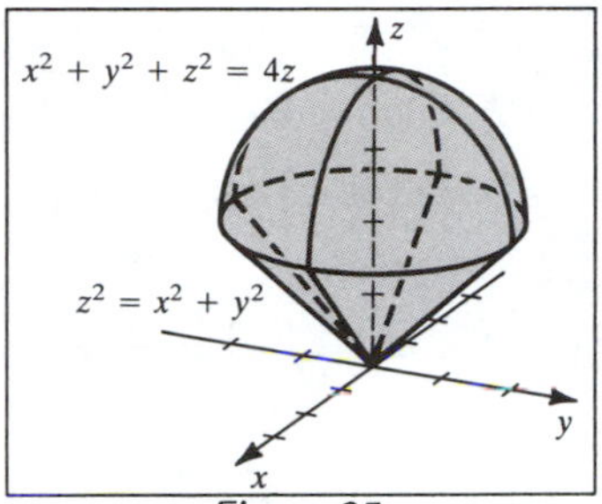

Figure 35

$$V = \int_0^{\pi/4}\int_0^{2\pi}\int_{2\sec\phi}^{4\cos\phi} (\rho^2\sin\phi)\,d\rho\,d\theta\,d\phi$$

$$= \tfrac{1}{3}\int_0^{\pi/4}\int_0^{2\pi} (64\cos^3\phi - 8\sec^3\phi)\sin\phi\,d\theta\,d\phi$$

$$= \tfrac{16\pi}{3}\int_0^{\pi/4}(8\cos^3\phi\sin\phi - \sec^2\phi\tan\phi)\,d\phi$$

$$= \tfrac{16\pi}{3}\Big[-2\cos^4\phi - \tfrac{1}{2}\tan^2\phi\Big]_0^{\pi/4} = \tfrac{16\pi}{3}.$$

Note: The solid is a hemisphere of radius 2, which has a volume of $\frac{2}{3}\pi r^3 = \frac{16\pi}{3}$.

[37] $\delta(x, y, z) = k(x^2 + y^2 + z^2) = k\rho^2$. By (17.32), $m = \iiint_Q \delta(\rho, \phi, \theta)\,dV =$

$$\int_0^{2\pi}\int_0^{\pi}\int_1^2 (k\rho^2)(\rho^2\sin\phi)\,d\rho\,d\phi\,d\theta = \tfrac{31}{5}k\int_0^{2\pi}\int_0^{\pi}\sin\phi\,d\phi\,d\theta = \tfrac{62}{5}k\int_0^{2\pi}d\theta = \tfrac{124}{5}k\pi.$$

[39] This is similar to Example 6 with $\rho = \sqrt{8}$ and $c = \frac{\pi}{4}$. From the innermost integral, $\sqrt{x^2 + y^2} \le z \le \sqrt{8 - x^2 - y^2} \Rightarrow x^2 + y^2 \le z^2 \le 8 - x^2 - y^2$. Hence, z varies between the cone $z^2 = x^2 + y^2$ and the sphere $x^2 + y^2 + z^2 = 8$. The cone can be described by $\phi = \frac{\pi}{4}$ and the sphere can be described by $\rho^2 = 8$, or $\rho = \sqrt{8}$. Thus,

$$\int_{-2}^{2}\int_{-\sqrt{4-x^2}}^{\sqrt{4-x^2}}\int_{\sqrt{x^2+y^2}}^{\sqrt{8-x^2-y^2}} (x^2 + y^2 + z^2)\,dz\,dy\,dx = \int_0^{2\pi}\int_0^{\pi/4}\int_0^{\sqrt{8}} (\rho^2)\,\rho^2\sin\phi\,d\rho\,d\phi\,d\theta$$

$$= \tfrac{128}{5}\sqrt{2}\int_0^{2\pi}\int_0^{\pi/4}\sin\phi\,d\phi\,d\theta$$

$$= \tfrac{128}{5}(\sqrt{2} - 1)\int_0^{2\pi}d\theta$$

$$= \tfrac{256\pi}{5}(\sqrt{2} - 1) \approx 66.63.$$

Exercises 17.9

1 (a) To find the u-curves and the v-curves, we will set u and v equal to the constants c and d, respectively. $u = c \Rightarrow 3x = c \Rightarrow x = \frac{1}{3}c$; vertical lines.

$v = d \Rightarrow 5y = d \Rightarrow y = \frac{1}{5}d$; horizontal lines.

(b) We must solve the given equations for x and y.

$u = 3x \Rightarrow x = \frac{1}{3}u$. $v = 5y \Rightarrow y = \frac{1}{5}v$.

3 (a) $u = c \Rightarrow x - y = c \Rightarrow y = x - c$; $v = d \Rightarrow 2x + 3y = d \Rightarrow y = -\frac{2}{3}x + \frac{1}{3}d$.

These are lines with slopes 1 and $-\frac{2}{3}$, respectively.

(b) Adding the equations $-2(x - y = u)$ and $2x + 3y = v$,

and then solving for y gives us $y = -\frac{2}{5}u + \frac{1}{5}v$.

Similarly, for x, $3(x - y = u)$ and $2x + 3y = v \Rightarrow x = \frac{3}{5}u + \frac{1}{5}v$.

5 (a) $u = c \Rightarrow x^3 = c \Rightarrow x = c^{1/3}$; vertical lines since c is simply a constant.

$v = d \Rightarrow x + y = d \Rightarrow y = -x + d$; lines with slope -1.

(b) $u = x^3 \Rightarrow x = u^{1/3}$. $v = x + y \Rightarrow y = v - x = v - u^{1/3}$.

9 (a) The vertices (0, 0), (0, 1), (2, 1), and (2, 0) are transformed into the vertices (0, 0), (0, 5), (6, 5), and (6, 0), respectively, by using the equations $u = 3x$ and $v = 5y$. The sides of the rectangle are segments of $x = 0$, $y = 1$, $x = 2$, and $y = 0$. In the uv-plane, these are $u = 0$, $v = 5$, $u = 6$, and $v = 0$, respectively. These equations also form a rectangle.

(b) $x^2 + y^2 = 1 \Rightarrow \left(\frac{u}{3}\right)^2 + \left(\frac{v}{5}\right)^2 = 1 \Rightarrow \frac{u^2}{9} + \frac{v^2}{25} = 1$, an ellipse.

11 (a) The vertices (0, 0), (0, 1), and (2, 0) are transformed into the vertices (0, 0), (−1, 3), and (2, 4), respectively by using the equations $u = x - y$ and $v = 2x + 3y$. The sides of the triangle are segments of $x = 0$, $y = 0$, and $y = -\frac{1}{2}x + 1$. From Exercise 3, we have $x = \frac{3}{5}u + \frac{1}{5}v$ and $y = -\frac{2}{5}u + \frac{1}{5}v$. In the uv-plane, $x = 0 \Rightarrow \frac{3}{5}u + \frac{1}{5}v = 0 \Rightarrow v = -3u$; $y = 0 \Rightarrow -\frac{2}{5}u + \frac{1}{5}v = 0 \Rightarrow v = 2u$; and $y = -\frac{1}{2}x + 1 \Rightarrow -\frac{2}{5}u + \frac{1}{5}v = -\frac{1}{2}(\frac{3}{5}u + \frac{1}{5}v) + 1 \Rightarrow v = \frac{1}{3}u + \frac{10}{3}$. These equations also form a triangle.

(b) $x + 2y = 1 \Rightarrow (\frac{3}{5}u + \frac{1}{5}v) + 2(-\frac{2}{5}u + \frac{1}{5}v) = 1 \Rightarrow -u + 3v = 5$, a line.

Note: Let J denote the indicated Jacobian.

13 By (17.34), $x = u^2 - v^2$ and $y = 2uv \Rightarrow$

$$J = \frac{\partial(x, y)}{\partial(u, v)} = \begin{vmatrix} \frac{\partial x}{\partial u} & \frac{\partial x}{\partial v} \\ \frac{\partial y}{\partial u} & \frac{\partial y}{\partial v} \end{vmatrix} = \begin{vmatrix} 2u & -2v \\ 2v & 2u \end{vmatrix} =$$

$$(2u)(2u) - (2v)(-2v) = 4u^2 + 4v^2.$$

[17] By (17.37), $x = 2u + 3v - w$, $y = v - 5w$, and $z = u + 4w \Rightarrow J = \dfrac{\partial(x, y, z)}{\partial(u, v, w)} =$

$$\begin{vmatrix} \dfrac{\partial x}{\partial u} & \dfrac{\partial x}{\partial v} & \dfrac{\partial x}{\partial w} \\ \dfrac{\partial y}{\partial u} & \dfrac{\partial y}{\partial v} & \dfrac{\partial y}{\partial w} \\ \dfrac{\partial z}{\partial u} & \dfrac{\partial z}{\partial v} & \dfrac{\partial z}{\partial w} \end{vmatrix} = \begin{vmatrix} 2 & 3 & -1 \\ 0 & 1 & -5 \\ 1 & 0 & 4 \end{vmatrix} = (2)(4) - (3)(5) + (-1)(-1) = -6.$$

[19] The boundary of R is triangular. Using the equations $x = u + v$ and $y = 2v$, we can transform the boundary of R into the boundary of S. $y = 0 \Rightarrow v = 0$; $x = 2 \Rightarrow u + v = 2$; $y = 2x \Rightarrow x = v \Rightarrow u = 0$. Thus, the boundary of S is given by $v = 0$, $u = 0$, and $u + v = 2$, which also forms a triangle. $x = u + v$ and $y = 2v \Rightarrow J = \dfrac{\partial(x, y)}{\partial(u, v)} = \begin{vmatrix} 1 & 1 \\ 0 & 2 \end{vmatrix} = 2$. As (u, v) traverses the vertices (0, 0), (2, 0), and (0, 2) in the positive direction, (x, y) traverses the vertices (0, 0), (2, 0), and (2, 4) in a positive direction.

Hence, using (17.35), $\iint_R (y - x)\, dx\, dy = \int_0^2 \int_0^{2-u} (v - u)(2)\, dv\, du$.

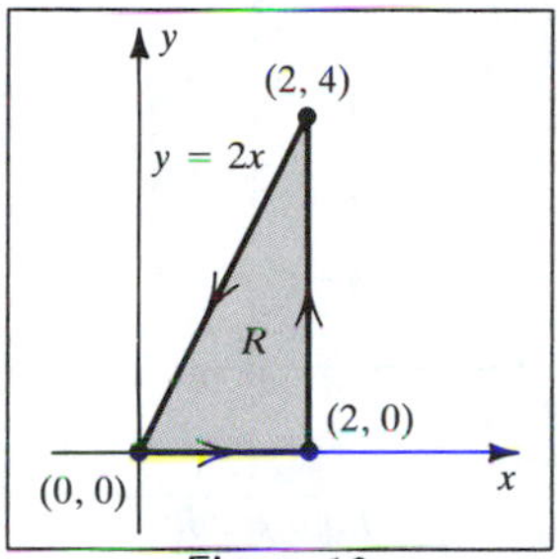

Figure 19a

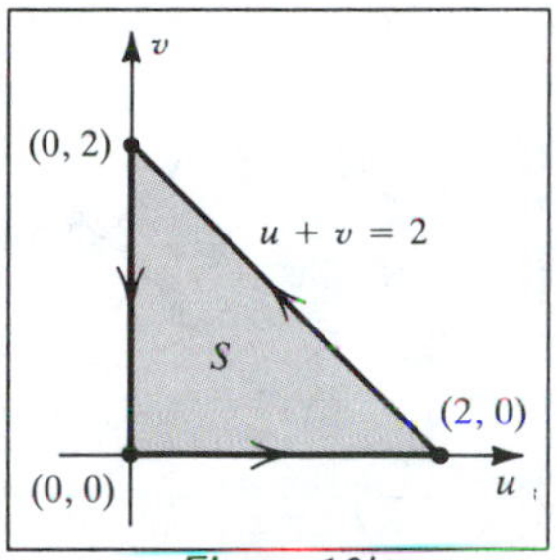

Figure 19b

[21] The boundary of R is an ellipse. $\frac{1}{4}x^2 + \frac{1}{9}y^2 = 1 \Rightarrow \frac{1}{4}(4u^2) + \frac{1}{9}(9v^2) = 1 \Rightarrow u^2 + v^2 = 1$, which is a circle. $x = 2u$ and $y = 3v \Rightarrow J = \dfrac{\partial(x, y)}{\partial(u, v)} = \begin{vmatrix} 2 & 0 \\ 0 & 3 \end{vmatrix} = 6$. As (u, v) traverses the points (1, 0), (0, 1), and (−1, 0) in the positive direction, (x, y) traverses the points (2, 0), (0, 3), and (−2, 0) in the positive direction.

Thus, $\iint_R (\frac{1}{4}x^2 + \frac{1}{9}y^2)\, dx\, dy = \int_{-1}^1 \int_{-\sqrt{1-v^2}}^{\sqrt{1-v^2}} (u^2 + v^2)(6)\, du\, dv$.

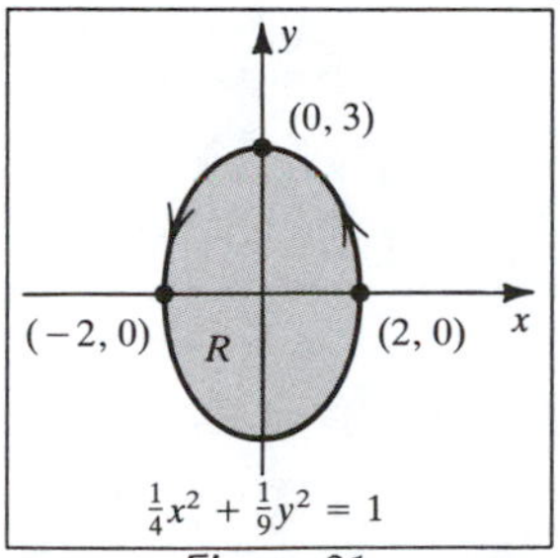

Figure 21a

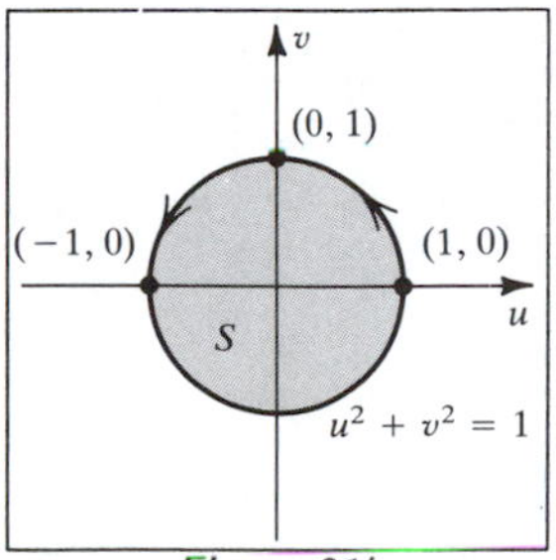

Figure 21b

[23] The vertices with rectangular coordinates (0, 1), (1, 0), (2, 1), and (1, 2) are transformed into the vertices (−1, 1), (1, 1), (1, 3), and (−1, 3) in a positive direction using the equations $u = x - y$ and $v = x + y$. Solving these equations for x and y gives us $x = \frac{1}{2}(u + v)$ and $y = \frac{1}{2}(v - u)$. The line segments in the xy-plane are transformed into the following boundaries in the uv-plane:

$y = -x + 1,\ 0 \le x \le 1 \Rightarrow \frac{1}{2}(v - u) = -\frac{1}{2}(u + v) + 1 \Rightarrow v = 1,\ -1 \le u \le 1.$

$y = x - 1,\ 1 \le x \le 2 \Rightarrow \frac{1}{2}(v - u) = \frac{1}{2}(u + v) - 1 \Rightarrow u = 1,\ 1 \le v \le 3.$

Similarly, $y = -x + 3,\ 1 \le x \le 2 \Rightarrow v = 3,\ -1 \le u \le 1.$

$y = x + 1,\ 0 \le x \le 1 \Rightarrow u = -1,\ 1 \le v \le 3.$ $J = \dfrac{\partial(x,\ y)}{\partial(u,\ v)} = \begin{vmatrix} \frac{1}{2} & \frac{1}{2} \\ -\frac{1}{2} & \frac{1}{2} \end{vmatrix} = \frac{1}{2}.$

Thus, $\iint\limits_R (x - y)^2 \cos^2(x + y)\,dx\,dy = \int_1^3 \int_{-1}^1 (u^2 \cos^2 v)(\frac{1}{2})\,du\,dv = \frac{1}{3}\int_1^3 \cos^2 v\,dv =$

$\frac{1}{3}\int_1^3 (\frac{1}{2} + \frac{1}{2}\cos 2v)\,dv = \frac{1}{3} + \frac{1}{12}\sin 6 - \frac{1}{12}\sin 2 \approx 0.23.$

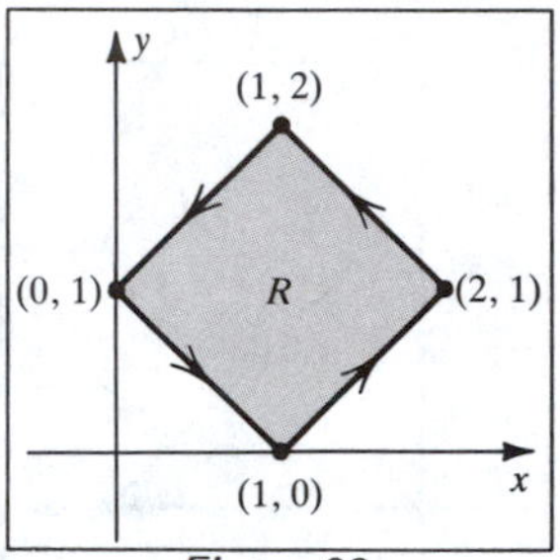

Figure 23a

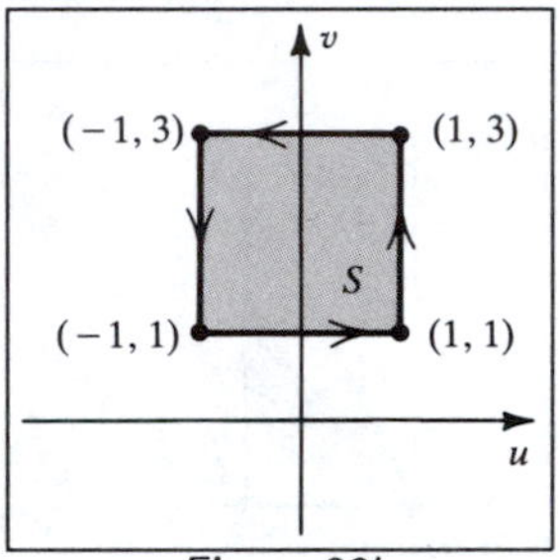

Figure 23b

[27] The vertices of the trapezoid $(-2,\ 0)$, $(-1,\ 0)$, $(0,\ 2)$, and $(0,\ 4)$ are transformed from a counterclockwise direction into the vertices $(4,\ -2)$, $(2,\ -1)$, $(2,\ 4)$, and $(4,\ 8)$ in a clockwise, or negative direction. Because of this change in direction, we will choose the minus sign in (17.35). The four line segments in the xy-plane are transformed into four line segments in the uv-plane connecting the vertices as shown in the figures. Since $x = \frac{1}{5}v - \frac{2}{5}u$ and $y = \frac{1}{5}u + \frac{2}{5}v$, $J = -\frac{1}{5}$. Notice that the Jacobian is negative. This indicates the change in direction produced by the transformation.

$$\text{Thus, } \iint_R \frac{2y + x}{y - 2x}\, dx\, dy = -\int_2^4 \int_{-u/2}^{2u} \left(\tfrac{v}{u}\right)\left(-\tfrac{1}{5}\right) dv\, du = \tfrac{1}{5}\int_2^4 \tfrac{15}{8}u\, du = \tfrac{9}{4}.$$

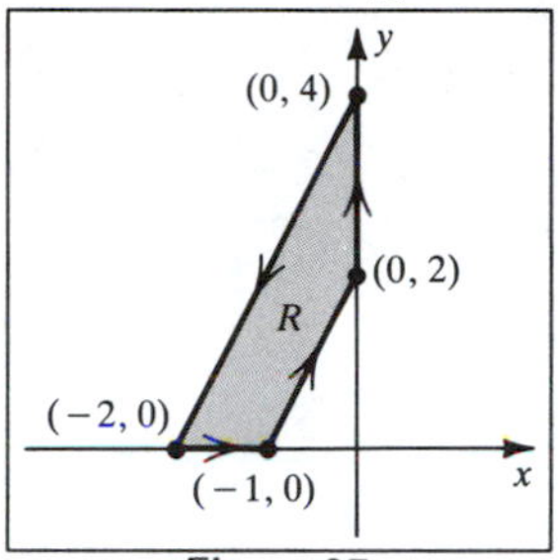

Figure 27a

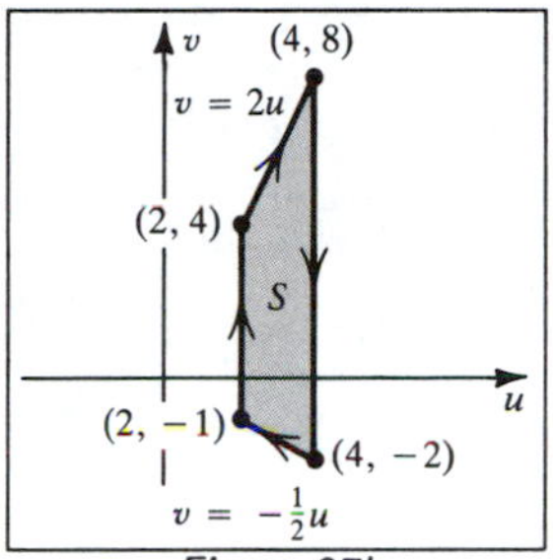

Figure 27b

[29] $\frac{x^2}{a^2} + \frac{y^2}{b^2} + \frac{z^2}{c^2} = 1$ $\{x = au,\ y = bv, \text{ and } z = cw\} \Rightarrow u^2 + v^2 + w^2 = 1,$

a sphere with radius 1. $J = \dfrac{\partial(x,\ y,\ z)}{\partial(u,\ v,\ w)} = \begin{vmatrix} a & 0 & 0 \\ 0 & b & 0 \\ 0 & 0 & c \end{vmatrix} = abc.$

Then by (17.38), $\iiint_R dx\, dy\, dz = \iiint_S (abc)\, du\, dv\, dw =$

$$abc\int_0^{2\pi}\int_0^{\pi}\int_0^1 (\rho^2 \sin\phi)\, d\rho\, d\phi\, d\theta = \tfrac{4}{3}\pi abc = \tfrac{4}{3}\pi(6378)^2(6356) \approx 1.08 \times 10^{12} \text{ km}^3.$$

[33] $x = \rho\sin\phi\cos\theta$, $y = \rho\sin\phi\sin\theta$, $z = \rho\cos\phi \Rightarrow$

$$J = \frac{\partial(x,\ y,\ z)}{\partial(\rho,\ \phi,\ \theta)} = \begin{vmatrix} \sin\phi\cos\theta & \rho\cos\phi\cos\theta & -\rho\sin\phi\sin\theta \\ \sin\phi\sin\theta & \rho\cos\phi\sin\theta & \rho\sin\phi\cos\theta \\ \cos\phi & -\rho\sin\phi & 0 \end{vmatrix}$$

Evaluating J along the third row gives us

$$J = \cos\phi\,(\rho^2\sin\phi\cos\phi\cos^2\theta + \rho^2\sin\phi\cos\phi\sin^2\theta) +$$
$$\rho\sin\phi\,(\rho\sin^2\phi\cos^2\theta + \rho\sin^2\phi\sin^2\theta)$$
$$= \rho^2\sin\phi\cos^2\phi\,(\cos^2\theta + \sin^2\theta) + \rho^2\sin\phi\sin^2\phi\,(\cos^2\theta + \sin^2\theta)$$
$$= \rho^2\sin\phi\,(\cos^2\phi + \sin^2\phi) = \rho^2\sin\phi.$$

35 $\dfrac{\partial(x,\, y)}{\partial(u,\, v)} \cdot \dfrac{\partial(u,\, v)}{\partial(x,\, y)} = \begin{vmatrix} f_u & f_v \\ g_u & g_v \end{vmatrix} \cdot \begin{vmatrix} u_x & u_y \\ v_x & v_y \end{vmatrix}$ {since $x = f(u,\, v)$ and $y = g(u,\, v)$}

$= \begin{vmatrix} f_u u_x + f_v v_x & f_u u_y + f_v v_y \\ g_u u_x + g_v v_x & g_u u_y + g_v v_y \end{vmatrix}$ {using the hint}

$= \begin{vmatrix} f_x & f_y \\ g_x & g_y \end{vmatrix}$ {by a chain rule, for example,

$f_x = \dfrac{\partial f}{\partial x} = \dfrac{\partial f}{\partial u}\dfrac{\partial u}{\partial x} + \dfrac{\partial f}{\partial v}\dfrac{\partial v}{\partial x} = f_u\, u_x + f_v\, v_x$}

$= \begin{vmatrix} 1 & 0 \\ 0 & 1 \end{vmatrix}$ {$f(u,\, v) = x \Rightarrow f_x = 1,\, f_y = 0;$

$g(u,\, v) = y \Rightarrow g_y = 1,\, g_x = 0$}

$= 1.$

17.10 Review Exercises

1 $\text{I} = \displaystyle\int_{-1}^{0}\int_{x+1}^{x^3} (x^2 - 2y)\, dy\, dx$

$= \displaystyle\int_{-1}^{0} \Big[\, x^2 y - y^2 \,\Big]_{x+1}^{x^3} dx$

$= \displaystyle\int_{-1}^{0} \Big[x^2(x^3 - x - 1) - x^6 + (x + 1)^2 \Big] dx$

$= \displaystyle\int_{-1}^{0} (x^5 - x^3 - x^6 + 2x + 1)\, dx = -\tfrac{5}{84}.$

5 $\text{I} = \displaystyle\int_{0}^{2}\int_{\sqrt{y}}^{1}\int_{z^2}^{y} xy^2z^3\, dx\, dz\, dy$

$= \displaystyle\int_{0}^{2}\int_{\sqrt{y}}^{1} \Big[\, \tfrac{1}{2}x^2y^2z^3 \,\Big]_{z^2}^{y} dz\, dy$

$= \tfrac{1}{2}\displaystyle\int_{0}^{2}\int_{\sqrt{y}}^{1} (y^4z^3 - y^2z^7)\, dz\, dy$

$= \tfrac{1}{2}\displaystyle\int_{0}^{2} \Big[\, \tfrac{1}{4}y^4z^4 - \tfrac{1}{8}y^2z^8 \,\Big]_{\sqrt{y}}^{1} dy = \tfrac{1}{16}\int_{0}^{2} (2y^4 - y^2 - y^6)\, dy = -\tfrac{107}{210}.$

7 R is the region bounded by the right branch of the hyperbola $(x^2/4) - (y^2/4) = 1$ and the vertical line $x = 4$. $x^2 - y^2 = 4$ and $x = 4 \Rightarrow y = \pm\sqrt{x^2 - 4}$ and $x \geq 2$.

Thus, $\iint_R f(x,\, y)\, dA = \displaystyle\int_{2}^{4}\int_{-\sqrt{x^2-4}}^{\sqrt{x^2-4}} f(x,\, y)\, dy\, dx.$

[11] The region is bounded by the graphs of $x = e^y$, $x = y^3$, $y = -1$, and $y = 1$.

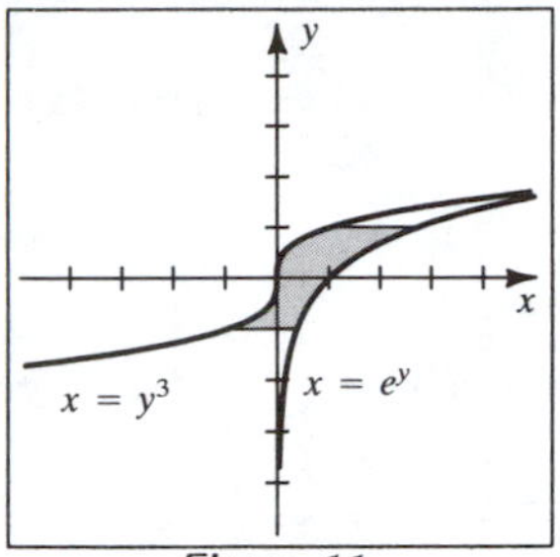

Figure 11

[15] R is the region bounded by $x = y^2$ $\{y = \pm\sqrt{x}\}$ and $x = 9$ from $y = 0$ to 3.

$$\int_0^3\int_{y^2}^9 ye^{-x^2}\,dx\,dy = \int_0^9\int_0^{\sqrt{x}} ye^{-x^2}\,dy\,dx = \tfrac{1}{2}\int_0^9 xe^{-x^2}\,dx = \tfrac{1}{4}(1 - e^{-81}) \approx 0.25.$$

[17] R is determined by $1 \le x \le 2$ and $1 \le y \le 3$. $z = f(x, y) = xy^2 \Rightarrow$

$$V = \iint_R f(x, y)\,dA = \int_1^2\int_1^3 xy^2\,dy\,dx = \tfrac{26}{3}\int_1^2 x\,dx = \tfrac{26}{3}\cdot\tfrac{3}{2} = 13.$$

[19] The trace of the surface described in the xy-plane is the circle $(x - 2)^2 + y^2 = 4$.

$$z = f(x, y) = (x^2 + y^2)^{1/2} \Rightarrow f_x = \frac{x}{(x^2 + y^2)^{1/2}} \text{ and } f_y = \frac{y}{(x^2 + y^2)^{1/2}}.$$

$$S = \iint_R \sqrt{f_x^2 + f_y^2 + 1}\,dA = \iint_R \sqrt{1 + 1}\,dA = \sqrt{2}\iint_R dA = 4\sqrt{2}\,\pi \approx 17.77$$

since the area of R (a circle with radius 2 in the xy-plane) is 4π.

[21] $e^\theta = r = 2$ when $\theta = \ln 2$. By (17.13),

$$A = \int_0^{\ln 2}\int_{e^\theta}^2 r\,dr\,d\theta = \tfrac{1}{2}\int_0^{\ln 2}(4 - e^{2\theta})\,d\theta = 2\ln 2 - \tfrac{3}{4} \approx 0.64.$$

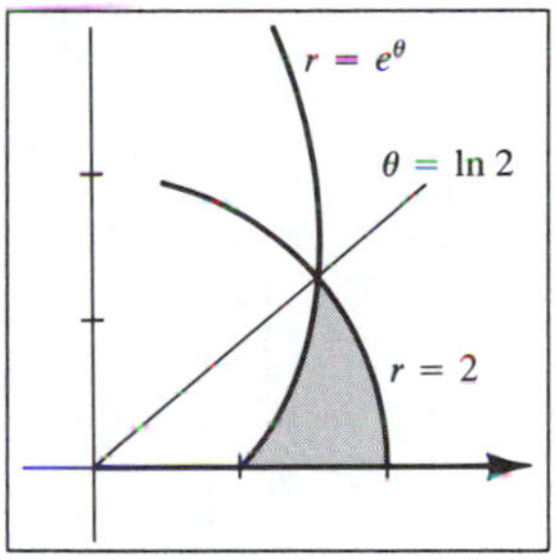

Figure 21

[23] $0 \le y \le \sqrt{16 - x^2}$ and $0 \le x \le 4 \Rightarrow R$ is the quarter of the circle in quadrant I whose boundary is $x^2 + y^2 = 16$. This integral represents the volume between the cone $z = \sqrt{x^2 + y^2}$ and the hemisphere with radius $\sqrt{32}$ and equation $z = \sqrt{32 - x^2 - y^2}$. These intersect when $\sqrt{x^2 + y^2} = \sqrt{32 - x^2 - y^2} \Rightarrow$ $x^2 + y^2 = 16$, or $z = \sqrt{x^2 + y^2} = 4$. By (17.32),

$$\int_0^4\int_0^{\sqrt{16-x^2}}\int_{\sqrt{x^2+y^2}}^{\sqrt{32-x^2-y^2}} \sqrt{x^2 + y^2 + z^2}\,dz\,dy\,dx = \int_0^{\pi/2}\int_0^{\pi/4}\int_0^{4\sqrt{2}} \rho(\rho^2\sin\phi)\,d\rho\,d\phi\,d\theta =$$

$$256\int_0^{\pi/2}\int_0^{\pi/4}\sin\phi\,d\phi\,d\theta = 256\left(1 - \tfrac{\sqrt{2}}{2}\right)\int_0^{\pi/2} d\theta = 64(2 - \sqrt{2})\pi \approx 117.78.$$

Note: Let k denote a constant of proportionality.

[25] $\delta(x, y) = k|x| = kx$ since $x \geq 0$. By (17.24), $m = \int_0^3\int_x^{2x}(kx)\,dy\,dx = k\int_0^3 x^2\,dx = 9k$.

$M_x = \int_0^3\int_x^{2x}(kx)\,y\,dy\,dx = \frac{3}{2}k\int_0^3 x^3\,dx = \frac{243}{8}k$.

$M_y = \int_0^3\int_x^{2x}(kx)\,x\,dy\,dx = k\int_0^3 x^3\,dx = \frac{81}{4}k$. Thus, $\bar{x} = \frac{M_y}{m} = \frac{9}{4}$ and $\bar{y} = \frac{M_x}{m} = \frac{27}{8}$.

[27] $\delta(r, \theta) = \frac{k}{r} \Rightarrow m = \int_0^{2\pi}\int_1^{2+\sin\theta}(\frac{k}{r})\,r\,dr\,d\theta = k\int_0^{2\pi}(1 + \sin\theta)\,d\theta = 2\pi k$.

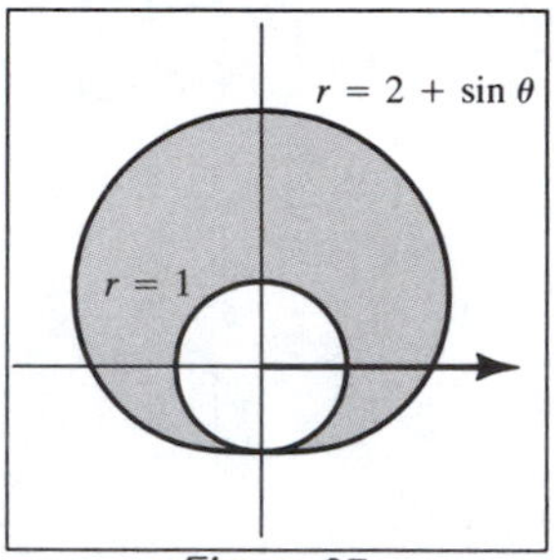

Figure 27

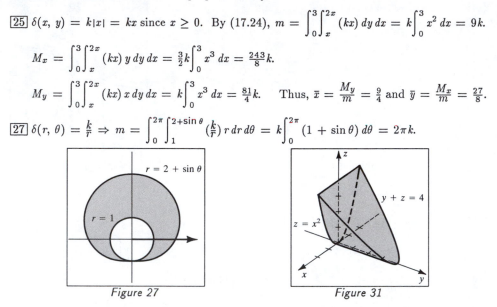

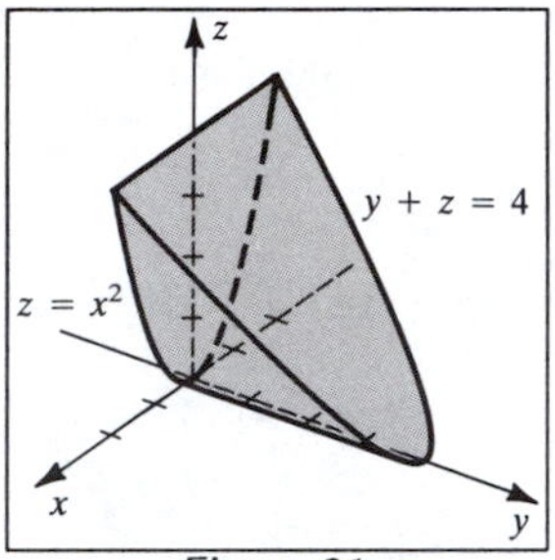

Figure 31

[31] Without loss of generality, let $\delta(x, y, z) = 1$. Let R be the region in the xz-plane bounded by $z = x^2$ and $z = 4$ $\{y + z = 4$ with $y = 0\}$. The solid is between the xz-plane and the plane $y = 4 - z$.

$m = V = \int_{-2}^2\int_{x^2}^4\int_0^{4-z} dy\,dz\,dx = \int_{-2}^2\int_{x^2}^4(4 - z)\,dz\,dx = \frac{1}{2}\int_{-2}^2(4 - x^2)^2\,dx = \frac{256}{15}$.

$M_{xz} = \int_{-2}^2\int_{x^2}^4\int_0^{4-z} y\,dy\,dz\,dx = \frac{1}{2}\int_{-2}^2\int_{x^2}^4(4 - z)^2\,dz\,dx$

$= \frac{1}{6}\int_{-2}^2(4 - x^2)^3\,dx = \frac{1}{6}\int_{-2}^2(64 - 48x^2 + 12x^4 - x^6)\,dx = \frac{2048}{105}$.

$M_{xy} = \int_{-2}^2\int_{x^2}^4\int_0^{4-z} z\,dy\,dz\,dx$

$= \int_{-2}^2\int_{x^2}^4(4z - z^2)\,dz\,dx = \int_{-2}^2(\frac{32}{3} - 2x^4 + \frac{1}{3}x^6)\,dx = \frac{1024}{35}$.

By symmetry, $\bar{x} = 0$. $\bar{y} = \frac{M_{xz}}{m} = \frac{8}{7}$ and $\bar{z} = \frac{M_{xy}}{m} = \frac{12}{7}$.

[35] Since $\delta(\rho, \theta, \phi) = k\rho$, by (17.32),

$m = \iiint_Q \delta(\rho, \phi, \theta)\,dV = \int_0^{2\pi}\int_0^{\pi}\int_0^a(k\rho)(\rho^2\sin\phi)\,d\rho\,d\phi\,d\theta =$

$\frac{1}{4}a^4k\int_0^{2\pi}\int_0^{\pi}\sin\phi\,d\phi\,d\theta = \frac{1}{2}a^4k\int_0^{2\pi}d\theta = \pi a^4k$.

[37] $\rho = 12$, $\phi = \frac{\pi}{6}$, and $\theta = \frac{3\pi}{4} \Rightarrow$

$x = \rho \sin\phi \cos\theta = 12\sin\frac{\pi}{6}\cos\frac{3\pi}{4} = 12(\frac{1}{2})(-\frac{\sqrt{2}}{2}) = -3\sqrt{2}$;

$y = \rho \sin\phi \sin\theta = 12\sin\frac{\pi}{6}\sin\frac{3\pi}{4} = 12(\frac{1}{2})(\frac{\sqrt{2}}{2}) = 3\sqrt{2}$; $z = \rho\cos\phi = 12\cos\frac{\pi}{6} = 12(\frac{\sqrt{3}}{2}) = 6\sqrt{3}$. Rectangular coordinates are $(-3\sqrt{2}, 3\sqrt{2}, 6\sqrt{3})$.

$r = \sqrt{(-3\sqrt{2})^2 + (3\sqrt{2})^2} = 6$. Cylindrical coordinates are $(6, \frac{3\pi}{4}, 6\sqrt{3})$.

[39] $z + 3r^2 = 9 \Rightarrow z = 9 - 3r^2 \Rightarrow z = 9 - 3x^2 - 3y^2$,

a paraboloid with vertex (0, 0, 9) and opening downward.

[41] Squaring both sides of $\rho\sin\phi = 4$ gives us $\rho^2\sin^2\phi = 16 \Rightarrow$

$\rho^2\sin^2\phi(\cos^2\theta + \sin^2\theta) = 16 \Rightarrow (\rho\sin\phi\cos\theta)^2 + (\rho\sin\phi\sin\theta)^2 = 16 \Rightarrow$

$x^2 + y^2 = 16$. This is a right circular cylinder of radius 4 with axis along the z-axis.

[45] (a) By (17.29), $z = x^2 - y^2 \Rightarrow$

$$z = r^2\cos^2\theta - r^2\sin^2\theta = r^2(\cos^2\theta - \sin^2\theta) = r^2\cos 2\theta.$$

(b) By (17.31), $z = x^2 - y^2 \Rightarrow \rho\cos\phi = \rho^2\sin^2\phi\cos^2\theta - \rho^2\sin^2\phi\sin^2\theta \Rightarrow$

$$\cos\phi = \rho\sin^2\phi\cos 2\theta.$$

[49] (a) On $0 \le x \le 4$, $0 \le y \le \sqrt{25 - x^2}$. $\int_0^4\int_0^{\sqrt{25-x^2}} dy\,dx$

(b) On $0 \le y \le 3$, $0 \le x \le 4$. On $3 \le y \le 5$, $0 \le x \le \sqrt{25 - y^2}$.

$$\int_0^3\int_0^4 dx\,dy + \int_3^5\int_0^{\sqrt{25-y^2}} dx\,dy$$

(c) For $0 \le \theta \le \arctan\frac{3}{4}$, $0 \le r \le 4\sec\theta$. For $\arctan\frac{3}{4} \le \theta \le \frac{\pi}{2}$, $0 \le r \le 5$.

$$\int_0^{\arctan(3/4)}\int_0^{4\sec\theta} r\,dr\,d\theta + \int_{\arctan(3/4)}^{\pi/2}\int_0^5 r\,dr\,d\theta$$

51 (a) We will subtract the volume above the plane $z = 3$ from the volume of the sphere $x^2 + y^2 + z^2 = 25$ (all in the first octant). The plane and the sphere intersect when $x^2 + y^2 + 3^2 = 25$, or $x^2 + y^2 = 16$.

$$V = \int_0^5 \int_0^{\sqrt{25-x^2}} \int_0^{\sqrt{25-x^2-y^2}} dz\,dy\,dx - \int_0^4 \int_0^{\sqrt{16-x^2}} \int_3^{\sqrt{25-x^2-y^2}} dz\,dy\,dx.$$

An alternate solution would be to treat R as the region in the xz-plane bounded by $0 \le x \le \sqrt{25 - z^2}$ and $0 \le z \le 3$. Then, $0 \le y \le \sqrt{25 - x^2 - z^2}$ and

$$V = \int_0^3 \int_0^{\sqrt{25-z^2}} \int_0^{\sqrt{25-x^2-z^2}} dy\,dx\,dz.$$

(b) We will add the volume inside the right circular cylinder $r = 4$ between $z = 0$ and $z = 3$ to the volume outside the cylinder and inside the hemisphere $z = \sqrt{25 - r^2}$.

$$\int_0^{\pi/2} \int_0^4 \int_0^3 r\,dz\,dr\,d\theta + \int_0^{\pi/2} \int_4^5 \int_0^{\sqrt{25-r^2}} r\,dz\,dr\,d\theta$$

(c) In the yz-plane, the plane intersects the sphere at (0, 4, 3).
Thus, $\tan\phi = \frac{4}{3}$ (*not* $\frac{3}{4}$), and $\phi = \arctan\frac{4}{3}$. We will add the volume bounded by $\rho = 3\sec\phi$ $\{z = 3\}$ to that bounded by $\rho = 5$.

$$\int_0^{\pi/2} \int_0^{\arctan(4/3)} \int_0^{3\sec\phi} (\rho^2 \sin\phi)\,d\rho\,d\phi\,d\theta + \int_0^{\pi/2} \int_{\arctan(4/3)}^{\pi/2} \int_0^5 (\rho^2 \sin\phi)\,d\rho\,d\phi\,d\theta$$

53 The region is bounded by $x + y = 2$, $x = y$, and $y = 0$, which forms a triangle with vertices (0, 0), (2, 0), and (1, 1). They are transformed into a triangle in the uv-plane with vertices (0, 0), (2, 2), and (0, 2) in a positive direction using the equations $u = x - y$ and $v = x + y$. Solving these equations for x and y gives us $x = \frac{1}{2}(u + v)$ and $y = -\frac{1}{2}(u - v) \Rightarrow J = \frac{1}{2}$.

$$\int_0^1 \int_y^{2-y} e^{(x-y)/(x+y)}\,dx\,dy = \int_0^2 \int_0^v e^{u/v}\left(\tfrac{1}{2}\right) du\,dv = \tfrac{1}{2}(e - 1)\int_0^2 v\,dv = e - 1 \approx 1.72.$$

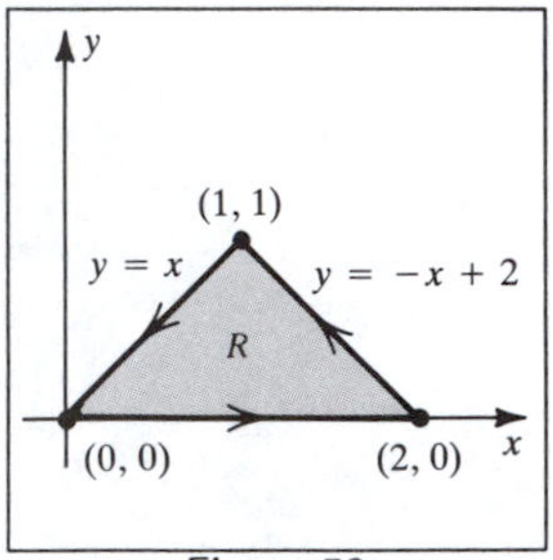

Figure 53a

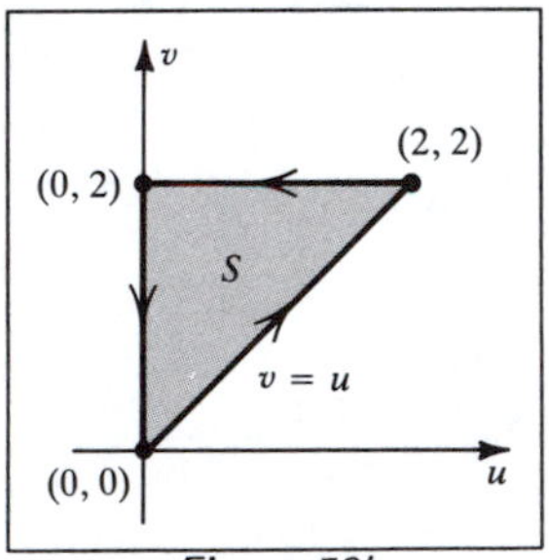

Figure 53b

Chapter 18: Vector Calculus

Exercises 18.1

[1] To obtain the vectors shown in *Figure 1*, substitute values for x and y into the function $\mathbf{F}(x, y) = x\mathbf{i} - y\mathbf{j}$. For example, $\mathbf{F}(1, 1) = 1\mathbf{i} - 1\mathbf{j} = \mathbf{i} - \mathbf{j}$. Sketch the vector $\mathbf{i} - \mathbf{j}$ with initial point $(1, 1)$ and terminal point $(1 + 1, 1 - 1) = (2, 0)$. $\mathbf{F}(2, -3) = 2\mathbf{i} - (-3)\mathbf{j} = 2\mathbf{i} + 3\mathbf{j}$. Sketch the vector $2\mathbf{i} + 3\mathbf{j}$ with initial point $(2, -3)$ and terminal point $(2 + 2, -3 - (-3)) = (4, 0)$.

In general, $\mathbf{F}(x, y) = x\mathbf{i} - y\mathbf{j}$ has initial point (x, y) and terminal point $(x + x, y - y) = (2x, 0)$. Hence, all terminal points are on the x-axis.

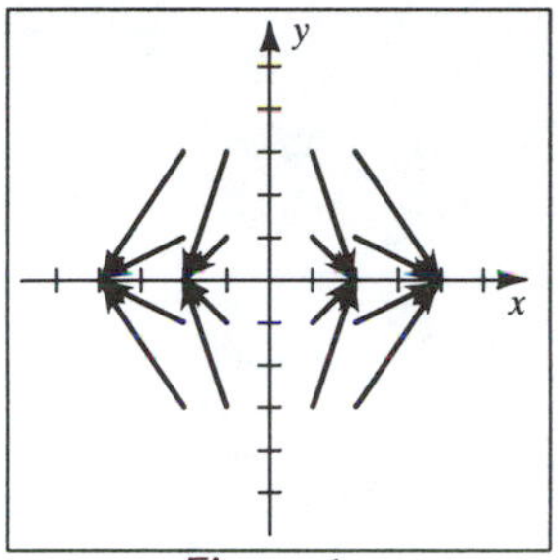

Figure 1

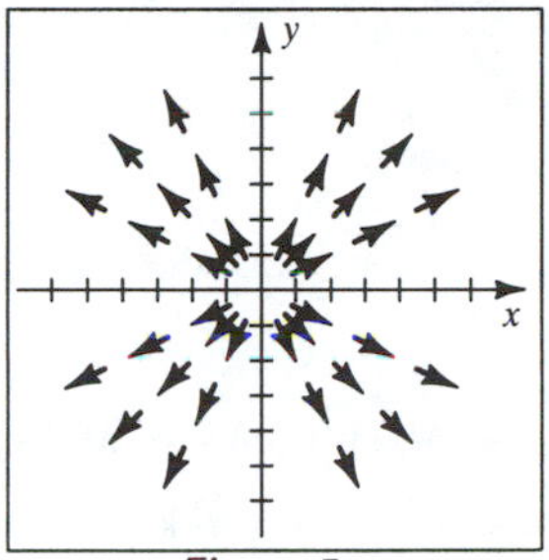

Figure 5

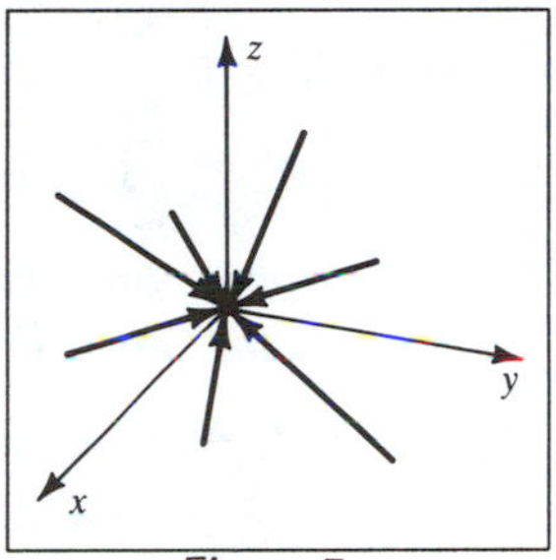

Figure 7

[5] $\mathbf{F}(x, y) = (x^2 + y^2)^{-1/2}(x\mathbf{i} + y\mathbf{j}) = \dfrac{x}{\sqrt{x^2 + y^2}}\mathbf{i} + \dfrac{y}{\sqrt{x^2 + y^2}}\mathbf{j} \Rightarrow$

$$\|\mathbf{F}(x, y)\| = \sqrt{\frac{x^2}{x^2 + y^2} + \frac{y^2}{x^2 + y^2}} = \sqrt{\frac{x^2 + y^2}{x^2 + y^2}} = \sqrt{1} = 1.$$

Hence, all vectors are unit vectors pointing away from the origin.

[7] The position vector $\mathbf{r} = x\mathbf{i} + y\mathbf{j} + z\mathbf{k}$ has initial point at the origin and terminal point at (x, y, z). Vectors of this form point away from the origin.

The vector field $\mathbf{F}(x, y, z) = -x\mathbf{i} - y\mathbf{j} - z\mathbf{k} = -(x\mathbf{i} + y\mathbf{j} + z\mathbf{k})$ includes all vectors with initial point (x, y, z) and terminal point at the origin. Vectors of this form point toward the origin.

[11] Using (18.3) with $f(x, y, z) = x^2 - 3y^2 + 4z^2$,

$$\begin{aligned}\mathbf{F}(x, y, z) &= \nabla f(x, y, z)\\ &= f_x(x, y, z)\,\mathbf{i} + f_y(x, y, z)\,\mathbf{j} + f_z(x, y, z)\,\mathbf{k}\\ &= 2x\mathbf{i} - 6y\mathbf{j} + 8z\mathbf{k}.\end{aligned}$$

$\mathbf{F}$ is a conservative vector field.

15 Using (18.5) with $M = x^2z$, $N = y^2x$, and $P = y + 2z$,

$\mathbf{F}(x, y, z) = x^2z\mathbf{i} + y^2x\mathbf{j} + (y + 2z)\mathbf{k} \Rightarrow$

$\operatorname{curl}\mathbf{F} = \nabla \times \mathbf{F}$

$$= \left(\frac{\partial P}{\partial y} - \frac{\partial N}{\partial z}\right)\mathbf{i} + \left(\frac{\partial M}{\partial z} - \frac{\partial P}{\partial x}\right)\mathbf{j} + \left(\frac{\partial N}{\partial x} - \frac{\partial M}{\partial y}\right)\mathbf{k}$$

$$= (1 - 0)\mathbf{i} + (x^2 - 0)\mathbf{j} + (y^2 - 0)\mathbf{k} = \mathbf{i} + x^2\mathbf{j} + y^2\mathbf{k}.$$

By (18.7), $\operatorname{div}\mathbf{F} = \nabla \cdot \mathbf{F} = \dfrac{\partial M}{\partial x} + \dfrac{\partial N}{\partial y} + \dfrac{\partial P}{\partial z} = 2xz + 2xy + 2.$

Note: A common mistake is to confuse vectors and scalars when working with curl and div. Remember, *curl* results in a *vector*, whereas *div* results in a *scalar*.

17 Using (18.6) with $M = 3xyz^2$, $N = y^2 \sin z$, and $P = xe^{2z}$,

$\mathbf{F}(x, y, z) = 3xyz^2\mathbf{i} + y^2 \sin z\mathbf{j} + xe^{2z}\mathbf{k} \Rightarrow$

$\operatorname{curl}\mathbf{F} = \nabla \times \mathbf{F}$

$$= \begin{vmatrix} \mathbf{i} & \mathbf{j} & \mathbf{k} \\ \dfrac{\partial}{\partial x} & \dfrac{\partial}{\partial y} & \dfrac{\partial}{\partial z} \\ M & N & P \end{vmatrix}$$

$$= (P_y - N_z)\mathbf{i} - (P_x - M_z)\mathbf{j} + (N_x - M_y)\mathbf{k}$$

$$= -y^2 \cos z\mathbf{i} + (6xyz - e^{2z})\mathbf{j} - 3xz^2\mathbf{k}.$$

$\operatorname{div}\mathbf{F} = \nabla \cdot \mathbf{F} = M_x + N_y + P_z = 3yz^2 + 2y \sin z + 2xe^{2z}.$

19 (a) $\mathbf{r} = x\mathbf{i} + y\mathbf{j} + z\mathbf{k}\ \{M = x,\ N = y,\ P = z\} \Rightarrow$

$$\nabla \cdot \mathbf{r} = M_x + N_y + P_z = 1 + 1 + 1 = 3.$$

(b) $\nabla \times \mathbf{r} = (P_y - N_z)\mathbf{i} - (P_x - M_z)\mathbf{j} + (N_x - M_y)\mathbf{k}$

$$= (0 - 0)\mathbf{i} - (0 - 0)\mathbf{j} + (0 - 0)\mathbf{k} = \mathbf{0}.$$

(c) $\|\mathbf{r}\| = (x^2 + y^2 + z^2)^{1/2} \Rightarrow \nabla\|\mathbf{r}\|$

$$= \frac{\partial}{\partial x}(x^2 + y^2 + z^2)^{1/2}\mathbf{i} + \frac{\partial}{\partial y}(x^2 + y^2 + z^2)^{1/2}\mathbf{j} + \frac{\partial}{\partial z}(x^2 + y^2 + z^2)^{1/2}\mathbf{k}$$

$$= \frac{x}{(x^2 + y^2 + z^2)^{1/2}}\mathbf{i} + \frac{y}{(x^2 + y^2 + z^2)^{1/2}}\mathbf{j} + \frac{z}{(x^2 + y^2 + z^2)^{1/2}}\mathbf{k}$$

$$= \frac{1}{\|\mathbf{r}\|}(x\mathbf{i} + y\mathbf{j} + z\mathbf{k}) = \frac{\mathbf{r}}{\|\mathbf{r}\|}.$$

Note: In Exercises 23-26, let $\mathbf{F} = M\mathbf{i} + N\mathbf{j} + P\mathbf{k}$ and $\mathbf{G} = Q\mathbf{i} + R\mathbf{j} + S\mathbf{k}$.

[23] $\nabla \times (\mathbf{F} + \mathbf{G}) = \nabla \times \left[(M+Q)\mathbf{i} + (N+R)\mathbf{j} + (P+S)\mathbf{k}\right]$

$$= \begin{vmatrix} \mathbf{i} & \mathbf{j} & \mathbf{k} \\ \dfrac{\partial}{\partial x} & \dfrac{\partial}{\partial y} & \dfrac{\partial}{\partial z} \\ M+Q & N+R & P+S \end{vmatrix}$$

$$= \left[(P+S)_y - (N+R)_z\right]\mathbf{i} - \left[(P+S)_x - (M+Q)_z\right]\mathbf{j} + \left[(N+R)_x - (M+Q)_y\right]\mathbf{k}$$

$$= \left[(P_y + S_y) - (N_z + R_z)\right]\mathbf{i} + \left[(M_z + Q_z) - (P_x + S_x)\right]\mathbf{j} + \left[(N_x + R_x) - (M_y + Q_y)\right]\mathbf{k}$$

$$= \left[(P_y - N_z)\mathbf{i} + (M_z - P_x)\mathbf{j} + (N_x - M_y)\mathbf{k}\right] + \left[(S_y - R_z)\mathbf{i} + (Q_z - S_x)\mathbf{j} + (R_x - Q_y)\mathbf{k}\right]$$

$$= \begin{vmatrix} \mathbf{i} & \mathbf{j} & \mathbf{k} \\ \dfrac{\partial}{\partial x} & \dfrac{\partial}{\partial y} & \dfrac{\partial}{\partial z} \\ M & N & P \end{vmatrix} + \begin{vmatrix} \mathbf{i} & \mathbf{j} & \mathbf{k} \\ \dfrac{\partial}{\partial x} & \dfrac{\partial}{\partial y} & \dfrac{\partial}{\partial z} \\ Q & R & S \end{vmatrix}$$

$$= \nabla \times \mathbf{F} + \nabla \times \mathbf{G}$$

When verifying identities, you may find it necessary to "look ahead" to the other side to determine your next step. In fact, you may want to work backwards to a common expression.

[25] $\nabla \times (f\mathbf{F}) = \nabla \times \left[(fM)\mathbf{i} + (fN)\mathbf{j} + (fP)\mathbf{k}\right]$

$$= \left[\frac{\partial}{\partial y}(fP) - \frac{\partial}{\partial z}(fN)\right]\mathbf{i} + \left[\frac{\partial}{\partial z}(fM) - \frac{\partial}{\partial x}(fP)\right]\mathbf{j} + \left[\frac{\partial}{\partial x}(fN) - \frac{\partial}{\partial y}(fM)\right]\mathbf{k}$$

$$= \left[(fP_y + f_y P) - (fN_z + f_z N)\right]\mathbf{i} + \left[(fM_z + f_z M) - (fP_x + f_x P)\right]\mathbf{j} + \left[(fN_x + f_x N) - (fM_y + f_y M)\right]\mathbf{k}$$

$$= f\left[(P_y - N_z)\mathbf{i} + (M_z - P_x)\mathbf{j} + (N_x - M_y)\mathbf{k}\right] + \left[(f_y P - f_z N)\mathbf{i} + (f_z M - f_x P)\mathbf{j} + (f_x N - f_y M)\mathbf{k}\right]$$

$$= f(\nabla \times \mathbf{F}) + (\nabla f) \times \mathbf{F}$$

[27] $\operatorname{curl} \operatorname{grad} f = \nabla \times (\nabla f)$

$$= \nabla \times (f_x\mathbf{i} + f_y\mathbf{j} + f_z\mathbf{k})$$

$$= \left[\frac{\partial}{\partial y}(f_z) - \frac{\partial}{\partial z}(f_y)\right]\mathbf{i} - \left[\frac{\partial}{\partial x}(f_z) - \frac{\partial}{\partial z}(f_x)\right]\mathbf{j} + \left[\frac{\partial}{\partial x}(f_y) - \frac{\partial}{\partial y}(f_x)\right]\mathbf{k}$$

$$= (f_{zy} - f_{yz})\mathbf{i} + (f_{xz} - f_{zx})\mathbf{j} + (f_{yx} - f_{xy})\mathbf{k}$$

$$= 0\mathbf{i} + 0\mathbf{j} + 0\mathbf{k} = \mathbf{0}.$$

[33] $f(x, y, z) = (x^2 + y^2 + z^2)^{-1/2} \Rightarrow$

$$f_x = -\tfrac{1}{2}(x^2 + y^2 + z^2)^{-3/2}(2x) = \frac{-x}{(x^2 + y^2 + z^2)^{3/2}} \Rightarrow$$

$$f_{xx} = \frac{(x^2 + y^2 + z^2)^{3/2}(-1) - (-x)(\frac{3}{2})(x^2 + y^2 + z^2)^{1/2}(2x)}{(x^2 + y^2 + z^2)^3}$$

$$= \frac{(x^2 + y^2 + z^2)^{1/2}\left[-(x^2 + y^2 + z^2)^1 + 3x^2\right]}{(x^2 + y^2 + z^2)^3}$$

$$= \frac{-(x^2 + y^2 + z^2) + 3x^2}{(x^2 + y^2 + z^2)^{5/2}} = \frac{2x^2 - y^2 - z^2}{(x^2 + y^2 + z^2)^{5/2}}.$$

Similarly, $f_{yy} = \dfrac{2y^2 - x^2 - z^2}{(x^2 + y^2 + z^2)^{5/2}}$ and $f_{zz} = \dfrac{2z^2 - x^2 - y^2}{(x^2 + y^2 + z^2)^{5/2}}$.

Thus, $f_{xx} + f_{yy} + f_{zz} = 0$, which is Laplace's equation, $\nabla^2 f = 0$.

[35] Define $\lim_{(x,y,z)\to(x_0,y_0,z_0)} \mathbf{F}(x, y, z) = \mathbf{a}$ to be $\left[\lim_{(x,y,z)\to(x_0,y_0,z_0)} M(x, y, z)\right]\mathbf{i} +$ $\left[\lim_{(x,y,z)\to(x_0,y_0,z_0)} N(x, y, z)\right]\mathbf{j} + \left[\lim_{(x,y,z)\to(x_0,y_0,z_0)} P(x, y, z)\right]\mathbf{k} =$ $u\mathbf{i} + v\mathbf{j} + w\mathbf{k}$, where $\mathbf{a} = u\mathbf{i} + v\mathbf{j} + w\mathbf{k}$. Thus, $\lim_{(x,y,z)\to(x_0,y_0,z_0)} \mathbf{F}(x, y, z) = \mathbf{a}$ means that for every $\epsilon > 0$, $\exists\delta > 0$, such that $\|\mathbf{F}(x, y, z) - \mathbf{a}\| < \epsilon$ whenever $0 < \|\langle x, y, z\rangle - \langle x_0, y_0, z_0\rangle\| = \sqrt{(x - x_0)^2 + (y - y_0)^2 + (z - z_0)^2} < \delta$. Geometrically this means that as the point (x, y, z) approaches (x_0, y_0, z_0), $\mathbf{F}(x, y, z)$ has nearly the same magnitude and direction as $\mathbf{a}$.

Exercises 18.2

[1] (1) Let $x = g(t) = 3t$ and $y = h(t) = t^3$ for $0 \le t \le 1$.

Then, $ds = \sqrt{[g'(t)]^2 + [h'(t)]^2}\,dt = \sqrt{3^2 + (3t^2)^2}\,dt = 3\sqrt{1 + t^4}\,dt$.

$x^3 + y = (3t)^3 + t^3 = 28t^3$.

By (18.9)(i), $\int_C f(x, y)\,ds = \int_0^1 (28t^3)(3\sqrt{1 + t^4})\,dt$ $\{u = 1 + t^4, \frac{1}{4}\,du = t^3\,dt\}$

$$= 84\left[\tfrac{1}{6}(1 + t^4)^{3/2}\right]_0^1 dt = 14(2^{3/2} - 1) \approx 25.60.$$

(2) $dx = g'(t)\,dt = 3\,dt$.

By (18.9)(ii), $\int_C f(x, y)\,dx = \int_0^1 (28t^3)\,3\,dt = 84\left[\tfrac{1}{4}t^4\right]_0^1 = 21$.

(3) $dy = h'(t)\,dt = 3t^2\,dt$.

By (18.9)(iii), $\int_C f(x, y)\,dy = \int_0^1 (28t^3)(3t^2)\,dt = 84\left[\tfrac{1}{6}t^6\right]_0^1 = 14$.

3 Let $y = g(x)$. Then, $y = x^3 + 1 \Rightarrow dy = g'(x)\,dx = 3x^2\,dx$ with $-1 \le x \le 1$.

We will bypass the parametric equations in t and write the line integral in terms of x.

$$\int_C 6x^2y\,dx + xy\,dy = \int_{-1}^{1}\left[6x^2(x^3+1) + x(x^3+1)(3x^2)\right]dx$$
$$= 3\int_{-1}^{1}(x^6 + 2x^5 + x^3 + 2x^2)\,dx \;\{2x^5 \text{ and } x^3 \text{ are odd}\}$$
$$= 6\int_0^1 (x^6 + 2x^2)\,dx = \tfrac{34}{7}.$$

Note: Let I denote the required line integral.

7 (a) First, write both line segments parametrically.

C_1: $x = t$, $y = 0$, $0 \le t \le 1 \Rightarrow dx = dt$ and $dy = 0\,dt$.

C_2: $x = 1$, $y = t$, $0 \le t \le 3 \Rightarrow dx = 0\,dt$ and $dy = dt$.

$$\int_C xy\,dx + (x+y)\,dy = \int_{C_1} xy\,dx + (x+y)\,dy + \int_{C_2} xy\,dx + (x+y)\,dy$$
$$= \int_0^1 (0 + t\cdot 0)\,dt + \int_0^3 \left[t\cdot 0 + (1+t)\right]dt$$
$$= \int_0^1 0\,dt + \int_0^3 (t+1)\,dt$$
$$= 0 + \tfrac{15}{2} = \tfrac{15}{2}.$$

(b) C_1: $x = 0$, $y = t$, $0 \le t \le 3 \Rightarrow dx = 0\,dt$ and $dy = dt$.

C_2: $x = t$, $y = 3$, $0 \le t \le 1 \Rightarrow dx = dt$ and $dy = 0\,dt$.

$$\text{I} = \int_0^3 \left[0 + (0+t)\right]dt + \int_0^1 \left[3t + (t+3)\cdot 0\right]dt = \tfrac{9}{2} + \tfrac{3}{2} = 6.$$

(c) The line segment from (0, 0) to (1, 3) can be described by $y = 3x$ $\{dy = 3\,dx\}$

with $0 \le x \le 1$. $\text{I} = \int_0^1 \left[x(3x) + (x+3x)\cdot 3\right]dx = 3\int_0^1 (x^2 + 4x)\,dx = 7.$

(d) $y = 3x^2 \Rightarrow dy = 6x\,dx$.

$$\text{I} = \int_0^1 \left[x(3x^2) + (x+3x^2)(6x)\right]dx = 3\int_0^1 (7x^3 + 2x^2)\,dx = \tfrac{29}{4}.$$

9 This exercise is similar to Exercise 3 except that it uses *three* functions of t.

We will convert the line integral to an integral involving only the variable t.

$x = e^t$, $y = e^{-t}$, $z = e^{2t} \Rightarrow dx = e^t\,dt$, $dy = -e^{-t}\,dt$, and $dz = 2e^{2t}\,dt \Rightarrow$

$$\int_C xz\,dx + (y+z)\,dy + x\,dz = \int_0^1 \left[e^{3t}(e^t) + (e^{-t} + e^{2t})(-e^{-t}) + e^t(2e^{2t})\right]dt =$$
$$\int_0^1 (e^{4t} - e^{-2t} - e^t + 2e^{3t})\,dt = \tfrac{1}{12}(3e^4 + 6e^{-2} - 12e + 8e^3 - 5) \approx 23.97.$$

11 (a) First, write each of the three line segments parametrically.

C_1: $x = t$, $y = z = 0 \Rightarrow dx = dt$, $dy = dz = 0\,dt$. $0 \le t \le 2$.

C_2: $x = 2$, $y = t$, $z = 0 \Rightarrow dy = dt$, $dx = dz = 0\,dt$. $0 \le t \le 3$.

C_3: $x = 2$, $y = 3$, $z = t \Rightarrow dz = dt$, $dx = dy = 0\,dt$. $0 \le t \le 4$.

$\int_C (x + y + z)\,dx + (x - 2y + 3z)\,dy + (2x + y - z)\,dz =$

$$\int_0^2 t\,dt + \int_0^3 (2 - 2t)\,dt + \int_0^4 (7 - t)\,dt = 2 - 3 + 20 = 19.$$

(b) C_1: $x = 0$, $y = 0$, $z = t \Rightarrow dz = dt$, $dx = dy = 0\,dt$. $0 \le t \le 4$.

C_2: $x = t$, $y = 0$, $z = 4 \Rightarrow dx = dt$, $dy = dz = 0\,dt$. $0 \le t \le 2$.

C_3: $x = 2$, $y = t$, $z = 4 \Rightarrow dy = dt$, $dx = dz = 0\,dt$. $0 \le t \le 3$.

$$\mathrm{I} = \int_0^4 (-t)\,dt + \int_0^2 (t + 4)\,dt + \int_0^3 (14 - 2t)\,dt = -8 + 10 + 33 = 35.$$

(c) A line segment C between $(0, 0, 0)$ and $(2, 3, 4)$ can be described parametrically by $x = 2t$, $y = 3t$, and $z = 4t$ with $0 \le t \le 1$. $\{ dx = 2\,dt$, $dy = 3\,dt$,

$$dz = 4\,dt\}\ \mathrm{I} = \int_0^1 \big[(9t)(2) + (8t)(3) + (3t)(4)\big]\,dt = \int_0^1 54t\,dt = 27.$$

15 Refer to Exercise 7 for the parametric equations.

For each part, $\int_C \mathbf{F} \cdot d\mathbf{r} = \int_C xy^2\,dx + x^2y\,dy$.

(a) $$W = \int_0^1 \Big[\, (t \cdot 0^2) + (t^2 \cdot 0)0 \,\Big]\,dt + \int_0^3 \Big[\, (1 \cdot t^2)0 + (1^2 \cdot t) \,\Big]\,dt$$

$$= \int_0^1 0\,dt + \int_0^3 t\,dt = 0 + \tfrac{9}{2} = \tfrac{9}{2}.$$

(b) Similarly, $W = \int_0^3 (0 + 0)\,dt + \int_0^1 (9t + 3t^2 \cdot 0)\,dt = \tfrac{9}{2}$.

(c) $W = \int_0^1 \Big[\, x(3x)^2 + x^2(3x)\,3 \,\Big]\,dx = \int_0^1 18x^3\,dx = \tfrac{9}{2}$.

(d) $W = \int_0^1 \Big[\, x(3x^2)^2 + x^2(3x^2)(6x) \,\Big]\,dx = \int_0^1 27x^5\,dx = \tfrac{9}{2}$.

19 Let $\mathbf{r} = x\mathbf{i} + y\mathbf{j} + z\mathbf{k}$ and then $d\mathbf{r} = dx\mathbf{i} + dy\mathbf{j} + dz\mathbf{k}$. By (18.12) with $\mathbf{F}(x, y, z) = y\mathbf{i} + z\mathbf{j} + x\mathbf{k}$, $W = \int_C \mathbf{F} \cdot d\mathbf{r} = \int_C y\,dx + z\,dy + x\,dz$. With $x = t$, $y = t^2$, and $z = t^3$, we have $dx = dt$, $dy = 2t\,dt$, and $dz = 3t^2\,dt$. Since $0 \le t \le 2$,

$$W = \int_0^2 \Big[\, (t^2)(1) + (t^3)(2t) + (t)(3t^2) \,\Big]\,dt = \int_0^2 (2t^4 + 3t^3 + t^2)\,dt = \tfrac{412}{15}.$$

[23] Let the wire be represented by a curve C as in Figure 18.10. Then the mass of the piece of wire between P_{k-1} and P_k can be considered concentrated at (u_k, v_k). Its mass is approximately $\delta(u_k, v_k)\Delta s_k$ and its moment with respect to the x-axis is approximately $v_k\,\delta(u_k, v_k)\Delta s_k$ and with respect to the y-axis is approximately $u_k\delta(u_k, v_k)\Delta s_k$. Using the limit of sums,

$$M_x = \lim_{\|P\|\to 0}\sum_k v_k\,\delta(u_k, v_k)\Delta s_k = \int_C y\,\delta(x, y)\,ds \text{ and}$$

$$M_y = \lim_{\|P\|\to 0}\sum_k u_k\,\delta(u_k, v_k)\Delta s_k = \int_C x\,\delta(x, y)\,ds.$$

From Example 3, $x = a\cos t$, $y = a\sin t$, $\delta(x, y) = ky$, and $ds = a\,dt$.

$$M_x = \int_C y\,\delta(x, y)\,ds = \int_0^\pi (a\sin t)(ka\sin t)\,a\,dt = \tfrac{1}{2}\pi ka^3 \text{ and}$$

$$M_y = \int_C x\,\delta(x, y)\,ds = \int_0^\pi (a\cos t)(ka\sin t)\,a\,dt = 0.$$

Thus, $\bar{x} = \dfrac{M_y}{m} = 0$ and $\bar{y} = \dfrac{M_x}{m}\ \{m = 2ka^2 \text{ from Example 3}\} = \tfrac{1}{4}\pi a$.

[27] In a manner similar to Exercise 23, define

$I_x = \int_C y^2\,\delta(x, y)\,ds$ and $I_y = \int_C x^2\,\delta(x, y)\,ds$. For the wire in Example 3,

$$I_x = \int_0^\pi (a\sin t)^2(ka\sin t)\,a\,dt = ka^4\int_0^\pi (1 - \cos^2 t)\sin t\,dt = \tfrac{4}{3}ka^4 \text{ and}$$

$$I_y = \int_0^\pi (a\cos t)^2(ka\sin t)\,a\,dt = ka^4\int_0^\pi \cos^2 t\sin t\,dt = \tfrac{2}{3}ka^4.$$

[29] If the density at (x, y, z) is $\delta(x, y, z)$, then the square of the distance from the x-axis is given by $y^2 + z^2$ and in a manner similar to Exercises 23 and 27,

$I_x = \int_C (y^2 + z^2)\,\delta(x, y, z)\,ds$.

Also, $I_y = \int_C (x^2 + z^2)\,\delta(x, y, z)\,ds$, and $I_z = \int_C (x^2 + y^2)\,\delta(x, y, z)\,ds$.

[31] $u_k = \frac{1}{10}k - \frac{1}{20} \Rightarrow v_k = (\frac{1}{10}k - \frac{1}{20})^4 \Rightarrow u_k v_k = (\frac{1}{10}k - \frac{1}{20})^5$.

$$\sum_{k=1}^{10}\sin(u_k v_k)\,\Delta x_k = \sum_{k=1}^{10}\sin(\tfrac{1}{10}k - \tfrac{1}{20})^5(\tfrac{1}{10}) \approx 0.1554.$$

Exercises 18.3

[1] By (18.13), $\int_C \mathbf{F}\cdot d\mathbf{r}$ is independent of path if and only if there exists a scalar function f such that $\nabla f = \mathbf{F}$, or, in this case,

$$f_x\,\mathbf{i} + f_y\,\mathbf{j} = (3x^2y + 2)\mathbf{i} + (x^3 + 4y^3)\mathbf{j}.$$

$f_x = 3x^2y + 2 \Rightarrow f = \int(3x^2y + 2)\,dx = x^3y + 2x + g(y)$, where $g(y)$ is a function of y only, and hence $D_x[g(y)] = 0$. Differentiating this expression for f with respect to y gives $f_y = x^3 + g'(y) = x^3 + 4y^3$ {j-component of F}. Thus, by comparing terms, we see that $g'(y) = 4y^3$ and $g(y) = \int 4y^3\,dy = y^4 + c$, where c is a scalar constant. So $f(x, y) = x^3y + 2x + y^4 + c$.

[3] $f_x\,\mathbf{i} + f_y\,\mathbf{j} = (2x\sin y + 4e^x)\mathbf{i} + x^2\cos y\,\mathbf{j} \Rightarrow f_x = 2x\sin y + 4e^x \Rightarrow$

$f = \int(2x\sin y + 4e^x)\,dx = x^2\sin y + 4e^x + g(y) \Rightarrow$

$f_y = x^2\cos y + g'(y) = x^2\cos y$ {j-component of $\mathbf{F}$}.

Thus, $g'(y) = 0$ and $g(y) = c$. So $f(x, y) = x^2\sin y + 4e^x + c$.

[7] $\nabla f(x, y, z) = \mathbf{F}(x, y, z) \Rightarrow f_x\,\mathbf{i} + f_y\,\mathbf{j} + f_z\,\mathbf{k} = 8xz\mathbf{i} + (1 - 6yz^3)\mathbf{j} + (4x^2 - 9y^2z^2)\mathbf{k}$ $\Rightarrow f_x = 8xz \Rightarrow f = \int 8xz\,dx = 4x^2z + g(y, z)$, where $g(y, z)$ is a function of only y and z, and hence, $D_x[g(y, z)] = 0$. Differentiating this expression for f with respect to y gives $f_y = g_y(y, z) = 1 - 6yz^3$ {j-component of $\mathbf{F}$}. Thus,

$g(y, z) = \int(1 - 6yz^3)\,dy = y - 3y^2z^3 + k(z)$, where $k(z)$ is a function of z only.

Hence, $f = 4x^2z + g(y, z) = 4x^2z + y - 3y^2z^3 + k(z)$.

$f_z = 4x^2 - 9y^2z^2 + k'(z) = 4x^2 - 9y^2z^2$ {k-component of $\mathbf{F}$} $\Rightarrow$

$k'(z) = 0$ and $k(z) = c$. So $f(x, y, z) = 4x^2z + y - 3y^2z^3 + c$.

Note: In Exercises 11-14, we will show that the line integral is independent of path by finding a scalar function f such that $\mathbf{F} = \nabla f$, and then use (18.14) to evaluate the given integral I.

[11] $\nabla f = \mathbf{F}(x, y) \Rightarrow f_x\,\mathbf{i} + f_y\,\mathbf{j} = (y^2 + 2xy)\mathbf{i} + (x^2 + 2xy)\mathbf{j} \Rightarrow f_x = y^2 + 2xy \Rightarrow$ $f = xy^2 + x^2y + g(y) \Rightarrow f_y = 2xy + x^2 + g'(y) = x^2 + 2xy$. So $g'(y) = 0$ and $g(y) = c$. Thus, let $f(x, y) = xy^2 + x^2y$ {we may omit "$+ c$"}. f is a scalar function such that $\mathbf{F} = \nabla f$. Since the line integral is independent of path, we only need to evaluate $f(x, y)$ at the given end points and subtract these values to find the value of the given integral. $\mathrm{I} = f(3, 1) - f(-1, 2) = 12 - (-2) = 14$.

[13] $f_x = 6xy^3 + 2z^2 \Rightarrow f = 3x^2y^3 + 2xz^2 + g(y, z) \Rightarrow$

$f_y = 9x^2y^2 + g_y(y, z) = 9x^2y^2$. So $g_y(y, z) = 0$ and $g(y, z) = k(z)$.

$f_z = 4xz + k'(z) = 4xz + 1 \Rightarrow k'(z) = 1$ and $k(z) = z + c$. Thus, let

$f(x, y, z) = 3x^2y^3 + 2xz^2 + z$. $\mathrm{I} = f(-2, 1, 3) - f(1, 0, 2) = -21 - 10 = -31$.

[15] $M = 4xy^3$ and $N = 2xy^3$. $\dfrac{\partial M}{\partial y} = \dfrac{\partial}{\partial y}(4xy^3) = 12xy^2$. $\dfrac{\partial N}{\partial x} = \dfrac{\partial}{\partial x}(2xy^3) = 2y^3$.

Since $\dfrac{\partial M}{\partial y} \neq \dfrac{\partial N}{\partial x}$, $\int_C \mathbf{F}\cdot d\mathbf{r}$ is not independent of path.

[19] If the line integral is path independent, then $\mathbf{F}(x, y, z) = \nabla f(x, y, z)$ for some scalar function f and $M = f_x$, $N = f_y$, and $P = f_z$. Consequently,

$$\frac{\partial M}{\partial y} = f_{xy} = f_{yx} = \frac{\partial N}{\partial x},\ \frac{\partial M}{\partial z} = f_{xz} = f_{zx} = \frac{\partial P}{\partial x},\text{ and } \frac{\partial N}{\partial z} = f_{yz} = f_{zy} = \frac{\partial P}{\partial y}.$$

[21] $M = 2xy$, $N = x^2 + z^2$, and $P = yz$. $\dfrac{\partial M}{\partial y} = 2x = \dfrac{\partial N}{\partial x}$, $\dfrac{\partial M}{\partial z} = 0 = \dfrac{\partial P}{\partial x}$,

but $\dfrac{\partial N}{\partial z} = 2z \neq z = \dfrac{\partial P}{\partial y}$. Thus, I is not independent of path.

23 Let $\mathbf{r} = x\mathbf{i} + y\mathbf{j} + z\mathbf{k}$. Then, $-\frac{\mathbf{r}}{\|\mathbf{r}\|}$ is a unit vector in the direction of the origin and $\|\mathbf{r}\| = \sqrt{x^2 + y^2 + z^2}$ is equal to the distance from the origin.

Since the magnitude of $\mathbf{F}$ is inversely proportional to the distance from the origin,

$$\|\mathbf{F}\| = \frac{c}{\|\mathbf{r}\|} = \frac{c}{(x^2 + y^2 + z^2)^{1/2}} = K(x, y, z),$$

where $c > 0$ and $K(x, y, z)$ is some scalar function representing the magnitude of $\mathbf{F}$.

Since $\mathbf{F}$ is directed toward the origin with magnitude K, $\mathbf{F}(x, y, z) = -K(x, y, z)\frac{\mathbf{r}}{\|\mathbf{r}\|}$.

Then, $f_x = \dfrac{-cx}{x^2 + y^2 + z^2}$, $f_y = \dfrac{-cy}{x^2 + y^2 + z^2}$, and $f_z = \dfrac{-cz}{x^2 + y^2 + z^2}$.

Now, $f = \displaystyle\int \frac{-cx}{x^2 + y^2 + z^2}\,dx = -\frac{1}{2}c\ln(x^2 + y^2 + z^2) + g(y, z) \Rightarrow$

$f_y = \dfrac{-cy}{x^2 + y^2 + z^2} + g_y(y, z)$. So $g_y(y, z) = 0$ and $g(y, z) = k(z)$.

$f_z = \dfrac{-cz}{x^2 + y^2 + z^2} + k'(z) \Rightarrow k'(z) = 0$ and $k(z) = d$.

Thus, $f(x, y, z) = -\frac{1}{2}c\ln(x^2 + y^2 + z^2) + d$, and $\mathbf{F}$ is conservative.

29 From the proof of (18.4), $f(x, y, z) = \dfrac{-c}{(x^2 + y^2 + z^2)^{1/2}} = \dfrac{-c}{\|\mathbf{r}\|}$, where $\mathbf{F}(x, y, z) = \nabla f(x, y, z)$. Since $\mathbf{F}$ is conservative, by (18.15), we can determine W by evaluating the potential function $f(x, y, z)$ at each end point and finding their difference. Since $\|\mathbf{r}\| = \sqrt{x^2 + y^2 + z^2}$, at P_1, $f(x, y, z) = \frac{-c}{d_1}$, and at P_2,

$$f(x, y, z) = \frac{-c}{d_2}. \text{ Thus, } W = \frac{-c}{d_2} - \frac{-c}{d_1} = c\left(\frac{1}{d_1} - \frac{1}{d_2}\right) = c\,\frac{d_2 - d_1}{d_1 d_2}.$$

Exercises 18.4

1 We will use Green's theorem (18.19) to convert the given line integral into a double integral over the region R bounded by C. This double integral will usually be easier to evaluate than the line integral. $M = x^2 + y$, $N = xy^2 \Rightarrow N_x = y^2$, $M_y = 1$.

From C, $y^2 = x$ and $y = -x \Rightarrow y^2 = -y \Rightarrow y^2 + y = 0 \Rightarrow y = 0, -1$.

It follows that R can be described by $y^2 \leq x \leq -y$ with $-1 \leq y \leq 0$.

Thus, $\displaystyle\oint_C M\,dx + N\,dy = \iint_R (N_x - M_y)\,dA \Rightarrow \oint_C (x^2 + y)\,dx + (xy^2)\,dy =$

$$\int_{-1}^{0}\int_{y^2}^{-y} (y^2 - 1)\,dx\,dy = \int_{-1}^{0} (y^2 - 1)(-y - y^2)\,dy = -\tfrac{7}{60}.$$

Note: In Exercises 5–14, let I denote the indicated line integral.

5 Since C is a circle, we will convert the given line integral into a double integral in polar coordinates. $M = xy$, $N = y + x \Rightarrow N_x = 1$, $M_y = x$. Thus,

$$I = \iint_R (N_x - M_y)\,dA = \iint_R (1 - x)\,dA =$$

$$\int_0^{2\pi}\int_0^1 (1 - r\cos\theta)\,r\,dr\,d\theta = \int_0^{2\pi} (\tfrac{1}{2} - \tfrac{1}{3}\cos\theta)\,d\theta = \pi.$$

7 $M = xy$, $N = \sin y \Rightarrow N_x = 0$, $M_y = x$.

Divide the triangle into the two regions R_1 and R_2 as shown in *Figure 7*. Thus,

$$I = \iint_{R_1} (0 - x)\,dA + \iint_{R_2} (0 - x)\,dA$$

$$= \int_1^2\int_{-x/2+3/2}^{x} (-x)\,dy\,dx + \int_2^3\int_{-x/2+3/2}^{-2x+6} (-x)\,dy\,dx$$

$$= \int_1^2 (-x)(\tfrac{3}{2}x - \tfrac{3}{2})\,dx + \int_2^3 (-x)(-\tfrac{3}{2}x + \tfrac{9}{2})\,dx$$

$$= -\tfrac{5}{4} + \tfrac{-7}{4} = -3.$$

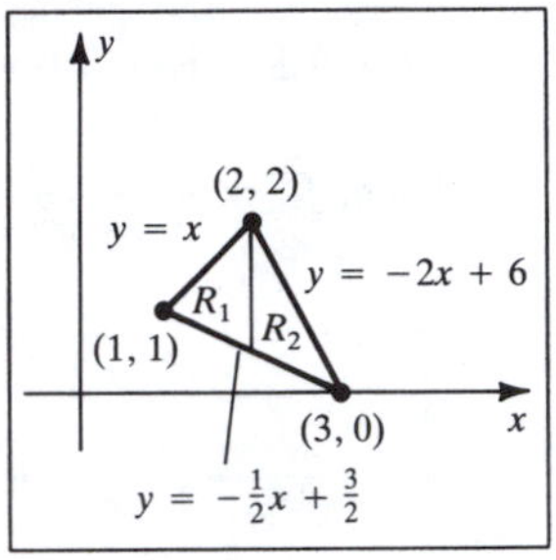

Figure 7

11 $M = x^4 + 4$, $N = xy \Rightarrow N_x = y$, $M_y = 0$.

Thus, $I = \iint_R y\,dA = \int_0^{2\pi}\int_0^{1+\cos\theta} (r\sin\theta)\,r\,dr\,d\theta = \int_0^{2\pi} \tfrac{1}{3}(1 + \cos\theta)^3 \sin\theta\,d\theta =$

$$-\tfrac{1}{3}\int_2^2 u^3\,du\ \{u = 1 + \cos\theta\} = 0.$$

15 A positive direction for the curve is shown in *Figure 15*.

C_1: $y = 4x^2$, $dy = 8x\,dx$ with $0 \le x \le 4$.

C_2: $y = 16x$, $dy = 16\,dx$ with $0 \le x \le 4$.

$$A = \oint x\,dy = \int_0^4 x(8x)\,dx + \int_4^0 x(16)\,dx$$

$$= \tfrac{512}{3} - 128 = \tfrac{128}{3}.$$

Note the limits of integration on the last integral. They are 4 to 0 because as C_2 is traversed in a positive direction, x decreases from 4 to 0.

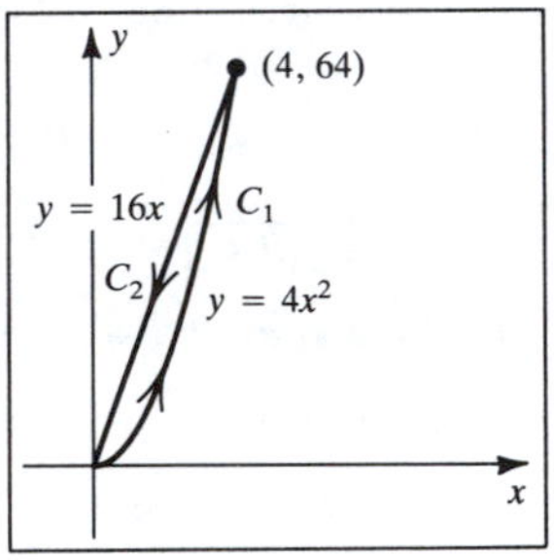

Figure 15

[19] $x = a\cos^3 t,\ y = a\sin^3 t \Rightarrow dx = -3a\cos^2 t\sin t\,dt,\ dy = 3a\sin^2 t\cos t\,dt.$

$$A = \tfrac{1}{2}\oint_C x\,dy - y\,dx$$
$$= \tfrac{1}{2}\int_0^{2\pi}\Big[(a\cos^3 t)(3a\sin^2 t\cos t) - (a\sin^3 t)(-3a\cos^2 t\sin t)\Big]\,dt$$
$$= \frac{3a^2}{2}\int_0^{2\pi}\sin^2 t\cos^2 t(\cos^2 t + \sin^2 t)\,dt$$
$$= \frac{3a^2}{2}\int_0^{2\pi}\Big[\tfrac{1}{2}(1 - \cos 2t)\cdot\tfrac{1}{2}(1 + \cos 2t)\Big]\,dt$$
$$= \frac{3a^2}{8}\int_0^{2\pi}(1 - \cos^2 2t)\,dt = \frac{3a^2}{8}\int_0^{2\pi}\Big[1 - \tfrac{1}{2}(1 + \cos 4t)\Big]\,dt = \tfrac{3}{8}\pi a^2.$$

[21] If $\mathbf{F}(x, y) = M(x, y)\mathbf{i} + N(x, y)\mathbf{j}$ is independent of path,

then by (18.16), $\dfrac{\partial M}{\partial y} = \dfrac{\partial N}{\partial x}$, and hence, by (18.19),

$$\oint_C \mathbf{F}\cdot d\mathbf{r} = \oint_C M\,dx + N\,dy = \iint_R\left(\frac{\partial N}{\partial x} - \frac{\partial M}{\partial y}\right)dA = \iint_R 0\,dA = 0.$$

[25] Without loss of generality, let $\delta(x, y) = 1$. Also, let M_x and M_y denote moments about the x- and y-axes, respectively. Since $\delta(x, y) = 1$, the area A of the region is equal to the mass m. By definition, $M_y = \iint_R x\,dA$. Let $\dfrac{\partial N}{\partial x} = x$ and $\dfrac{\partial M}{\partial y} = 0$. Then, $\left(\dfrac{\partial N}{\partial x} - \dfrac{\partial M}{\partial y}\right) = x$ and $N = \tfrac{1}{2}x^2 + C_1$, $M = C_2$. Without loss of generality, let $C_1 = C_2 = 0$. It now follows by Green's theorem, $\oint_C M\,dx + N\,dy = \iint_R\left(\dfrac{\partial N}{\partial x} - \dfrac{\partial M}{\partial y}\right)dA \Rightarrow \oint_C 0\cdot dx + \tfrac{1}{2}x^2\,dy = \iint_R x\,dA = M_y.$

Now let $\dfrac{\partial N}{\partial x} = 0$ and $\dfrac{\partial M}{\partial y} = -y$. Then, $\left(\dfrac{\partial N}{\partial x} - \dfrac{\partial M}{\partial y}\right) = y$ and $N = C_1$, $M = -\tfrac{1}{2}y^2 + C_2$. Let $C_1 = C_2 = 0$. Then, $\oint_C M\,dx + N\,dy = \iint_R\left(\dfrac{\partial N}{\partial x} - \dfrac{\partial M}{\partial y}\right)dA \Rightarrow \oint_C -\tfrac{1}{2}y^2\,dx + 0\cdot dy = \iint_R y\,dA = M_x$. We now have $M_y = \dfrac{1}{2}\oint_C x^2\,dy$ and $M_x = -\dfrac{1}{2}\oint_C y^2\,dx$. So, $\bar{x} = \dfrac{M_y}{A} = \dfrac{1}{2A}\oint_C x^2\,dy$ and $\bar{y} = \dfrac{M_x}{A} = -\dfrac{1}{2A}\oint_C y^2\,dx.$

[27] Place the semicircle as shown in *Figure 27.*

On C_1: $x = a\cos t,\ y = a\sin t,\ 0 \le t \le \pi \Rightarrow$

$$dx = -a\sin t\,dt,\ dy = a\cos t\,dt.$$

On C_2: $x = t,\ y = 0,\ -a \le t \le a \Rightarrow$

$$dx = dt,\ dy = 0\,dt.$$

The area A of the semicircle is $\frac{1}{2}\pi a^2$

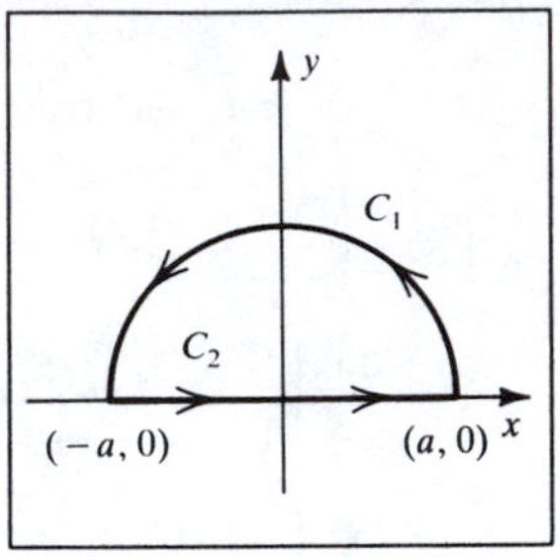

Figure 27

$$\bar{x} = \frac{1}{2A}\oint_C x^2\,dy$$

$$= \frac{1}{\pi a^2}\oint_C x^2\,dy = \frac{1}{\pi a^2}\left(\oint_{C_1} x^2\,dy + \oint_{C_2} x^2\,dy\right)$$

$$= \frac{1}{\pi a^2}\left(\int_0^\pi (a^2\cos^2 t)(a\cos t)\,dt + \int_{-a}^{a} t^2\cdot 0\,dt\right)$$

$$= \tfrac{a}{\pi}\int_0^\pi \cos^3 t\,dt = \tfrac{a}{\pi}\int_0^\pi (1 - \sin^2 t)\cos t\,dt = \tfrac{a}{\pi}\int_0^0 (1 - u^2)\,du\ \{u = \sin t\} = 0.$$

{This can also be seen by symmetry.}

$$\bar{y} = -\frac{1}{2A}\oint_C y^2\,dx = -\frac{1}{\pi a^2}\left(\oint_{C_1} y^2\,dx + \oint_{C_2} y^2\,dx\right)$$

$$= -\frac{1}{\pi a^2}\left(\int_0^\pi (a^2\sin^2 t)(-a\sin t)\,dt + \int_{-a}^{a} 0\,dt\right)$$

$$= \tfrac{a}{\pi}\int_0^\pi \sin^3 t\,dt = \tfrac{a}{\pi}\int_0^\pi (1 - \cos^2 t)\sin t\,dt$$

$$= \tfrac{2a}{\pi}\int_0^1 (1 - u^2)\,du\ \{u = \cos t\} = \tfrac{2a}{\pi}\cdot\tfrac{2}{3} = \tfrac{4a}{3\pi}.$$

Exercises 18.5

[1] $x^2 + y^2 + z^2 = a^2\ \{z \ge 0\} \Rightarrow z = f(x, y) = (a^2 - x^2 - y^2)^{1/2} \Rightarrow$

$$f_x = \frac{-x}{(a^2 - x^2 - y^2)^{1/2}} \text{ and } f_y = \frac{-y}{(a^2 - x^2 - y^2)^{1/2}} \Rightarrow$$

$$\sqrt{f_x^2 + f_y^2 + 1} = \frac{a}{(a^2 - x^2 - y^2)^{1/2}}.$$

R_{xy} is the region inside the circle $x^2 + y^2 = a^2\ \{z = 0\}$. By (18.23)(i),

$$\mathrm{I} = \iint_{R_{xy}} g(x, y, f(x, y))\sqrt{[f_x(x, y)]^2 + [f_y(x, y)]^2 + 1}\,dA$$ (continued)

$$= \iint_{R_{xy}} x^2 \left[\frac{x^2}{a^2 - x^2 - y^2} + \frac{y^2}{a^2 - x^2 - y^2} + \frac{a^2 - x^2 - y^2}{a^2 - x^2 - y^2} \right]^{1/2} dA$$

$$= \iint_{R_{xy}} x^2 \frac{a}{(a^2 - x^2 - y^2)^{1/2}} \, dA$$

$$= \int_0^{2\pi} \int_0^a r^2 \cos^2\theta \left(\frac{a}{(a^2 - r^2)^{1/2}} \right) r \, dr \, d\theta \ \{\text{polar coor. over the circle } x^2 + y^2 = a^2\}$$

$$= a \int_0^{2\pi} \cos^2\theta \left[\int_0^a \frac{r^3}{(a^2 - r^2)^{1/2}} \, dr \right] d\theta$$

Since $\cos^2\theta$ is constant with respect to r, it can be factored out of the inside iterated integral. To integrate with respect to r, use the trig substitution $r = a \sin t$, $(a^2 - r^2)^{1/2} = a \cos t$, and $dr = a \cos t$. Thus,

$$\text{I} = a \int_0^{2\pi} \cos^2\theta \left[\int_0^{\pi/2} a^3 \sin^3 t \, dt \right] d\theta$$

$$= a \int_0^{2\pi} \cos^2\theta \left(\tfrac{2}{3}a^3\right) d\theta = \tfrac{2}{3}a^4 \int_0^{2\pi} \tfrac{1}{2}(1 + \cos 2\theta) \, d\theta = \tfrac{2}{3}\pi a^4.$$

3 $2x + 3y + z = 6 \Rightarrow z = f(x, y) = 6 - 2x - 3y \Rightarrow f_x = -2$ and $f_y = -3 \Rightarrow \sqrt{f_x^2 + f_y^2 + 1} = \sqrt{14}$. R_{xy} is the region bounded by $x = 0$, $y = 0$, and $2x + 3y = 6$, which is the trace of the plane in the xy-plane $\{z = 0\}$. By (18.23)(i),

$$\text{I} = \iint_{R_{xy}} g(x, y, f(x, y)) \sqrt{[f_x(x, y)]^2 + [f_y(x, y)]^2 + 1} \, dA$$

$$= \iint_{R_{xy}} (x + y)\sqrt{14} \, dA = \sqrt{14} \int_0^2 \int_0^{(6-3y)/2} (x + y) \, dx \, dy$$

$$= \sqrt{14} \int_0^2 \left[\tfrac{9}{8}(2 - y)^2 + \tfrac{3}{2}(2y - y^2) \right] dy = \sqrt{14} \int_0^2 \left(\tfrac{9}{2} - \tfrac{3}{2}y - \tfrac{3}{8}y^2\right) dy = 5\sqrt{14}.$$

5 (a) $2x + 3y + 4z = 12 \Rightarrow x = k(y, z) = \frac{1}{2}(12 - 3y - 4z) \Rightarrow$

$$\sqrt{k_y^2 + k_z^2 + 1} = \sqrt{(-\tfrac{3}{2})^2 + (-\tfrac{4}{2})^2 + 1} = \sqrt{\tfrac{29}{4}}.$$

R_{yz} is bounded by $y = 0$, $z = 0$, and $3y + 4z = 12$, which is the trace of the plane in the yz-plane $\{x = 0\}$.

By (18.23)(iii) with $g(x, y, z) = g(k(y, z), y, z) = xy^2z^3 = \frac{1}{2}(12 - 3y - 4z)y^2z^3$,

$$\text{I} = \iint_{R_{yz}} g(k(y, z), y, z) \sqrt{[k_y(y, z)]^2 + [k_z(y, z)]^2 + 1} \, dA$$

$$= \int_0^4 \int_0^{(12-3y)/4} \tfrac{1}{2}(12 - 3y - 4z)y^2z^3 \left(\tfrac{1}{2}\sqrt{29}\right) dz \, dy.$$

(b) Similarly, $y = h(x, z) = \frac{1}{3}(12 - 2x - 4z) \Rightarrow \sqrt{h_x^2 + h_z^2 + 1} = \sqrt{\frac{29}{9}}$.

R_{xz} is bounded by $x = 0$, $z = 0$, and $2x + 4z = 12$. By (18.23)(ii) with

$g(x, y, z) = g(x, h(x, z), z) = xy^2z^3 = x\left[\frac{1}{3}(12 - 2x - 4z)\right]^2 z^3$,

$$\text{I} = \int_0^3 \int_0^{6-2z} x\left[\tfrac{1}{3}(12 - 2x - 4z)\right]^2 z^3 \left(\tfrac{1}{3}\sqrt{29}\right) dx \, dz.$$

11 $z = f(x, y) = \sqrt{a^2 - x^2 - y^2} \Rightarrow$

$$f_x = \frac{-x}{(a^2 - x^2 - y^2)^{1/2}} \text{ and } f_y = \frac{-y}{(a^2 - x^2 - y^2)^{1/2}}.$$

Using the formula for n that appears below (18.24), $\mathbf{n} = \dfrac{-f_x\mathbf{i} - f_y\mathbf{j} + \mathbf{k}}{\sqrt{f_x^2 + f_y^2 + 1}}$.

$$\begin{aligned}
\mathrm{I} &= \iint_S \mathbf{F}\cdot\mathbf{n}\,dS = \iint_{R_{xy}} \frac{-xf_x - yf_y + z}{\sqrt{f_x^2 + f_y^2 + 1}}\sqrt{f_x^2 + f_y^2 + 1}\,dA \\
&= \iint_{R_{xy}} (-xf_x - yf_y + z)\,dA \\
&= \iint_{R_{xy}} \left[\frac{x^2 + y^2}{(a^2 - x^2 - y^2)^{1/2}} + (a^2 - x^2 - y^2)^{1/2}\right] dA \\
&= \iint_{R_{xy}} \frac{a^2}{(a^2 - x^2 - y^2)^{1/2}}\,dA.
\end{aligned}$$

Since R_{xy} is the interior of the circle $x^2 + y^2 = a^2$,

$$\begin{aligned}
\mathrm{I} &= \int_0^{2\pi}\int_0^a a^2(a^2 - r^2)^{-1/2}\,r\,dr\,d\theta \\
&= a^3\int_0^{2\pi} d\theta \;\{u = a^2 - r^2,\ -\tfrac{1}{2}\,du = r\,dr\} = 2\pi a^3.
\end{aligned}$$

Note: Since the radical obtained in changing from dS to dA is the same as the radical in the denominator of n, we will only compute $\mathbf{F}\cdot\langle -f_x, -f_y, 1\rangle$,

and denote this value by $\mathbb{A}$.

13 $z = f(x, y) = (x^2 + y^2)^{1/2} \Rightarrow$

$$f_x = \frac{x}{(x^2 + y^2)^{1/2}} \text{ and } f_y = \frac{y}{(x^2 + y^2)^{1/2}}.$$

Let $\mathbb{A} = \mathbf{F}\cdot\langle -f_x, -f_y, 1\rangle$

$$= \frac{-2x}{(x^2 + y^2)^{1/2}} + \frac{-5y}{(x^2 + y^2)^{1/2}} + 3.$$

$$\mathrm{I} = \iint_S \mathbf{F}\cdot\mathbf{n}\,dS = \iint_{R_{xy}} \mathbb{A}\,dA.$$

R_{xy} is the interior of the circle $x^2 + y^2 = 1$.

Then, using polar coordinates,

$$\begin{aligned}
\mathrm{I} &= \int_0^1\int_0^{2\pi}\left(\frac{-2r\cos\theta}{r} + \frac{-5r\sin\theta}{r} + 3\right) r\,d\theta\,dr \\
&= \int_0^1\int_0^{2\pi}(-2\cos\theta - 5\sin\theta + 3)\,r\,d\theta\,dr = 6\pi\int_0^1 r\,dr = 3\pi.
\end{aligned}$$

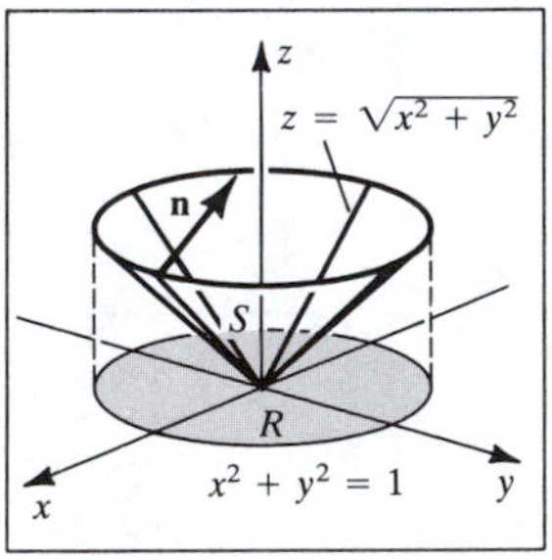

Figure 13

[15] $z = f(x, y) = 6 - 2x - 3y \Rightarrow f_x = -2$ and $f_y = -3$.

$\mathbb{A} = \mathbf{F} \cdot \langle -f_x, -f_y, 1 \rangle = 2x + 3y + z$, which is 6 on $S\{2x + 3y + z = 6\}$.

By (18.25), the flux of F through S is

$$\iint_S \mathbf{F} \cdot \mathbf{n}\, dS = \iint_{R_{xy}} \mathbb{A}\, dA = \int_0^3 \int_0^{(6-2x)/3} (6)\, dy\, dx = \int_0^3 (12 - 4x)\, dx = 18.$$

[17] The cube has six surfaces. The unit outer normal vectors are $\pm\mathbf{i}$, $\pm\mathbf{j}$, and $\pm\mathbf{k}$.

$$\iint_S \mathbf{F} \cdot \mathbf{n}\, dS = \iint_{S_1} \mathbf{F} \cdot \mathbf{i}\, dS + \iint_{S_2} \mathbf{F} \cdot (-\mathbf{i})\, dS + \iint_{S_3} \mathbf{F} \cdot \mathbf{j}\, dS + \iint_{S_4} \mathbf{F} \cdot (-\mathbf{j})\, dS +$$
$$\iint_{S_5} \mathbf{F} \cdot \mathbf{k}\, dS + \iint_{S_6} \mathbf{F} \cdot (-\mathbf{k})\, dS.$$

Let the six integrals be I_1 through I_6. $\mathbf{F}(x, y, z) = (x + y)\mathbf{i} + z\mathbf{j} + xz\mathbf{k}$.

(1) $x = 1$, $\mathbf{n} = \mathbf{i}$, and $\mathbf{F} \cdot \mathbf{n} = (x + y)(1) = x + y = 1 + y\,\{x = 1\}$.

$$\mathrm{I}_1 = \int_{-1}^1 \int_{-1}^1 (1 + y)\, dy\, dz = 4.$$

(2) $x = -1$, $\mathbf{n} = -\mathbf{i}$, and $\mathbf{F} \cdot \mathbf{n} = (x + y)(-1) = -x - y = 1 - y\,\{x = -1\}$.

$$\mathrm{I}_2 = \int_{-1}^1 \int_{-1}^1 (1 - y)\, dy\, dz = 4.$$

(3) $y = 1$, $\mathbf{n} = \mathbf{j}$, and $\mathbf{F} \cdot \mathbf{n} = z(1) = z$. $\quad \mathrm{I}_3 = \int_{-1}^1 \int_{-1}^1 z\, dx\, dz = 0.$

(4) $y = -1$, $\mathbf{n} = -\mathbf{j}$, and $\mathbf{F} \cdot \mathbf{n} = z(-1) = -z$. $\quad \mathrm{I}_4 = -\mathrm{I}_3 = 0.$

(5) $z = 1$, $\mathbf{n} = \mathbf{k}$, and $\mathbf{F} \cdot \mathbf{n} = xz(1) = x\,\{z = 1\}$. $\quad \mathrm{I}_5 = \int_{-1}^1 \int_{-1}^1 x\, dx\, dy = 0.$

(6) $z = -1$, $\mathbf{n} = -\mathbf{k}$, and $\mathbf{F} \cdot \mathbf{n} = xz(-1) = x\,\{z = -1\}$. $\quad \mathrm{I}_6 = \mathrm{I}_5 = 0.$

Thus, $\iint_S \mathbf{F} \cdot \mathbf{n}\, dS = 4 + 4 + 0 + 0 + 0 + 0 = 8.$

[19] If $x = k(y, z)$, then $g(x, y, z) = x - k(y, z) = 0$ is the surface S,

and $\mathbf{n}$ is given by $\mathbf{n} = \dfrac{\nabla g(x, y, z)}{\|\nabla g(x, y, z)\|} = \dfrac{\mathbf{i} - k_y\,\mathbf{j} - k_z\,\mathbf{k}}{\sqrt{k_y^2 + k_z^2 + 1}}$.

So, by (18.23)(iii), $\displaystyle\iint_S \mathbf{F} \cdot \mathbf{n}\, dS = \iint_{R_{yz}} \frac{M - Nk_y - Pk_z}{\sqrt{k_y^2 + k_z^2 + 1}} \sqrt{k_y^2 + k_z^2 + 1}\, dA =$

$$\iint_{R_{yz}} \left[M - Nk_y(y, z) - Pk_z(y, z)\right] dy\, dz.$$

Exercises 18.6

[1] By (18.26) with $\mathbf{F} = y \sin x\,\mathbf{i} + y^2 z\,\mathbf{j} + (x + 3z)\,\mathbf{k}$,

$$\begin{aligned}
\iint_S \mathbf{F} \cdot \mathbf{n}\, dS &= \iiint_Q \nabla \cdot \mathbf{F}\, dV = \iiint_Q \operatorname{div} \mathbf{F}\, dV \\
&= \iiint_Q \left[\frac{\partial}{\partial x}(y \sin x) + \frac{\partial}{\partial y}(y^2 z) + \frac{\partial}{\partial z}(x + 3z)\right] dV \\
&= \int_{-1}^1 \int_{-1}^1 \int_{-1}^1 (y \cos x + 2yz + 3)\, dz\, dy\, dx \\
&= \int_{-1}^1 \int_{-1}^1 (2y \cos x + 6)\, dy\, dx = \int_{-1}^1 12\, dx = 24.
\end{aligned}$$

3 $\mathbf{F} = (x^2 + \sin yz)\,\mathbf{i} + (y - xe^{-z})\,\mathbf{j} + z^2\,\mathbf{k} \Rightarrow$

$\nabla \cdot \mathbf{F} = \operatorname{div}\mathbf{F} = \frac{\partial}{\partial x}(x^2 + \sin yz) + \frac{\partial}{\partial y}(y - xe^{-z}) + \frac{\partial}{\partial z}(z^2) = 2x + 1 + 2z.$

Using cylindrical coordinates, $x^2 + y^2 = 4 \Rightarrow r^2 = 4 \Rightarrow r = 2$ and $z = 2 - x \Rightarrow z = 2 - r\cos\theta$. Thus, Q is the region between the planes $z = 2 - r\cos\theta$ and $z = 0$ bounded by the cylinder $r = 2$.

$$\iint_S \mathbf{F}\cdot\mathbf{n}\,dS = \iiint_Q \operatorname{div}\mathbf{F}\,dV = \iiint_Q (2x + 1 + 2z)\,dV$$
$$= \int_0^{2\pi}\int_0^2\int_0^{2-r\cos\theta} (2r\cos\theta + 1 + 2z)\,r\,dz\,dr\,d\theta$$
$$= \int_0^{2\pi}\int_0^2 \Big[\, 2r\cos\theta\; z + z + z^2 \,\Big]_0^{2-r\cos\theta} r\,dr\,d\theta$$
$$= \int_0^{2\pi}\int_0^2 (6r - r^2\cos\theta - r^3\cos^2\theta)\,dr\,d\theta$$
$$= \int_0^{2\pi} (12 - \tfrac{8}{3}\cos\theta - 4\cos^2\theta)\,d\theta = 20\pi.$$

7 By (18.24), the flux of $\mathbf{F}$ over S is $\iint_S \mathbf{F}\cdot\mathbf{n}\,dS$. Using the divergence theorem, we will convert this double integral into a triple integral.

$\mathbf{F} = 3x\mathbf{i} + xz\mathbf{j} + z^2\mathbf{k} \Rightarrow \operatorname{div}\mathbf{F} = \frac{\partial}{\partial x}(3x) + \frac{\partial}{\partial y}(xz) + \frac{\partial}{\partial z}(z^2) = 3 + 2z.$

The trace of the paraboloid $z = 4 - x^2 - y^2$ in the xy-plane is $x^2 + y^2 = 4$.

Using cylindrical coordinates, $\iint_S \mathbf{F}\cdot\mathbf{n}\,dS = \iiint_Q \operatorname{div}\mathbf{F}\,dV =$

$$\int_0^{2\pi}\int_0^2\int_0^{4-r^2} (3 + 2z)\,r\,dz\,dr\,d\theta = \int_0^{2\pi}\int_0^2 (28r - 11r^3 + r^5)\,dr\,d\theta = \int_0^{2\pi} \tfrac{68}{3}\,d\theta = \tfrac{136\pi}{3}.$$

9 $\mathbf{F} = 2xz\mathbf{i} + xyz\mathbf{j} + yz\mathbf{k} \Rightarrow \operatorname{div}\mathbf{F} = 2z + xz + y.$

S is the surface of the wedge shape shown in *Figure 9* and Q is its interior. Thus, Q is the region between the planes $z = 0$ and $z = (4 - x)/2$ and bounded by $0 \le x \le 4$ and $0 \le y \le 2$.

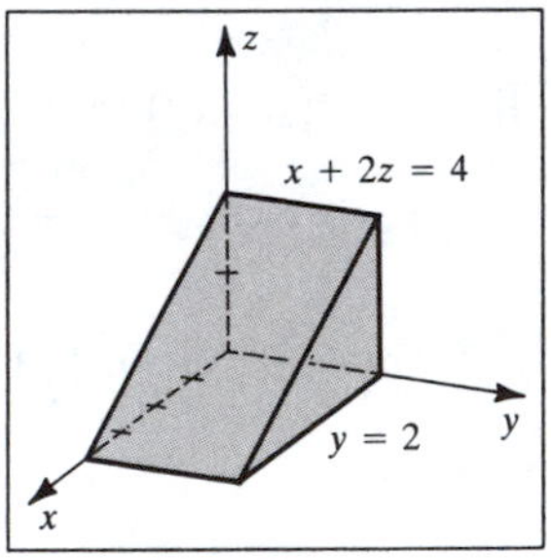

Figure 9

$$\iint_S \mathbf{F}\cdot\mathbf{n}\,dS = \iiint_Q \operatorname{div}\mathbf{F}\,dV$$
$$= \int_0^2\int_0^4\int_0^{(4-x)/2} (2z + xz + y)\,dz\,dx\,dy$$
$$= \tfrac{1}{8}\int_0^2\int_0^4 (32 - 6x^2 + x^3 + 16y - 4xy)\,dx\,dy$$
$$= 4\int_0^2 (2 + y)\,dy = 24.$$

[11] (1) The unit outer normal for a sphere is $\mathbf{n} = \frac{\mathbf{r}}{\|\mathbf{r}\|} = \frac{x\mathbf{i} + y\mathbf{j} + z\mathbf{k}}{\sqrt{x^2 + y^2 + z^2}}$ and $\mathbf{F} \cdot \mathbf{n} = \sqrt{x^2 + y^2 + z^2} = a$ since $x^2 + y^2 + z^2 = a^2$ on S. The upper half of the sphere is given by $z = \sqrt{a^2 - x^2 - y^2}$ and the lower half is given by $z = -\sqrt{a^2 - x^2 - y^2}$. $z_x = \mp \frac{x}{\sqrt{a^2 - x^2 - y^2}}$ and $z_y = \mp \frac{y}{\sqrt{a^2 - x^2 - y^2}}$ on the upper and lower halves of the sphere, respectively. Thus,

on both the upper and lower halves, $\sqrt{z_x^2 + z_y^2 + 1} = \frac{a}{(a^2 - x^2 - y^2)^{1/2}}$.

$$\begin{aligned}
\iint_S \mathbf{F} \cdot \mathbf{n}\, dS &= \iint_{S(\text{upper})} \mathbf{F} \cdot \mathbf{n}\, dS + \iint_{S(\text{lower})} \mathbf{F} \cdot \mathbf{n}\, dS \\
&= 2 \iint_{R_{xy}} a\sqrt{z_x^2 + z_y^2 + 1}\, dA \\
&= 2 \iint_{R_{xy}} \frac{a^2}{(a^2 - x^2 - y^2)^{1/2}}\, dA \\
&= 2(2\pi a^3) \;\{\text{Exercise 11, §18.5}\} = 4\pi a^3.
\end{aligned}$$

(2) $\operatorname{div} \mathbf{F} = 1 + 1 + 1 = 3$.

$$\iiint_Q 3\, dV = 3(\tfrac{4}{3}\pi a^3) \;\{\text{since the volume of a sphere is } \tfrac{4}{3}\pi r^3\} = 4\pi a^3.$$

13 (1) The region Q has 3 surfaces as shown in *Figure 13a*.

On S_1, $\mathbf{n} = -\mathbf{i}$ and $\mathbf{F}\cdot\mathbf{n} = -(x + z) = -z$ since $x = 0$ on S_1.

Now S_1 is R_{yz} and so $\iint_{S_1} \mathbf{F}\cdot\mathbf{n}\, dS = -\int_0^1\int_0^{2\pi} (r\sin\theta)\, r\, d\theta\, dr = 0.$

(Note that we have used polar coordinates where $y = r\cos\theta$ and $z = r\sin\theta$.)

On S_3, $\mathbf{n} = \mathbf{i}$ and $\mathbf{F}\cdot\mathbf{n} = x + z = 2 + z$ since $x = 2$ on S_3.

$\iint_{S_3} \mathbf{F}\cdot\mathbf{n}\, dS = \iint_{S_3} (2 + z)\, dS = \iint_{S_3} 2\, dS + \iint_{S_3} z\, dS = 2\left[\pi(1)^2\right] + 0 = 2\pi,$
where the first integral is twice the area of S_3,

and the value of the second integral was already shown to be zero.

On S_2, $\mathbf{n} = \dfrac{\nabla g(x, y, z)}{\|\nabla g(x, y, z)\|} = \dfrac{2y\mathbf{j} + 2z\mathbf{k}}{\sqrt{4y^2 + 4z^2}} = y\mathbf{j} + z\mathbf{k}$ since $y^2 + z^2 = 1$ on S_2.

$\mathbf{F}\cdot\mathbf{n} = y(y + z) + z(x + y) = y^2 + 2yz + xz$. The region R_{xy}, which the upper and lower halves of S_2 project onto, is the rectangle $0 \le x \le 2$, $-1 \le y \le 1$ as shown in *Figure 13b*. For the upper half $z = \sqrt{1 - y^2}$ and the lower half $z = -\sqrt{1 - y^2}$. In both cases, $\sqrt{z_x^2 + z_y^2 + 1} = \dfrac{1}{\sqrt{1 - y^2}}$. Now,

$$\begin{aligned}
\iint_{S_2} \mathbf{F}\cdot\mathbf{n}\, dS &= \iint_{S_2\text{(upper)}} \mathbf{F}\cdot\mathbf{n}\, dS + \iint_{S_2\text{(lower)}} \mathbf{F}\cdot\mathbf{n}\, dS \\
&= \int_{-1}^{1}\int_0^2 \left(y^2 + 2y\sqrt{1 - y^2} + x\sqrt{1 - y^2}\right)\cdot\frac{1}{\sqrt{1 - y^2}}\, dx\, dy \;+ \\
&\qquad \int_{-1}^{1}\int_0^2 \left(y^2 - 2y\sqrt{1 - y^2} - x\sqrt{1 - y^2}\right)\cdot\frac{1}{\sqrt{1 - y^2}}\, dx\, dy \\
&= 2\int_{-1}^{1}\int_0^2 \frac{y^2}{\sqrt{1 - y^2}}\, dx\, dy = 8\int_0^1 \frac{y^2}{\sqrt{1 - y^2}}\, dy = 8\int_0^{\pi/2} \sin^2\theta\, d\theta = 2\pi,
\end{aligned}$$

where the substitution $y = \sin\theta$ has been used.

Thus, $\iint_S \mathbf{F}\cdot\mathbf{n}\, dS = 0 + 2\pi + 2\pi = 4\pi$.

(2) $\operatorname{div}\mathbf{F} = 1 + 1 + 0 = 2$.

$\iiint_Q 2\, dV = 2\left[\pi(1)^2(2)\right] = 4\pi$ since the volume of a cylinder is $\pi r^2 h$.

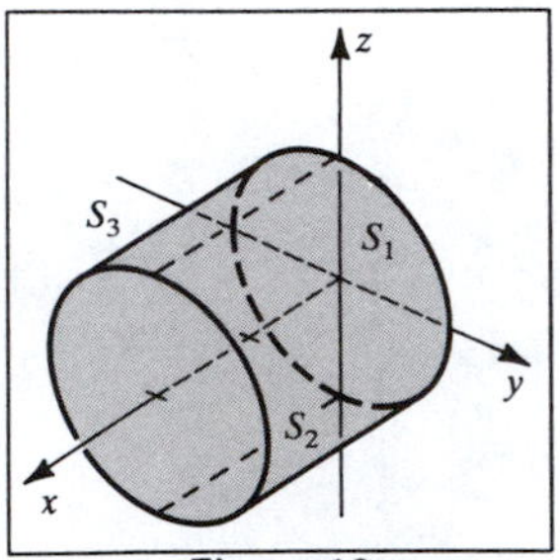

Figure 13a

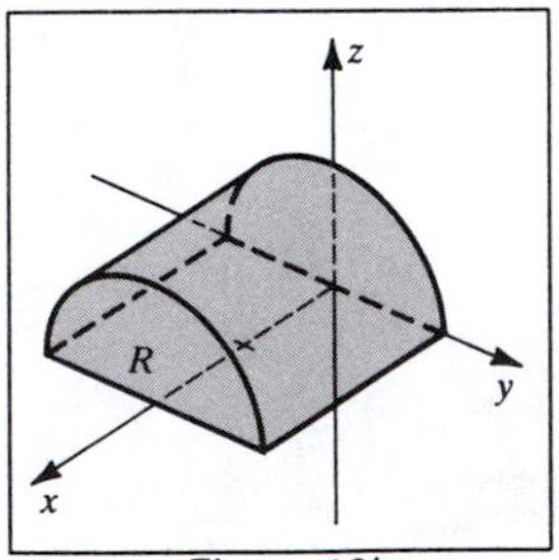

Figure 13b

[17] Let $\mathbf{F} = f\nabla g$. Then,

$$\begin{aligned} \text{RHS} &= \iint_S (f\nabla g)\cdot \mathbf{n}\, dS \\ &= \iint_S \mathbf{F}\cdot\mathbf{n}\, dS \ \{\text{hint}\} \\ &= \iiint_Q \nabla\cdot\mathbf{F}\, dV \ \{(18.26)\} \\ &= \iiint_Q \nabla\cdot(f\nabla g)\, dV \\ &= \iiint_Q \big[f[\nabla\cdot(\nabla g)] + (\nabla f)\cdot(\nabla g)\big]\, dV \ \{\text{Example 5, §18.1}\} \\ &= \iiint_Q (f\nabla^2 g + \nabla f\cdot\nabla g)\, dV = \text{LHS}. \end{aligned}$$

[19] $$\begin{aligned} \text{RHS} = \iint_S f\mathbf{F}\cdot\mathbf{n}\, dS &= \iiint_Q \operatorname{div} f\mathbf{F}\, dV \ \{(18.26)\} \\ &= \iiint_Q (f\operatorname{div}\mathbf{F} + \nabla f\cdot\mathbf{F})\, dV \ \{\text{Example 5, §18.1}\} \\ &= \iiint_Q \mathbf{F}\cdot\mathbf{F}\, dV \ \{\operatorname{div}\mathbf{F} = 0 \text{ and } \mathbf{F} = \nabla f\} = \text{LHS}. \end{aligned}$$

[21] $\iint_S \operatorname{curl}\mathbf{F}\cdot\mathbf{n}\, dS = \iiint_Q \operatorname{div}\operatorname{curl}\mathbf{F}\, dV = 0$ since $\operatorname{div}\operatorname{curl}\mathbf{F} = 0$.

The following can be used to show that $\operatorname{div}\operatorname{curl}\mathbf{F} = 0$ with $\mathbf{F} = M\mathbf{i} + N\mathbf{j} + P\mathbf{k}$.

$$\begin{aligned} \operatorname{div}\operatorname{curl}\mathbf{F} = \nabla\cdot(\nabla\times\mathbf{F}) &= \nabla\cdot\big[(P_y - N_z)\mathbf{i} + (M_z - P_x)\mathbf{j} + (N_x - M_y)\mathbf{k}\big] \\ &= \frac{\partial}{\partial x}(P_y - N_z) + \frac{\partial}{\partial y}(M_z - P_x) + \frac{\partial}{\partial z}(N_x - M_y) \\ &= P_{yx} - N_{zx} + M_{zy} - P_{xy} + N_{xz} - M_{yz} \\ &= (P_{yx} - P_{xy}) + (N_{zx} - N_{xz}) + (M_{zy} - M_{yz}) = 0. \end{aligned}$$

Note: In Exercises 23–24, Exercise 50 in §15.2 is used.

[23] By (14.33(v)), $(\mathbf{c}\times\mathbf{F})\cdot\mathbf{n} = \mathbf{c}\cdot(\mathbf{F}\times\mathbf{n})$ and by Exercise 26, §18.1, $\nabla\cdot(\mathbf{c}\times\mathbf{F}) = (\nabla\times\mathbf{c})\cdot\mathbf{F} - (\nabla\times\mathbf{F})\cdot\mathbf{c} = -\mathbf{c}\cdot(\nabla\times\mathbf{F})$ since $\mathbf{c}$ is constant and $\nabla\times\mathbf{c} = \mathbf{0}$. Now by substituting $\mathbf{c}\times\mathbf{F}$ into the divergence theorem we have the following equalities. $\mathbf{c}\cdot\iint_S \mathbf{F}\times\mathbf{n}\, dS = \iint_S \mathbf{c}\cdot(\mathbf{F}\times\mathbf{n})\, dS = \iint_S (\mathbf{c}\times\mathbf{F})\cdot\mathbf{n}\, dS =$ $\iiint_Q \nabla\cdot(\mathbf{c}\times\mathbf{F})\, dV = \iiint_Q -\mathbf{c}\cdot(\nabla\times\mathbf{F})\, dV = -\mathbf{c}\cdot\iiint_Q \nabla\times\mathbf{F}\, dV$. Since $\mathbf{c}$ is an arbitrary vector, $\iint_S \mathbf{F}\times\mathbf{n}\, dS = -\iiint_Q \nabla\times\mathbf{F}\, dV$.

(To see this, let $\mathbf{c} = \langle 1, 0, 0\rangle$, $\langle 0, 1, 0\rangle$, and $\langle 0, 0, 1\rangle$ and compare components.)

[27] Let $f = p$ in Exercise 24. Then, $-\iint_S p\mathbf{n}\, dS = -\iiint_Q \nabla p\, dV = -\iiint_Q (-62.5\,\mathbf{k})\, dV =$ $(62.5\iiint_Q dV)\mathbf{k} = 62.5\big[\pi(1)^2(10)\big]\mathbf{k}$ {volume of the cylindrical tank} $= (625\pi \text{ lb})\mathbf{k}$.

The force is directed upward since it is in the direction of the unit vector $\mathbf{k}$.

Exercises 18.7

1 (1) $\mathbf{F} = y^2\mathbf{i} + z^2\mathbf{j} + x^2\mathbf{k} \Rightarrow$

$\text{curl}\,\mathbf{F} = -2z\mathbf{i} - 2x\mathbf{j} - 2y\mathbf{k}$

$\{(18.5)$ with $M = y^2$, $N = z^2$, and $P = x^2\}$.

A unit normal to S ($g(x, y, z) = x + y + z - 1$),

pointing into the first octant,

is $\dfrac{\nabla g(x, y, z)}{\|\nabla g(x, y, z)\|} = \frac{1}{\sqrt{3}}(\mathbf{i} + \mathbf{j} + \mathbf{k})$.

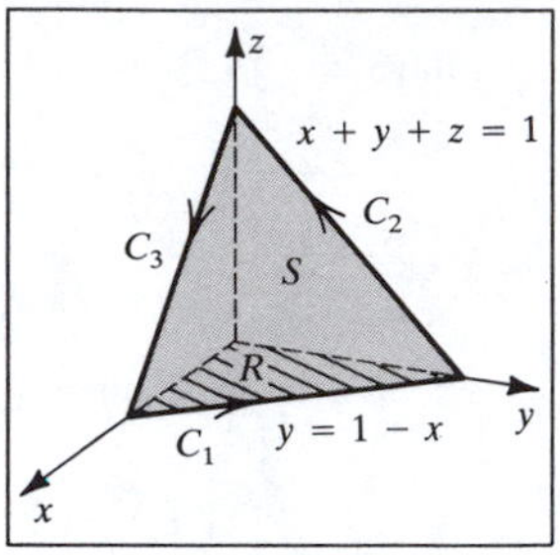

Figure 1

$\text{curl}\,\mathbf{F} \cdot \mathbf{n} = -\frac{2}{\sqrt{3}}(z + x + y) = -\frac{2}{\sqrt{3}}$ since $x + y + z = 1$ on S. If we let $z = 1 - x - y$, then $\sqrt{z_x^2 + z_y^2 + 1} = \sqrt{3}$.

Thus, $\iint\limits_S \text{curl}\,\mathbf{F} \cdot \mathbf{n}\, dS = \iint\limits_S -\frac{2}{\sqrt{3}}\, dS = \iint\limits_{R_{xy}} -\frac{2}{\sqrt{3}} \cdot \sqrt{3}\, dA = -2 \iint\limits_{R_{xy}} dA =$
$-2(\frac{1}{2}) = -1$ since R_{xy} is a triangular region with area $\frac{1}{2}$.

(2) To calculate $\oint_C \mathbf{F} \cdot \mathbf{T}\, ds = \oint_C \mathbf{F} \cdot d\mathbf{r} = \oint_C y^2\, dx + z^2\, dy + x^2\, dz$

$\{d\mathbf{r} = dx\mathbf{i} + dy\mathbf{j} + dz\mathbf{k}\}$, we express $C = C_1 \cup C_2 \cup C_3$ as shown in *Figure 1*.

The line integral is taken on the boundary of S in the first octant.

We must express each curve parametrically.

C_1: $x = 1 - t$, $y = t$, and $z = 0$ with $0 \le t \le 1$

$\{dx = -dt,\ dy = dt,\ dz = 0\, dt\}$.

$$\int_{C_1} \mathbf{F} \cdot d\mathbf{r} = \oint_{C_1} (t^2)(-dt) + 0^2\, dt + (1 - t)^2(0\, dt) = \int_0^1 -t^2\, dt = -\tfrac{1}{3}.$$

We can compute $\int_{C_2} \mathbf{F} \cdot d\mathbf{r}$ and $\int_{C_3} \mathbf{F} \cdot d\mathbf{r}$ similarly as follows.

C_2: $x = 0$, $y = 1 - t$, $z = t$, $0 \le t \le 1$, and $\int_{C_2} \mathbf{F} \cdot d\mathbf{r} = \int_0^1 -t^2\, dt = -\frac{1}{3}$.

C_3: $x = t$, $y = 0$, $z = 1 - t$, $0 \le t \le 1$, and $\int_{C_3} \mathbf{F} \cdot d\mathbf{r} = \int_0^1 -t^2\, dt = -\frac{1}{3}$.

Their sum is -1.

3 (1) $\mathbf{F} = z\mathbf{i} + x\mathbf{j} + y\mathbf{k} \Rightarrow \text{curl}\,\mathbf{F} = \mathbf{i} + \mathbf{j} + \mathbf{k}$. $z = f(x, y) = \sqrt{a^2 - x^2 - y^2} \Rightarrow$

$f_x = \dfrac{-x}{\sqrt{a^2 - x^2 - y^2}}$, $f_y = \dfrac{-y}{\sqrt{a^2 - x^2 - y^2}}$, and $\mathbf{n} = \dfrac{-f_x\mathbf{i} - f_y\mathbf{j} + \mathbf{k}}{\sqrt{f_x^2 + f_y^2 + 1}}$. Then,

$$\begin{aligned}
\iint_S \text{curl}\,\mathbf{F}\cdot\mathbf{n}\,dS &= \iint_{R_{xy}} \frac{-f_x - f_y + 1}{\sqrt{f_x^2 + f_y^2 + 1}}\sqrt{f_x^2 + f_y^2 + 1}\,dA \\
&= \iint_{R_{xy}} (-f_x - f_y + 1)\,dA \\
&= \iint_{R_{xy}} \left[\frac{x}{\sqrt{a^2 - x^2 - y^2}} + \frac{y}{\sqrt{a^2 - x^2 - y^2}} + \frac{\sqrt{a^2 - x^2 - y^2}}{\sqrt{a^2 - x^2 - y^2}}\right] dA \\
&= \iint_{R_{xy}} \frac{x + y + \sqrt{a^2 - x^2 - y^2}}{\sqrt{a^2 - x^2 - y^2}}\,dA \\
&= \int_0^{2\pi}\int_0^a \frac{r\cos\theta + r\sin\theta + \sqrt{a^2 - r^2}}{\sqrt{a^2 - r^2}}\,r\,dr\,d\theta \\
&= \int_0^a\int_0^{2\pi} \frac{r^2(\cos\theta + \sin\theta)}{\sqrt{a^2 - r^2}}\,d\theta\,dr + \int_0^{2\pi}\int_0^a r\,dr\,d\theta \\
&= \int_0^a 0\,dr + \int_0^{2\pi} \tfrac{1}{2}a^2\,d\theta = \pi a^2.
\end{aligned}$$

(2) The curve C is the boundary of S. C is the trace of the hemisphere $z = (a^2 - x^2 - y^2)^{1/2}$ in the xy-plane, which is the circle $x^2 + y^2 = a^2$. To calculate $\oint_C \mathbf{F}\cdot d\mathbf{r} = \oint_C z\,dx + x\,dy + y\,dz$, we express C as $x = a\cos t$, $y = a\sin t$, $z = 0$, $0 \le t \le 2\pi$. Then, $\oint_C \mathbf{F}\cdot d\mathbf{r} =$

$\int_0^{2\pi} x\,dy = \pi a^2$ since by (18.20), the integral equals the area of the circle.

5 Since only C is specified in the exercise, and not S, we will need to choose S ourselves. S must have C as its boundary and there are *many* choices for S. We will try to pick S so the surface integral is as simple as possible to work with. Let the surface S be the disk given by $z = 1$ bounded by C with $\mathbf{n} = \mathbf{k}$.

$\mathbf{F} = (3z - \sin x)\mathbf{i} + (x^2 + e^y)\mathbf{j} + (y^3 - \cos z)\mathbf{k} \Rightarrow$

$\text{curl}\,\mathbf{F} = 3y^2\mathbf{i} + 3\mathbf{j} + 2x\mathbf{k}$ and $z = f(x, y) = 1 \Rightarrow \sqrt{f_x^2 + f_y^2 + 1} = 1$.

Then, $\oint_C \mathbf{F}\cdot d\mathbf{r} = \iint_S \text{curl}\,\mathbf{F}\cdot\mathbf{n}\,dS = \iint_S (3y^2\mathbf{i} + 3\mathbf{j} + 2x\mathbf{k})\cdot\mathbf{k}\,dS = \iint_S 2x\,dS =$

$$\iint_{R_{xy}} 2x(1)\,dA = \int_0^1\int_0^{2\pi} (2r\cos\theta)\,r\,d\theta\,dr = \int_0^1 0\,dr = 0.$$

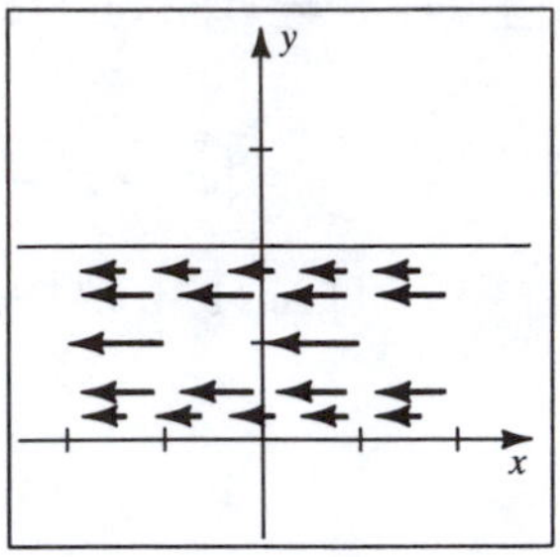

Figure 9

[9] $\mathbf{F} = (y^2 - 2y)\mathbf{i} + 0\mathbf{j} + 0\mathbf{k} \Rightarrow \text{curl}\,\mathbf{F} = 2(1 - y)\mathbf{k}$.
curl F is a function of y only and is independent of x and z.
Refer to *Figure 9*. The curl meter rotates counterclockwise for $0 < y < 1$ and clockwise for $1 < y < 2$.
There is no rotation if $y = 1$. $\text{curl}\,\mathbf{F} = 2(1 - y)\mathbf{k}$;
$|(\text{curl}\,\mathbf{F}) \cdot \mathbf{k}| = |2(1 - y)|$ has a maximum value 2 at $y = 0$ and $y = 2$ and a minimum value 0 at $y = 1$.
See Figure 18.57 and Example 2.

[13] To show that F is irrotational, we must show that $\text{curl}\,\mathbf{F} = \mathbf{0}$ (18.31)(iv).
$\mathbf{F}(x, y, z) = y\mathbf{i} + (x + e^z)\mathbf{j} + (1 + ye^z)\mathbf{k} \Rightarrow$
$$\text{curl}\,\mathbf{F} = (e^z - e^z)\mathbf{i} - (0 - 0)\mathbf{j} + (1 - 1)\mathbf{k} = \mathbf{0} \Rightarrow \mathbf{F} \text{ is irrotational.}$$

[15] (a) By (18.26), $\iint_S \text{curl}\,\mathbf{F} \cdot \mathbf{n}\, dS = \iiint_Q \nabla \cdot \text{curl}\,\mathbf{F}\, dV = \iiint_Q \text{div}\,\text{curl}\,\mathbf{F}\, dV = 0$
since $\text{div}\,\text{curl}\,\mathbf{F} = 0$ {see Exercise 21, §18.6}.

(b) We may assume that S is a sphere of radius a with center at the origin.
Let S_1 be the upper hemisphere ($z \geq 0$) and S_2 be the lower hemisphere ($z \leq 0$).
Let C be the circle $x^2 + y^2 = a^2$ in the xy-plane. To apply Stokes' theorem, we let N denote the unit upper normal. Then, $\mathbf{N} = \mathbf{n}$ on S_1, and $\mathbf{N} = -\mathbf{n}$ on S_2.
Thus, $\iint_S \text{curl}\,\mathbf{F} \cdot \mathbf{n}\, dS = \iint_{S_1} \text{curl}\,\mathbf{F} \cdot \mathbf{N}\, dS + \iint_{S_2} \text{curl}\,\mathbf{F} \cdot \mathbf{N}\, dS =$
$\iint_{S_1} \text{curl}\,\mathbf{F} \cdot \mathbf{n}\, dS + \iint_{S_2} \text{curl}\,\mathbf{F} \cdot (-\mathbf{n})\, dS = \oint_C \mathbf{F} \cdot d\mathbf{r} - \oint_C \mathbf{F} \cdot d\mathbf{r} = 0.$
Note that on S_1, we traverse C in a counterclockwise direction, and on S_2, we traverse C in the opposite direction.

[17] By Exercise 25, §18.1, with $\mathbf{F} = \nabla g$,
we have $\nabla \times (f\nabla g) = f(\nabla \times \nabla g) + (\nabla f \times \nabla g)$.
By Exercise 27, §18.1, with $f = g$, $\nabla \times \nabla g = \mathbf{0}$, so $\nabla \times (f\nabla g) = \nabla f \times \nabla g$.
Now by Stokes' theorem, $\oint_C f\nabla g \cdot d\mathbf{r} = \iint_S \nabla \times (f\nabla g) \cdot \mathbf{n}\, dS = \iint_S (\nabla f \times \nabla g) \cdot \mathbf{n}\, dS$.

18.8 Review Exercises

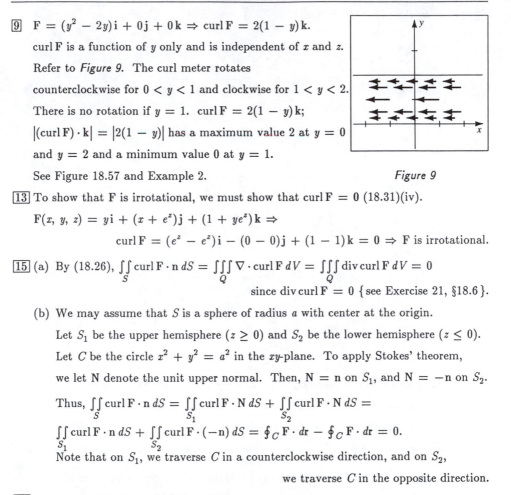

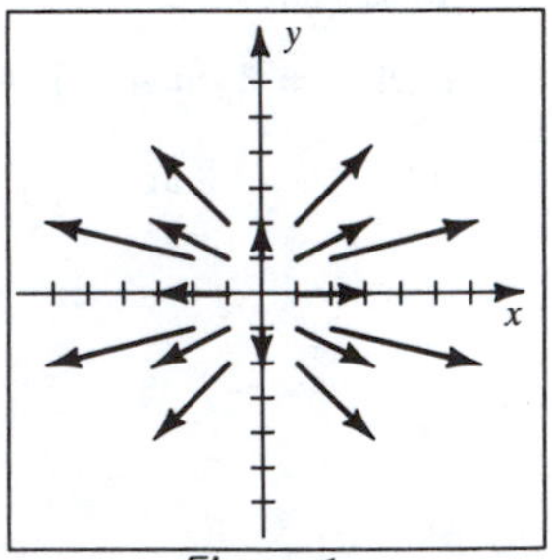

Figure 1

[1] $\mathbf{F}(x, y) = 2x\mathbf{i} + y\mathbf{j} \Rightarrow \mathbf{F}(0, 1) = \mathbf{j}$, $\mathbf{F}(0, -1) = -\mathbf{j}$,
$\mathbf{F}(1, 1) = 2\mathbf{i} + \mathbf{j}$, etc. Plot several vectors as shown in *Figure 1*.

[5] By (18.3), $f(x, y) = y^2 \tan x \Rightarrow$

$$\mathbf{F}(x, y) = \nabla f(x, y) = f_x\,\mathbf{i} + f_y\,\mathbf{j} = (y^2 \sec^2 x)\,\mathbf{i} + (2y \tan x)\,\mathbf{j}.$$

[7] Let C be divided into two parts and represent each part parametrically.

On C_1: $x = t$, $y = 0$, $-1 \le t \le 1$. On C_2: $x = -1$, $y = t$, $0 \le t \le 4$.

$$\text{Thus, } \int_C y^2\,dx + xy\,dy = \int_1^{-1} 0\,dt + \int_0^4 (-t)\,dt = -8.$$

[13] The parametric equations for C are $x = 2t$, $y = 4t$, and $z = 8t$ with $0 \le t \le 1$.

$dx = 2\,dt$, $dy = 4\,dt$, and $dz = 8\,dt \Rightarrow$

$$\mathrm{I} = \int_C x\,dx + (x + y)\,dy + (x + y + z)\,dz$$

$$= \int_0^1 \Big[(2t)(2) + (2t + 4t)(4) + (2t + 4t + 8t)(8)\Big]\,dt = \int_0^1 140t\,dt = 70.$$

[17] By (18.3), we must find a function f such that $\mathbf{F} = \nabla f$. $f_x = x + y \Rightarrow$

$f = \frac{1}{2}x^2 + xy + g(y) \Rightarrow f_y = x + g'(y) = x + y \Rightarrow g'(y) = y$ and $g(y) = \frac{1}{2}y^2 + c$.

If we let $c = 0$, then $f(x, y) = \frac{1}{2}x^2 + xy + \frac{1}{2}y^2$,

and the line integral is independent of path.

By (18.14), its value is $f(2, 3) - f(1, -1) = \frac{25}{2} - 0 = \frac{25}{2}$.

[19] $f_x = 2xe^{2y} \Rightarrow f = x^2e^{2y} + g(y, z) \Rightarrow f_y = 2x^2e^{2y} + g_y(y, z) = 2x^2e^{2y} + 2y \cot z$

$\Rightarrow g_y(y, z) = 2y \cot z$ and $g(y, z) = y^2 \cot z + k(z)$.

$f_z = g_z(y, z) = -y^2 \csc^2 z + k'(z) = -y^2 \csc^2 z \Rightarrow k'(z) = 0$ and $k(z) = c$. Thus,

$f(x, y, z) = x^2e^{2y} + y^2 \cot z + c$, and by (18.31), $\int_C \mathbf{F} \cdot d\mathbf{r}$ is independent of path.

[21] $\oint_C M\,dx + N\,dy = \iint_R (N_x - M_y)\,dA \Rightarrow$

$$\oint_C xy\,dx + (x^2 + y^2)\,dy = \iint_R (2x - x)\,dA = \iint_R x\,dA$$

$$= \int_0^1\int_0^{1-x} x\,dy\,dx = \int_0^1 (x - x^2)\,dx = \Big[\tfrac{1}{2}x^2 - \tfrac{1}{3}x^3\Big]_0^1 = \tfrac{1}{6}.$$

[25] $z = f(x, y) = x + y \Rightarrow \sqrt{f_x^2 + f_y^2 + 1} = \sqrt{3}$.

R_{xy} is the triangular region bounded by $x = 0$, $y = 0$, and $y = 2 - 2x$. Thus,

$$\iint_S xyz\,dS = \iint_{R_{xy}} xyz\sqrt{3}\,dA = \sqrt{3}\int_0^1\int_0^{2-2x} xy(x + y)\,dy\,dx$$

$$= \sqrt{3}\int_0^1 \Big[\tfrac{1}{2}x^2(2 - 2x)^2 + \tfrac{1}{3}x(2 - 2x)^3\Big]\,dx$$

$$= \tfrac{1}{3}\sqrt{3}\int_0^1 (-2x^4 + 12x^3 - 18x^2 + 8x)\,dx = \tfrac{1}{5}\sqrt{3}.$$

27 $\mathbf{F} = x^3\mathbf{i} + y^3\mathbf{j} + z^3\mathbf{k} \Rightarrow \operatorname{div}\mathbf{F} = 3(x^2 + y^2 + z^2)$. Using cylindrical coordinates,

$$\iint_S \mathbf{F}\cdot\mathbf{n}\,dS = \iiint_Q \operatorname{div}\mathbf{F}\,dV = \iiint_Q 3(x^2 + y^2 + z^2)\,dV =$$

$$3\int_0^{2\pi}\int_0^1\int_0^1 (r^2 + z^2)\,r\,dz\,dr\,d\theta = 3\int_0^{2\pi}\int_0^1 (r^3 + \tfrac{1}{3}r)\,dr\,d\theta = 3\cdot\tfrac{5}{12}\int_0^{2\pi} d\theta = \tfrac{5\pi}{2}.$$

29 (1) $\mathbf{F} = y^2\mathbf{i} + 2x\mathbf{j} + 5y\mathbf{k} \Rightarrow \operatorname{curl}\mathbf{F} = 5\mathbf{i} + (2 - 2y)\mathbf{k}$.

$z = f(x, y) = (4 - x^2 - y^2)^{1/2} \Rightarrow$

$$f_x = \frac{-x}{(4 - x^2 - y^2)^{1/2}},\ f_y = \frac{-y}{(4 - x^2 - y^2)^{1/2}},\ \text{and } \mathbf{n} = \frac{-f_x\mathbf{i} - f_y\mathbf{j} + \mathbf{k}}{\sqrt{f_x^2 + f_y^2 + 1}}.$$

Then, $\displaystyle\iint_S \operatorname{curl}\mathbf{F}\cdot\mathbf{n}\,dS = \iint_{R_{xy}}\left[\frac{5x}{(4 - x^2 - y^2)^{1/2}} + (2 - 2y)\right]dA$

{where R_{xy} is the region bounded by $x^2 + y^2 = 4$}

$$= \int_0^2\int_0^{2\pi}\left[\frac{5r\cos\theta}{(4 - r^2)^{1/2}} + (2 - 2r\sin\theta)\right]r\,d\theta\,dr = \int_0^2 4\pi r\,dr = 8\pi.$$

(2) The boundary C of the surface S is the circle $x^2 + y^2 = 4$.

To calculate $\oint_C \mathbf{F}\cdot d\mathbf{r} = \oint_C y^2\,dx + 2x\,dy + 5y\,dz$, we express C as $x = 2\cos t$, $y = 2\sin t$, and $z = 0$ for $0 \le t \le 2\pi$ with $dx = -2\sin t\,dt$, $dy = 2\cos t\,dt$, and $dz = 0\,dt$. Then,

$$\oint_C \mathbf{F}\cdot d\mathbf{r} = \int_0^{2\pi}(-8\sin^3 t + 8\cos^2 t)\,dt$$

$$= 8\int_0^{2\pi}\left[-(1 - \cos^2 t)\sin t + \tfrac{1}{2}(1 + \cos 2t)\right]dt$$

$$= 8\int_1^1 (1 - u^2)\,du\ \{u = \cos t\} + 4\left[t + \tfrac{1}{2}\sin 2t\right]_0^{2\pi}$$

$$= 0 + 4(2\pi) = 8\pi.$$

Chapter 19: Differential Equations

Exercises 19.1

1 (a) $y' = 3x^2 \Rightarrow y = \int 3x^2\,dx = x^3 + C.$

The solutions are vertical translations of $y = x^3$ with y-intercept C.

(b) $y = 2$ when $x = 0 \Rightarrow 2 = 0^3 + C \Rightarrow C = 2,$ and hence, $y = x^3 + 2.$

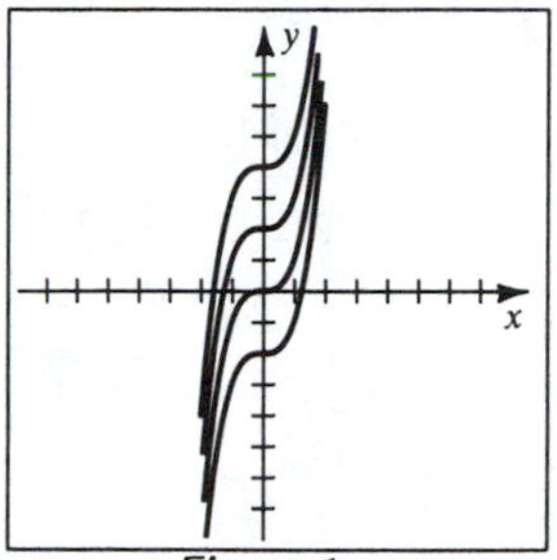

Figure 1

5 $y = C_1 e^x + C_2 e^{2x} \Rightarrow y' = C_1 e^x + 2C_2 e^{2x}$ and $y'' = C_1 e^x + 4C_2 e^{2x}.$

$$y'' - 3y' + 2y = C_1 e^x + 4C_2 e^{2x} - 3C_1 e^x - 6C_2 e^{2x} + 2C_1 e^x + 2C_2 e^{2x} = 0.$$

9 Differentiating $(y^2 - x^2 - xy = c)$ implicitly yields $2yy' - 2x - y - xy' = 0 \Rightarrow$ $(2y - x)y' - (2x + y) = 0 \Rightarrow -(2y - x)y' + (2x + y) = 0 \Rightarrow$

$$(x - 2y)y' + 2x + y = 0.$$

Note: Let k and C denote constants in the following exercises.

11 $\sec x\,dy - 2y\,dx = 0 \Rightarrow \frac{1}{2y}\,dy = \cos x\,dx \Rightarrow \frac{1}{2}\int \frac{1}{y}\,dy = \int \cos x\,dx \Rightarrow$

$\frac{1}{2}\ln|y| = \sin x + k \Rightarrow \ln|y| = 2\sin x + 2k \Rightarrow |y| = e^{2\sin x + 2k} \Rightarrow$

$$y = \pm\, e^{2k}\, e^{2\sin x} \Rightarrow y = Ce^{2\sin x}, \text{ where } C = \pm\, e^{2k} \text{ is a constant.}$$

15 $3y\,dx + (xy + 5x)\,dy = 0 \Rightarrow \frac{3}{x}\,dx + \left(1 + \frac{5}{y}\right)dy = 0$ {divide by xy} $\Rightarrow$

$3\int \frac{1}{x}\,dx + \int\left(1 + \frac{5}{y}\right)dy = \int 0\,dx \Rightarrow 3\ln|x| + y + 5\ln|y| = k \Rightarrow$

$\ln|x^3| + y + \ln|y^5| = k \Rightarrow e^{\ln|x^3|}\,e^y\,e^{\ln|y^5|} = e^k \Rightarrow$

$|x^3| \cdot e^y \cdot |y^5| = e^k \Rightarrow x^3 e^y y^5 = C.$

Assume $x \neq 0$, but note that $y = 0$ is a solution to the original equation.

19 $e^{x+2y}\,dx - e^{2x-y}\,dy = 0 \Rightarrow e^x e^{2y}\,dx - e^{2x}e^{-y}\,dy = 0 \Rightarrow e^{-x}\,dx = e^{-3y}\,dy \Rightarrow$

$-e^{-x} = -\frac{1}{3}e^{-3y} + k \Rightarrow e^{-3y} = 3e^{-x} - 3k \Rightarrow -3y = \ln(C + 3e^{-x}) \Rightarrow$

$$y = -\tfrac{1}{3}\ln(C + 3e^{-x}), \text{ where } C = -3k.$$

21 $y(1 + x^3)\,y' + x^2(1 + y^2) = 0 \Rightarrow y(1 + x^3)\frac{dy}{dx} + x^2(1 + y^2) = 0 \Rightarrow$

$\frac{y}{1 + y^2}\,dy + \frac{x^2}{1 + x^3}\,dx = 0$ {multiply by dx} $\Rightarrow$

$\frac{1}{2}\ln(1 + y^2) + \frac{1}{3}\ln|1 + x^3| = k \Rightarrow (1 + y^2)^{1/2}|1 + x^3|^{1/3} = e^k \Rightarrow$

$$(1 + y^2)^{1/2} = \pm\, e^k(1 + x^3)^{-1/3} \Rightarrow y^2 = C(1 + x^3)^{-2/3} - 1, \text{ where } C = e^{2k}.$$

[25] $e^y \sin x\,dx - \cos^2 x\,dy = 0 \Rightarrow e^y \sin x\,dx - \frac{\cos x}{\sec x}\,dy = 0 \Rightarrow$

$\sec x \tan x\,dx - e^{-y}\,dy = 0 \left\{\text{multiply by } \frac{\sec x}{\cos x} \cdot e^{-y}\right\} \Rightarrow \sec x + e^{-y} = C.$

[27] $2y^2 y' = 3y - y' \Rightarrow 2y^2 \frac{dy}{dx} = 3y - \frac{dy}{dx} \Rightarrow (2y^2 + 1)\,dy - 3y\,dx = 0 \Rightarrow$

$\left(2y + \frac{1}{y}\right) dy - 3\,dx = 0 \Rightarrow y^2 + \ln|y| - 3x = C.$

Letting $y = 1$ and $x = 3 \Rightarrow C = -8$. Thus,

$y^2 + \ln y = 3x - 8$ for $y > 0$. (We have $y > 0$ since the condition has $y = 1 > 0$.)

[31] $(xy + x)\,dx + \sqrt{4 + x^2}\,dy = 0 \Rightarrow$

$\frac{x}{(4 + x^2)^{1/2}}\,dx + \frac{1}{y + 1}\,dy = 0 \left\{\text{divide by } \frac{1}{(y + 1)\sqrt{4 + x^2}}\right\} \Rightarrow$

$\sqrt{4 + x^2} + \ln|y + 1| = C.$ Letting $y = 1$ and $x = 0 \Rightarrow C = 2 + \ln 2$.

Thus, $\ln|y + 1| = 2 + \ln 2 - \sqrt{4 + x^2} \Rightarrow |y + 1| = 2e^{2-\sqrt{4+x^2}} \Rightarrow$

$y = 2e^{2-\sqrt{4+x^2}} - 1$ for $y > -1$.

(We pick the positive solution since the condition has $y = 1$, and hence, $y + 1 > 0$.)

[33] $\cot x\,dy - (1 + y^2)\,dx = 0 \Rightarrow \frac{1}{1 + y^2}\,dy - \tan x\,dx = 0 \Rightarrow \tan^{-1} y - \ln|\sec x| = C.$

Letting $y = 1$ and $x = 0 \Rightarrow C = \frac{\pi}{4}$. Thus, $\tan^{-1} y - \ln \sec x = \frac{\pi}{4}$.

[35] Differentiating $(x^2 - y^2 = c)$ implicitly yields $2x - 2yy' = 0 \Rightarrow y' = \frac{x}{y}$.

Orthogonal trajectories will have slopes $y' = -\frac{y}{x} \Rightarrow \frac{1}{y}\,dy + \frac{1}{x}\,dx = 0 \Rightarrow$

$\ln|y| + \ln|x| = C \Rightarrow |xy| = e^C \Rightarrow xy = k$; hyperbolas.

Note: In Exercises 37–40, it is easiest to solve for c first, thereby eliminating c upon differentiation.

[39] Differentiating implicitly, $\frac{y^2}{x^3} = c \Rightarrow \frac{(x^3)(2y\,y') - (y^2)(3x^2)}{x^6} = 0 \Rightarrow$

$2yx^3 y' = 3x^2 y^2 \Rightarrow y' = \frac{3y}{2x}$. Orthogonal trajectories will have slopes $y' = -\frac{2x}{3y} \Rightarrow$

$\frac{dy}{dx} = -\frac{2x}{3y} \Rightarrow 3y\,dy + 2x\,dx = 0 \Rightarrow 3y^2 + 2x^2 = k$; ellipses.

[41] (a) Step 1: Let $f(x, y) = xy$, $k = 0$, $a = x_0 = 0$, $b = 1$, and $n = 8$.

Hence, $h = \frac{b - a}{n} = \frac{1 - 0}{8} = \frac{1}{8}$. Step 2: Let $y_0 = 1$.

Step 3: $y_1 = y_0 + hf(x_0, y_0) = 1 + \frac{1}{8}(0)(1) = 1$. Step 4: $x_1 \not\approx b$, so go back to Step 3 with $x_1 = \frac{1}{8}$. $y_2 = y_1 + hf(x_1, y_1) = 1 + \frac{1}{8}(\frac{1}{8})(1) = 1.015625$.

In a similar manner, $y_3 \approx 1.0474$, $y_4 = 1.0965$, $y_5 = 1.1650$,

$y_6 = 1.2560$, $y_7 \approx 1.3738$, and $y_8 \approx 1.5240$.

(b) $y' = xy \Rightarrow \frac{1}{y}\,dy = x\,dx \Rightarrow \ln y = \frac{1}{2}x^2 + C$ { no absolute value since $y > 0$ } $\Rightarrow$

$y = e^{x^2/2 + C}$. $y = 1$ at $x = 0 \Rightarrow C = 0$.

Thus, $y = e^{x^2/2}$ and $y = e^{1/2} \approx 1.648721$ at $x = 1$.

[43] $y_1 = y_0 + \frac{1}{2}h\big[f(x_0, y_0) + f(x_0 + h, y_0 + hf(x_0, y_0))\big] =$

$1 + \frac{1}{16}\big[f(0, 1) + f(0 + \frac{1}{8}, 1 + \frac{1}{8}f(0, 1))\big] = 1 + \frac{1}{16}\big[0 + f(\frac{1}{8}, 1)\big] = 1 + \frac{1}{16}\cdot\frac{1}{8} =$

1.0078125. In a similar manner, $y_2 \approx 1.031679$, $y_3 \approx 1.072735$, $y_4 \approx 1.132971$,

$y_5 \approx 1.215399$, $y_6 \approx 1.324299$, $y_7 \approx 1.465587$, and $y_8 \approx 1.647355$.

Exercises 19.2

Note: Let *IF* denote the integrating factor.

[1] The form in (19.1) is $y' + P(x)y = Q(x)$. $y' + 2y = e^{2x} \Rightarrow P(x) = 2$ and

$IF = e^{\int 2\,dx} = e^{2x}$. Multiplying both sides by *IF* gives us $y'e^{2x} + 2e^{2x}y = e^{4x}$.

Since the left side of the last equation equals $D_x(ye^{2x})$, we have

$D_x(ye^{2x}) = e^{4x} \Rightarrow ye^{2x} = \frac{1}{4}e^{4x} + C \Rightarrow y = \frac{1}{4}e^{2x} + Ce^{-2x}$.

[3] $y' - \frac{3}{x}y = x^4 \Rightarrow P = -\frac{3}{x}$, $Q = x^4$, $IF = e^{\int(-3/x)\,dx} = e^{-3\ln|x|} = |x|^{-3}$.

$x^{-3}y' - 3x^{-4}y = x \Rightarrow D_x(yx^{-3}) = x \Rightarrow x^{-3}y = \frac{1}{2}x^2 + C \Rightarrow y = \frac{1}{2}x^5 + Cx^3$.

[7] $x^2\,dy + (2xy - e^x)\,dx = 0 \Rightarrow x^2y' + (2xy - e^x) = 0 \Rightarrow$

$y' + \frac{2}{x}y = \frac{1}{x^2}e^x \Rightarrow IF = e^{\int(2/x)\,dx} = e^{2\ln|x|} = x^2$.

$x^2y' + 2xy = e^x \Rightarrow x^2y = e^x + C \Rightarrow y = (e^x + C)/x^2$.

[11] $(y\sin x - 2)\,dx + \cos x\,dy = 0 \Rightarrow y'\cos x + y\sin x = 2 \Rightarrow y' + y\tan x = 2\sec x \Rightarrow$

$IF = e^{\int\tan x\,dx} = e^{\ln|\sec x|} = |\sec x|$. $y'\sec x + y\sec x\tan x = 2\sec^2 x \Rightarrow$

$y\sec x = 2\tan x + C \Rightarrow y = 2\sin x + C\cos x$.

[15] $xy' + (2 + 3x)y = xe^{-3x} \Rightarrow y' + \left(\frac{2}{x} + 3\right)y = e^{-3x} \Rightarrow$

$IF = e^{\int[(2/x) + 3]\,dx} = e^{2\ln|x|}e^{3x} = x^2e^{3x}$.

$x^2e^{3x}y' + (2xe^{3x} + 3x^2e^{3x})y = x^2 \Rightarrow D_x(x^2e^{3x}y) = \int x^2\,dx \Rightarrow$

$x^2e^{3x}y = \frac{1}{3}x^3 + C \Rightarrow y = \frac{1}{3}xe^{-3x} + Cx^{-2}e^{-3x} = \left(\frac{1}{3}x + \frac{C}{x^2}\right)e^{-3x}$.

[19] $\tan x\,dy + (y - \sin x)\,dx = 0 \Rightarrow y'\tan x + y = \sin x \Rightarrow y' + y\cot x = \cos x \Rightarrow$

$IF = e^{\int\cot x\,dx} = |\sin x|$. $y'\sin x + y\cos x = \sin x\cos x \Rightarrow$

$D_x(y\sin x) = \int\sin x\cos x\,dx \Rightarrow y\sin x = \frac{1}{2}\sin^2 x + C \Rightarrow y = \frac{1}{2}\sin x + C\csc x$.

[23] $xy' - y = x^2 + x \Rightarrow y' - \frac{1}{x}y = x + 1 \Rightarrow IF = e^{\int(-1/x)\,dx} = |x|^{-1}$.

$x^{-1}y' - x^{-2}y = 1 + x^{-1} \Rightarrow x^{-1}y = x + \ln|x| + C \Rightarrow y = x(x + \ln|x| + C)$.

Letting $y = 2$ and $x = 1 \Rightarrow C = 1$. Thus, $y = x(x + \ln x + 1)$, for $x > 0$.

25 $xy' + y + xy = e^{-x} \Rightarrow y' + \left(\frac{1}{x} + 1\right)y = \frac{1}{x}e^{-x} \Rightarrow$

$IF = e^{\int[(1/x)+1]\,dx} = e^{\ln|x|}e^x = xe^x.$ $xe^x y' + (e^x + xe^x)\,y = 1 \Rightarrow$

$D_x(xe^x y) = \int 1\,dx \Rightarrow xe^x y = x + C \Rightarrow y = e^{-x} + Cx^{-1}e^{-x}.$

Letting $y = 0$ and $x = 1 \Rightarrow C = -1$. Thus, $y = e^{-x}(1 - x^{-1})$.

27 (a) $R\frac{dQ}{dt} + \frac{Q}{C} = V \Rightarrow \frac{dQ}{dt} + \frac{1}{RC}Q = \frac{V}{R} \Rightarrow IF = e^{\int(1/(RC))\,dt} = e^{t/RC}.$

$e^{t/RC}\frac{dQ}{dt} + \frac{1}{RC}e^{t/RC}Q = e^{t/RC}\left(\frac{V}{R}\right) \Rightarrow Qe^{t/RC} = CVe^{t/RC} + k \Rightarrow$

$Q = CV + ke^{-t/RC}.$ $Q(0) = 0 \Rightarrow k = -CV$. Thus, $Q = CV(1 - e^{-t/RC})$.

(b) $R\frac{dQ}{dt} + \frac{Q}{C} = V \Rightarrow R\,dQ + \left(\frac{Q}{C} - V\right)dt = 0 \Rightarrow$

$\frac{C}{Q - CV}dQ + \frac{1}{R}dt = 0 \left\{\text{divide by } \frac{Q - CV}{CR}\right\} \Rightarrow$

$C\ln(Q - CV) + \frac{t}{R} = k \Rightarrow \ln(Q - CV) = \frac{k}{C} - \frac{t}{RC} \Rightarrow$

$Q - CV = e^{k/C}e^{-t/RC} \Rightarrow Q = CV + me^{-t/RC}\ \{m = e^{k/C}\}.$

$Q(0) = 0 \Rightarrow m = -CV$. Thus, $Q = CV(1 - e^{-t/RC})$.

29 The rate at which salt comes into the tank is 6 gal/min $\times\ \frac{1}{3}$ lb/gal $= 2$ lb/min.

Since $f(t)$ denotes the total amount of salt in the tank, the rate at which salt leaves the tank is 6 gal/min $\times\ \frac{1}{80}f(t)$ lb/gal $= \frac{3}{40}f(t)$ lb/gal.

The net rate of change is $f'(t) = 2 - \frac{3}{40}f(t) \Rightarrow f'(t) + 0.075f(t) = 2 \Rightarrow$

$IF = e^{0.075t}$, $f(t)e^{0.075t} = \frac{2}{0.075}e^{0.075t} + C$, and $f(t) = \frac{80}{3} + Ce^{-0.075t}$.

$f(0) = K \Rightarrow C = K - \frac{80}{3}$. Thus, $f(t) = \frac{80}{3}(1 - e^{-0.075t}) + Ke^{-0.075t}$ lb.

31 (a) Since $f'(t)$ is proportional to $M - f(t)$, $f'(t) = k\big[M - f(t)\big]$ $\{k \text{ is a constant}\} \Rightarrow$

$f'(t) + kf(t) = kM.$ $IF = e^{\int k\,dt} = e^{kt}.$ $e^{kt}f'(t) + ke^{kt}f(t) = ke^{kt}M \Rightarrow$

$f(t)\,e^{kt} = Me^{kt} + C$, and $f(t) = M + Ce^{-kt}$. $f(1) = A \Rightarrow$

$C = (A - M)\,e^k$. Thus, $f(t) = M + (A - M)\,e^{k(1-t)}$.

(b) $M = 30,\ A = 5 \Rightarrow f(t) = 30 - 25e^{k(1-t)}$. $f(2) = 8 \Rightarrow 8 = 30 - 25e^{-k} \Rightarrow$

$e^k = \frac{25}{22}$. So $f(t) = 30 - 25(\frac{25}{22})^{1-t}$ and $f(20) \approx 27.8 \approx 28$ items.

35 (a) Since the rate of elimination is proportional to y, $y' = -ky$, $k > 0 \Rightarrow$

$y' + ky = 0 \Rightarrow IF = e^{kt}$, $e^{kt}y' + ke^{kt}y = 0 \Rightarrow ye^{kt} = C$, and $y = Ce^{-kt}$.

$y(0) = y_0 \Rightarrow y = y_0e^{-kt}$.

(b) $y' + ky = I \Rightarrow IF = e^{\int k\,dt} = e^{kt} \Rightarrow ye^{kt} = \frac{I}{k}e^{kt} + C \Rightarrow y = \frac{I}{k} + Ce^{-kt}$.

$y(0) = 0 \Rightarrow C = -\frac{I}{k}$ so $y = \frac{I}{k}(1 - e^{-kt})$. As $t \to \infty$, $y = \frac{I}{k}$.

(c) $y(0) = \frac{I}{k}$ and a half-life of 2 hr $\Rightarrow$ $y(2) = \frac{1}{2} \cdot \frac{I}{k} = \frac{I}{k}(1 - e^{-2k}) \Rightarrow$

$\frac{1}{2} = (1 - e^{-2k}) \Rightarrow k = \ln\sqrt{2}$. Since the long-term amount is

$$\frac{I}{k},\ y = 100 = \frac{I}{\ln\sqrt{2}} \Rightarrow I = 100\ln\sqrt{2} \approx 34.7 \text{ mg/hr} \approx 0.58 \text{ mg/min}.$$

[39] $h = \frac{1}{8}$, $x_0 = 0$, and $y_0 = 0.1 \Rightarrow K_1 = 0.01$, $K_2 \approx 0.014032$, $K_3 \approx 0.014082$, $K_4 \approx 0.025980 \Rightarrow y_1 = y_0 + \frac{1}{48}(K_1 + 2K_2 + 2K_3 + K_4) \approx 0.101921$.

Similarly, $y_2 \approx 0.107841$, $y_3 \approx 0.121837$, and $y_4 \approx 0.148170$.

Exercises 19.3

[1] By (19.5), the auxiliary equation of $y'' - 5y' + 6y = 0$ is $m^2 - 5m + 6$.

$m^2 - 5m + 6 = (m - 2)(m - 3) = 0 \Rightarrow m = 2, 3$.

By (19.6), the general solution is $y = C_1 e^{m_1 x} + C_2 e^{m_2 x} = C_1 e^{2x} + C_2 e^{3x}$.

[5] The auxiliary equation is $m^2 + 4m + 4 = (m + 2)^2 = 0 \Rightarrow$

$m = -2$ is a double root. By (19.7), the general solution is

$$y = C_1 e^{mx} + C_2 x e^{mx} = C_1 e^{-2x} + C_2 x e^{-2x} = e^{-2x}(C_1 + C_2 x).$$

[7] The auxiliary equation is $m^2 - 4m + 1 = 0 \Rightarrow m = \dfrac{4 \pm \sqrt{12}}{2} = 2 \pm \sqrt{3}$.

By (19.6), $y = C_1 e^{(2+\sqrt{3})x} + C_2 e^{(2-\sqrt{3})x}$.

[13] The auxiliary equation is $9m^2 - 24m + 16 = (3m - 4)^2 = 0 \Rightarrow$

$m = \frac{4}{3}$ is a double root. By (19.7), $y = C_1 e^{4x/3} + C_2 x e^{4x/3} = e^{4x/3}(C_1 + C_2 x)$.

[17] The auxiliary equation is $m^2 - 2m + 2 = 0 \Rightarrow m = \dfrac{2 \pm \sqrt{-4}}{2} = 1 \pm i$,

which are distinct complex solutions.

By (19.10), with $s = t = 1$ $\{s \pm ti = 1 \pm 1i\}$, $y = e^x(C_1 \cos x + C_2 \sin x)$.

[19] $m^2 - 4m + 13 = 0 \Rightarrow m = \dfrac{4 \pm \sqrt{-36}}{2} = 2 \pm 3i$.

By (19.10) with $s \pm ti = 2 \pm 3i$, $y = e^{2x}(C_1 \cos 3x + C_2 \sin 3x)$.

[21] $m^2 + 6m + 2 = 0 \Rightarrow m = \dfrac{-6 \pm \sqrt{28}}{2} = -3 \pm \sqrt{7}$.

By (19.6), $y = C_1 e^{(-3+\sqrt{7})x} + C_2 e^{(-3-\sqrt{7})x}$.

[23] $m^2 - 3m + 2 = 0 \Rightarrow m = 1, 2$, so $y = C_1 e^x + C_2 e^{2x}$. $x = 0$, $y = 0 \Rightarrow$

$0 = C_1 e^0 + C_2 e^0 \Rightarrow C_1 + C_2 = 0$. $y' = C_1 e^x + 2C_2 e^{2x}$ and $x = 0$, $y' = 2 \Rightarrow$

$C_1 + 2C_2 = 2$. Thus, $C_2 = 2$, $C_1 = -2$, and $y = -2e^x + 2e^{2x}$.

[25] $m^2 + 1 = 0 \Rightarrow m = \pm i$, so by (19.10), $y = C_1 \cos x + C_2 \sin x$. $x = 0$, $y = 1 \Rightarrow$

$1 = C_1 \cos 0 + C_2 \sin 0 \Rightarrow C_1 = 1$. $y' = C_2 \cos x - \sin x$ and $x = 0$, $y' = 2 \Rightarrow$

$2 = C_2 \cos 0 - \sin 0 \Rightarrow C_2 = 2$. Thus, $y = \cos x + 2\sin x$.

27 $m^2 + 8m + 16 = 0 \Rightarrow m = -4$ is a double root, so $y = C_1 e^{-4x} + C_2 x e^{-4x}$.

$x = 0,\ y = 2 \Rightarrow C_1 = 2.\ \ y = 2e^{-4x} + C_2 x e^{-4x} \Rightarrow$

$y' = -8e^{-4x} + C_2(e^{-4x} - 4xe^{-4x})$ and $x = 0,\ y' = 1 \Rightarrow -8 + C_2 = 1 \Rightarrow$

$C_2 = 9$. Thus, $y = 2e^{-4x} + 9xe^{-4x} = e^{-4x}(2 + 9x)$.

Exercises 19.4

1 The auxiliary equation for $y'' + y = 0$ is $m^2 + 1 = 0 \Rightarrow m = \pm\, i$.

By (19.10), the complementary solution is $y_c = C_1 \cos x + C_2 \sin x$.

To find a particular solution y_p, we use (19.13) with $y_1 = \cos x$ and $y_2 = \sin x$.

$$\begin{cases} u' \cos x + v' \sin x = 0 \\ -u' \sin x + v' \cos x = \tan x \end{cases}$$

Solving this as a system of equations in the unknowns u' and v',

we multiply the first equation by $\sin x$ and the second by $\cos x$, yielding the system

$$\begin{cases} u' \cos x \sin x + v' \sin^2 x = 0 \\ -u' \cos x \sin x + v' \cos^2 x = \sin x \end{cases}$$

Adding equations yields $v'(\sin^2 x + \cos^2 x) = \sin x \Rightarrow v' = \sin x$ and then substituting

into the first equation gives us $u' = -\dfrac{\sin^2 x}{\cos x} = -\dfrac{1 - \cos^2 x}{\cos x} = -\sec x + \cos x$.

Integrating produces $u = -\ln|\sec x + \tan x| + \sin x$ and $v = -\cos x$.

Thus, $y_p = u\,y_1 + v\,y_2 = -\cos x \ln|\sec x + \tan x| + \sin x \cos x - \cos x \sin x$.

The general solution is $y = y_c + y_p = C_1 \cos x + C_2 \sin x - \cos x \ln|\sec x + \tan x|$.

5 The auxiliary equation is $m^2 - 1 = 0 \Rightarrow m = \pm 1$.

By (19.6), $y_c = C_1 e^x + C_2 e^{-x}$. Using (19.13),

$$\begin{cases} u' e^x + v' e^{-x} = 0 \\ u' e^x - v' e^{-x} = e^x \cos x \end{cases} \Rightarrow \begin{cases} u' = \tfrac{1}{2}\cos x \\ v' = -\tfrac{1}{2} e^{2x} \cos x \end{cases}$$

Adding equations yields $2u' e^x = e^x \cos x$, or $u' = \frac{1}{2}\cos x$. Substituting $\frac{1}{2}\cos x$ for

u' into the first equation gives us $v' e^{-x} = -\frac{1}{2}e^x \cos x$, or $v' = -\frac{1}{2} e^{2x} \cos x$.

Integrating produces $u = \frac{1}{2}\sin x$ and $v = -\frac{1}{5} e^{2x} \cos x - \frac{1}{10} e^{2x} \sin x$.

The general solution is $y = C_1 e^x + C_2 e^{-x} + \frac{2}{5} e^x \sin x - \frac{1}{5} e^x \cos x$.

9 $m^2 - 3m - 4 = 0 \Rightarrow m = -1, 4.\ \ y_c = C_1 e^{-x} + C_2 e^{4x}$. Using (19.13),

$$\begin{cases} u' e^{-x} + v' e^{4x} = 0 \\ -u' e^{-x} + v'(4e^{4x}) = 2 \end{cases} \Rightarrow \begin{cases} v' = \tfrac{2}{5} e^{-4x} \\ u' = -\tfrac{2}{5} e^x \end{cases}$$

Integrating gives us $u = -\frac{2}{5}e^x$ and $v = -\frac{1}{10}e^{-4x}$.

The general solution is $y = C_1 e^{-x} + C_2 e^{4x} - \frac{1}{2}$.

[11] $m^2 - 3m + 2 = 0 \Rightarrow m = 1, 2.$ $y_c = C_1 e^x + C_2 e^{2x}$.

Since -1 is not a root of the auxiliary equation, by (19.14)(i), let $y_p = Ae^{-x}$.

Then, substituting the expression for y_p into the differential equation,

$y_p'' - 3y_p' + 2y_p = (A + 3A + 2A)e^{-x} = 4e^{-x} \Rightarrow A = \frac{2}{3}$.

The general solution is $y = C_1 e^x + C_2 e^{2x} + \frac{2}{3}e^{-x}$.

[15] $m^2 - 1 = 0 \Rightarrow m = \pm 1.$ $y_c = C_1 e^x + C_2 e^{-x}$. By (19.14)(ii), let

$y_p = (A + Bx)e^{2x}$. $y_p' = e^{2x}(2Bx + 2A + B)$ and $y_p'' = 4e^{2x}(Bx + A + B)$.

Then, $y_p'' - y_p = (4B + 4A)e^{2x} + 4Bxe^{2x} - (A + Bx)e^{2x} =$

$(4B + 3A)e^{2x} + 3Bxe^{2x} = xe^{2x} \Rightarrow B = \frac{1}{3}$ and $A = -\frac{4}{9}$.

The general solution is $y = C_1 e^x + C_2 e^{-x} + \frac{1}{9}(-4 + 3x)e^{2x}$.

[17] $m^2 - 6m + 13 = 0 \Rightarrow m = 3 \pm 2i$. By (19.10), $y_c = e^{3x}(C_1 \cos 2x + C_2 \sin 2x)$.

Let $y_p = Ae^x \cos x + Be^x \sin x$. $y_p' = e^x\left[(A + B)\cos x + (B - A)\sin x\right]$ and

$y_p'' = e^x(2B\cos x - 2A\sin x)$. Then, $y_p'' - 6y_p' + 13y_p$

$= \left[2B - 6(A + B) + 13A\right]e^x \cos x + \left[-2A - 6(B - A) + 13B\right]e^x \sin x$

$= (7A - 4B)\, e^x \cos x + (4A + 7B)\, e^x \sin x = e^x \cos x \Rightarrow A = \frac{7}{65}$ and $B = -\frac{4}{65}$.

The general solution is $y = e^{3x}(C_1 \cos 2x + C_2 \sin 2x) + \frac{1}{65}e^x(7\cos x - 4\sin x)$.

[19] $L(Cy) = (D^2 + bD + c)(Cy)$

$= D^2(Cy) + bD(Cy) + c(Cy)$

$= CD^2y + CbDy + Ccy$

$= C(D^2y + bDy + cy) = CL(y)$.

[21] Since we are approximating y at $x = \frac{1}{2}$, let $b = \frac{1}{2}$.

$x_0 = 0$, $y_0 = 1$, and $h = (b - a)/n = (\frac{1}{2} - 0)/4 = \frac{1}{8} \Rightarrow$

$y_1 = 2y_0 - y_{-1} + h^2 f(x_0, y_0) = 2(1) - (0.984496) + (\frac{1}{8})^2(-2)(1) = 0.984254$.

In a similar manner, $y_2 \approx 0.938711$, $y_3 \approx 0.867501$, and $y_4 \approx 0.776805$.

Exercises 19.5

[1] From Hooke's law, $F = ky \Rightarrow 5 = k(6 \text{ in.} = \frac{1}{2} \text{ ft}) \Rightarrow k = 10$. $W = mg \Rightarrow$

$m = \frac{W}{g} = \frac{5}{32}$. Also, $\omega^2 = \frac{k}{m} = 64 \Rightarrow \omega = 8$. By Example 1,

$y = C_1 \cos 8t + C_2 \sin 8t$. $y = -\frac{1}{3}$ when $t = 0 \Rightarrow -\frac{1}{3} = C_1 \cos 0 + C_2 \sin 0 \Rightarrow$

$C_1 = -\frac{1}{3}$. $y' = -8C_1 \sin 8t + 8C_2 \cos 8t = 0$ when $t = 0 \Rightarrow$

$0 = -8C_1 \sin 0 + 8C_2 \cos 0 \Rightarrow C_2 = 0$. Thus, $y = -\frac{1}{3}\cos 8t$.

[3] $ky = F \Rightarrow k(1) = 10 \Rightarrow k = 10$ and $m = \frac{W}{g} = \frac{10}{32}$. Also, $\omega^2 = \frac{k}{m} = 32$.

Using (19.16) with $c = 5$ and $2p = \frac{c}{m} = 16$ or $p = 8 \Rightarrow p^2 - \omega^2 = 32 > 0$,

and the motion is overdamped. By Case 1, $y = e^{-8t}(C_1 e^{\sqrt{32}\,t} + C_2 e^{-\sqrt{32}\,t})$.

$y = 0$ when $t = 0 \Rightarrow C_1 + C_2 = 0$ and

$y' = e^{-8t}\left[C_1 e^{\sqrt{32}\,t}(\sqrt{32} - 8) - C_2 e^{-\sqrt{32}\,t}(\sqrt{32} + 8)\right] = 2$ when $t = 0 \Rightarrow$

$(\sqrt{32} - 8)\,C_1 + (-\sqrt{32} - 8)\,C_2 = 2$.

So $C_1 = \frac{1}{8}\sqrt{2}$ and $C_2 = -\frac{1}{8}\sqrt{2}$. Thus, $y = \frac{1}{8}\sqrt{2}\,e^{-8t}(e^{4\sqrt{2}\,t} - e^{-4\sqrt{2}\,t})$.

[5] $ky = F \Rightarrow k(\frac{1}{4}) = 4 \Rightarrow k = 16$ and $m = \frac{W}{g} = \frac{4}{32}$. Also, $\omega^2 = \frac{k}{m} = 128$.

Using (19.16) with $c = 2$ and $2p = \frac{c}{m} = 16$ or $p = 8 \Rightarrow p^2 - \omega^2 = -64 < 0$,

and the motion is underdamped. The roots of the auxiliary equation

$m^2 + 2pm + \omega^2 = 0$ are $a \pm bi = -8 \pm 8i$.

By Case 3, $y = e^{-8t}(C_1 \cos 8t + C_2 \sin 8t)$. $y = \frac{1}{3}$ when $t = 0 \Rightarrow C_1 = \frac{1}{3}$ and

$y' = -e^{-8t}\left[8(C_1 - C_2)\cos 8t + 8(C_1 + C_2)\sin 8t\right] = 0$ when $t = 0 \Rightarrow$

$-8(\frac{1}{3} - C_2) = 0$, or $C_2 = \frac{1}{3}$. Thus, $y = \frac{1}{3}e^{-8t}(\cos 8t + \sin 8t)$.

[9] By Case 2, critical damping will occur when $p^2 - \omega^2 = 0$.

Let the damping force be $-c\frac{dy}{dt}$. $2p = \frac{c}{m} = \frac{c}{3/4} = \frac{4}{3}c \Rightarrow p = \frac{2}{3}c$ and $\omega^2 = 32$.

$p^2 - \omega^2 = \frac{4}{9}c^2 - 32 = 0$ when $c = 6\sqrt{2}$.

Exercises 19.6

[1] Substituting $y = \sum_{n=0}^{\infty} a_n x^n$ and $y'' = \sum_{n=2}^{\infty} n(n-1)a_n x^{n-2}$, we obtain

$y'' + y = \sum_{n=2}^{\infty} n(n-1)a_n x^{n-2} + \sum_{n=0}^{\infty} a_n x^n = 0$. We replace n with $n + 2$ in the

first series so that both series have the same powers of x. Hence

$$\sum_{n=0}^{\infty} (n+2)(n+1)a_{n+2} x^n + \sum_{n=0}^{\infty} a_n x^n = \sum_{n=0}^{\infty}\left[(n+2)(n+1)a_{n+2} + a_n\right]x^n = 0.$$

Since the coefficients of x are equal to 0 for every n, $a_{n+2} = -\dfrac{1}{(n+2)(n+1)}a_n$.

The even-numbered terms corresponding to $n = 0, 2, 4, \ldots$, are $a_2 = -\frac{1}{2\cdot 1}a_0$,

$a_4 = -\frac{1}{4\cdot 3}a_2 = \frac{1}{4!}a_0$, $a_6 = -\frac{1}{6\cdot 5}a_4 = -\frac{1}{6!}a_0$, …, and, in general,

$$a_{2k} = (-1)^k \frac{1}{(2k)!}a_0.$$

The odd-numbered terms corresponding to $n = 1, 3, 5, \ldots$, are $a_3 = -\frac{1}{3\cdot 2}a_1$,

$a_5 = -\frac{1}{5\cdot 4}a_3 = \frac{1}{5!}a_1$, $a_7 = -\frac{1}{7\cdot 6}a_5 = -\frac{1}{7!}a_1$, …, and, in general,

$$a_{2k+1} = (-1)^k \frac{1}{(2k+1)!}a_1.$$

(continued)

As in Example 2, $y = \sum_{n=0}^{\infty} a_n x^n$ can be represented as the sum of two series.

Thus, $y = a_0 \sum_{n=0}^{\infty} \frac{(-1)^n}{(2n)!} x^{2n} + a_1 \sum_{n=0}^{\infty} \frac{(-1)^n}{(2n+1)!} x^{2n+1} = a_0 \cos x + a_1 \sin x.$

5 Using the same substitutions as in Example 4,

$$y'' - xy' + 2y = \sum_{n=2}^{\infty} n(n-1) a_n x^{n-2} - x \sum_{n=1}^{\infty} n a_n x^{n-1} + 2 \sum_{n=0}^{\infty} a_n x^n$$

$$= \sum_{n=0}^{\infty} (n+2)(n+1) a_{n+2} x^n - \sum_{n=1}^{\infty} n a_n x^n + 2 \sum_{n=0}^{\infty} a_n x^n$$

{change n to $n + 2$ in the first summation}

$$= (2a_2 + 2a_0) + \sum_{n=1}^{\infty} \Big[(n+2)(n+1) a_{n+2} - n a_n + 2 a_n\Big] x^n$$

$$= 0 \Rightarrow a_2 = -a_0 \text{ and } a_{n+2} = \frac{n-2}{(n+2)(n+1)} a_n.$$

When $n = 2$, $a_4 = 0$, and hence $a_6 = a_8 = a_{10} = \cdots = 0$.

The odd-numbered terms are $a_3 = \frac{-1}{3 \cdot 2} a_1 = -\frac{1}{3!} a_1$, $a_5 = \frac{1}{5 \cdot 4} a_3 = -\frac{1}{5!} a_1$,

$a_7 = \frac{3}{7 \cdot 6} a_5 = -\frac{3 \cdot 1}{7!} a_1, \ldots,$ and $a_{2k+1} = -\frac{(2k-3)(2k-5)\cdots 5 \cdot 3 \cdot 1}{(2k+1)!} a_1$, for $k \geq 2$.

$y = \sum_{n=0}^{\infty} a_n x^n = a_0 + a_1 x + a_2 x^2 + a_3 x^3 + \sum_{n=5}^{\infty} a_n x^n.$

Since $a_2 = -a_0$ and $a_3 = -\frac{1}{6} a_1$, we have

$$y = a_0(1 - x^2) + a_1 \left[x - \tfrac{1}{6} x^3 - \sum_{n=2}^{\infty} \frac{(2n-3)(2n-5)\cdots 5 \cdot 3 \cdot 1}{(2n+1)!} x^{2n+1} \right].$$

7 $$xy' + y' - 3y = x \sum_{n=1}^{\infty} n a_n x^{n-1} + \sum_{n=1}^{\infty} n a_n x^{n-1} - 3 \sum_{n=0}^{\infty} a_n x^n$$

$$= \sum_{n=1}^{\infty} n a_n x^n + \sum_{n=0}^{\infty} (n+1) a_{n+1} x^n - 3 \sum_{n=0}^{\infty} a_n x^n$$

$$= \sum_{n=1}^{\infty} n a_n x^n + a_1 + \sum_{n=1}^{\infty} (n+1) a_{n+1} x^n - 3a_0 - 3 \sum_{n=1}^{\infty} a_n x^n$$

$$= -3a_0 + a_1 + \sum_{n=1}^{\infty} \Big[n a_n + (n+1) a_{n+1} - 3 a_n \Big] x^n$$

$$= 0 \Rightarrow a_1 = 3a_0 \text{ and } a_{n+1} = \frac{3-n}{n+1} a_n.$$

Hence, $a_2 = \frac{2}{2} a_1 = 3a_0$, $a_3 = \frac{1}{3} a_2 = a_0$, $a_4 = 0$,

and since every other a_i is $\frac{3-n}{n+1}$ times its predecessor, the rest of the terms are 0.

Thus, $y = \sum_{n=0}^{3} a_n x^n = a_0 + 3a_0 x + 3a_0 x^2 + a_0 x^3 = a_0 (x+1)^3.$

[9] $y'' - y - 5x = \sum_{n=2}^{\infty} n(n-1)a_n x^{n-2} - \sum_{n=0}^{\infty} a_n x^n - 5x$

$= \sum_{n=0}^{\infty} (n+2)(n+1)a_{n+2} x^n - \sum_{n=0}^{\infty} a_n x^n - 5x$

$= -5x + \sum_{n=0}^{\infty} \left[(n+2)(n+1)a_{n+2} - a_n\right] x^n$

{let $n = 0$, 1 to determine the coefficient of x}

$= (2a_2 - a_0) + (-5 + 6a_3 - a_1)x + \sum_{n=2}^{\infty} \left[(n+2)(n+1)a_{n+2} - a_n\right] x^n$

$= 0 \Rightarrow a_2 = \frac{1}{2\cdot 1}a_0,\ a_3 = \frac{a_1+5}{3\cdot 2}$, and $a_{n+2} = \frac{1}{(n+2)(n+1)}a_n$.

For $n = 2, 4, 6, \ldots$: $a_4 = \frac{1}{4\cdot 3}a_2 = \frac{1}{4!}a_0,\ a_6 = \frac{1}{6\cdot 5}a_4 = \frac{1}{6!}a_0, \ldots,$

and $a_{2k} = \frac{1}{(2k)!}a_0$.

For $n = 3, 5, 7, \ldots$: $a_5 = \frac{1}{5\cdot 4}a_3 = \frac{a_1+5}{5!},\ a_7 = \frac{1}{7\cdot 6}a_5 = \frac{a_1+5}{7!},$

and $a_{2k+1} = \frac{a_1+5}{(2k+1)!}$.

Thus, $y = -5x + a_0 \sum_{n=0}^{\infty} \frac{1}{(2n)!} x^{2n} + (a_1+5) \sum_{n=0}^{\infty} \frac{1}{(2n+1)!} x^{2n+1}$

$= -5x + a_0 \cosh x + (a_1+5)\sinh x$

$= -5x + a_0\left(\frac{e^x + e^{-x}}{2}\right) + (a_1+5)\left(\frac{e^x - e^{-x}}{2}\right)$

$= -5x + (\frac{1}{2}a_0 + \frac{1}{2}a_1 + \frac{5}{2})e^x + (\frac{1}{2}a_0 - \frac{1}{2}a_1 - \frac{5}{2})e^{-x}$

19.7 Review Exercises

[1] $xe^y\,dx - \csc x\,dy = 0 \Rightarrow x\sin x\,dx - e^{-y}\,dy = 0 \Rightarrow$

$\int x\sin x\,dx - \int e^{-y}\,dy = 0$ {Formula 82} $\Rightarrow \sin x - x\cos x + e^{-y} = C.$

[5] $y' + (\sec x)\,y = 2\cos x \Rightarrow IF = e^{\int \sec x\,dx} = |\sec x + \tan x|.$

$y'(\sec x + \tan x) + y\sec x(\sec x + \tan x) = 2\cos x(\sec x + \tan x) = 2 + 2\sin x \Rightarrow$

$D_x\left[y(\sec x + \tan x)\right] = 2 + 2\sin x \Rightarrow$

$y(\sec x + \tan x) = 2x - 2\cos x + C \Rightarrow y = \frac{2x - 2\cos x + C}{\sec x + \tan x}.$

[9] $y\sqrt{1-x^2}\,\frac{dy}{dx} = \sqrt{1-y^2} \Rightarrow \frac{y}{\sqrt{1-y^2}}\,dy = \frac{1}{\sqrt{1-x^2}}\,dx \Rightarrow$

$-\sqrt{1-y^2} = \sin^{-1}x + C \Rightarrow \sqrt{1-y^2} + \sin^{-1}x = C.$

[13] $y' + (2\cos x)\,y = \cos x \Rightarrow IF = e^{\int 2\cos x\,dx} = e^{2\sin x}.$

$y'\,e^{2\sin x} + 2y\cos x\,e^{2\sin x} = \cos x\,e^{2\sin x} \Rightarrow$

$\int D_x(y\,e^{2\sin x})\,dx = \int \cos x\,e^{2\sin x}\,dx$ $\{u = 2\sin x,\ \frac{1}{2}\,du = \cos x\,dx\} \Rightarrow$

$y\,e^{2\sin x} = \frac{1}{2}e^{2\sin x} + C \Rightarrow y = \frac{1}{2} + Ce^{-2\sin x}.$

[15] The auxiliary equation is $m^2 - 8m + 16 = (m - 4)^2 = 0 \Rightarrow$

$m = 4$ is a double root. Thus, by (19.7), $y = C_1 e^{4x} + C_2 x e^{4x} = e^{4x}(C_1 + C_2 x)$.

[19] The auxiliary equation is $m^2 - 1 = 0 \Rightarrow m = \pm 1$.

By (19.6), $y_c = C_1 e^{-x} + C_2 e^{x}$. By (19.14)(iii), let $y_p = Ae^x \cos x + Be^x \sin x$.

Then, $y_p'' - y_p = (2B - A)e^x \cos x + (-2A - B)e^x \sin x = e^x \sin x \Rightarrow A = -\frac{2}{5}$

and $B = -\frac{1}{5}$. The general solution is $y = C_1 e^{-x} + C_2 e^x - \frac{1}{5}e^x(2\cos x + \sin x)$.

[23] The auxiliary equation is $m^2 - 3m + 2 = 0 \Rightarrow m = 1, 2$.

By (19.6), $y_c = C_1 e^x + C_2 e^{2x}$. Let $y_p = Ae^{5x}$.

Then, $y_p'' - 3y_p' + 2y_p = (25A - 15A + 2A)e^{5x} = e^{5x} \Rightarrow A = \frac{1}{12}$.

The general solution is $y = C_1 e^x + C_2 e^{2x} + \frac{1}{12}e^{5x}$.

[25] Since $D_x(xy) = xy' + y$, $xy' + y = (x - 2)^2 \Rightarrow$

$$xy = \tfrac{1}{3}(x - 2)^3 + C \Rightarrow y = \frac{(x-2)^3}{3x} + \frac{C}{x}.$$

[27] $m^2 + 5m + 7 = 0 \Rightarrow m = \dfrac{-5 \pm \sqrt{-3}}{2} = -\frac{5}{2} \pm \frac{1}{2}\sqrt{3}\,i$.

By (19.10) with $s \pm ti = -\frac{5}{2} \pm \frac{1}{2}\sqrt{3}\,i$, $y = e^{-5x/2}\left[C_1 \cos\left(\tfrac{1}{2}\sqrt{3}\,x\right) + C_2 \sin\left(\tfrac{1}{2}\sqrt{3}\,x\right)\right]$.

[31] $\cot x\, dy = (y - \cos x)\, dx \Rightarrow \cot x\left(\dfrac{dy}{dx}\right) = y - \cos x \Rightarrow y' = y \tan x - \sin x \Rightarrow$

$y' - y \tan x = -\sin x \Rightarrow IF = e^{\int -\tan x\, dx} = |\cos x|$.

$y' \cos x - y \sin x = -\sin x \cos x \Rightarrow y \cos x = \frac{1}{2}\cos^2 x + C \Rightarrow y = \frac{1}{2}\cos x + C \sec x$.

[35] $y' - (3 \sin 2\pi t)y = 2000 \sin 2\pi t \Rightarrow IF = e^{\int -3 \sin 2\pi t\, dt} = e^{(3/(2\pi)) \cos 2\pi t}$.

$y' e^{(3/(2\pi)) \cos 2\pi t} - (3 \sin 2\pi t)\, e^{(3/(2\pi)) \cos 2\pi t} = (2000 \sin 2\pi t)\, e^{(3/(2\pi)) \cos 2\pi t} \Rightarrow$

$ye^{(3/(2\pi)) \cos 2\pi t} = -\frac{2000}{3}\, e^{(3/(2\pi)) \cos 2\pi t} + C \Rightarrow y = -\frac{2000}{3} + Ce^{-(3/(2\pi)) \cos 2\pi t}$.

$y = 500$ when $t = 0 \Rightarrow 500 = -\frac{2000}{3} + Ce^{-3/(2\pi)} \Rightarrow C = \frac{3500}{3}\, e^{3/(2\pi)}$.

$y = \frac{3500}{3}\, e^{(3/(2\pi))(1 - \cos 2\pi t)} - \frac{2000}{3}$.

The maximum value of y will occur when the exponent is maximum, that is,

when $\cos 2\pi t = -1$. Thus, the maximum is $\frac{3500}{3}\, e^{3/\pi} - \frac{2000}{3} \approx 2365$.

37 $\frac{dy}{dt} = k(a - y)(b - y) \Rightarrow \frac{dy}{(a - y)(b - y)} = k\,dt$. Using partial fractions and integrating, we have $\int\left[\frac{1}{b - a}\left(\frac{1}{a - y} - \frac{1}{b - y}\right)\right]dy = \int k\,dt \Rightarrow$

$\frac{1}{b - a}\ln\left(\frac{b - y}{a - y}\right) = kt + C$. $y = 0$ when $t = 0 \Rightarrow C = \frac{\ln(b/a)}{b - a}$. Thus,

$\frac{1}{b - a}\ln\left(\frac{b - y}{a - y}\right) = kt + \frac{\ln(b/a)}{b - a} \Rightarrow \frac{1}{b - a}\left[\ln\left(\frac{b - y}{a - y}\right) + \ln\left(\frac{a}{b}\right)\right] = kt \Rightarrow$

$\frac{1}{b - a}\ln\left[\frac{a}{b}\left(\frac{b - y}{a - y}\right)\right] = kt \Rightarrow \frac{a}{b}\left(\frac{b - y}{a - y}\right) = e^{k(b-a)t} \Rightarrow a(b - y) = b(a - y)e^{k(b-a)t}$

$\Rightarrow bye^{k(b-a)t} - ay = abe^{k(b-a)t} - ab \Rightarrow y = f(t) = \frac{ab\left[e^{k(b-a)t} - 1\right]}{be^{k(b-a)t} - a}$.

39 The slope of the line connecting the origin with the point $P(x, y)$ is $m = \frac{y}{x}$.

If $y = f(x)$ has a tangent line at $P(x, y)$ perpendicular to this line, then $\frac{dy}{dx} = -\frac{x}{y}$.

Thus, $y\,dy + x\,dx = 0 \Rightarrow y^2 + x^2 = C$, which is a circle with center at the origin.

Appendix I: Mathematical Induction

Note: P_n is the statement in the text for Exercises 1–22.

[1] (i) P_1 is true, since $2(1) = 1(1+1) = 2$.

(ii) Assume P_k is true:

$2 + 4 + 6 + \cdots + 2k = k(k+1)$. Hence

$$\begin{aligned} 2 + 4 + 6 + \cdots + 2k + 2(k+1) &= k(k+1) + 2(k+1) \\ &= (k+1)(k+2) = (k+1)(k+1+1). \end{aligned}$$

Thus, P_{k+1} is true and the proof is complete.

[3] (i) P_1 is true, since $2(1) - 1 = (1)^2 = 1$.

(ii) Assume P_k is true:

$1 + 3 + 5 + \cdots + (2k-1) = k^2$. Hence

$$\begin{aligned} 1 + 3 + 5 + \cdots + (2k-1) + 2(k+1) - 1 &= k^2 + 2(k+1) - 1 \\ &= k^2 + 2k + 1 \\ &= (k+1)^2. \end{aligned}$$

Thus, P_{k+1} is true and the proof is complete.

[5] (i) P_1 is true, since $5(1) - 3 = \frac{1}{2}(1)[5(1) - 1] = 2$.

(ii) Assume P_k is true:

$2 + 7 + 12 + \cdots + (5k-3) = \frac{1}{2}k(5k-1)$. Hence

$2 + 7 + 12 + \cdots + (5k-3) + 5(k+1) - 3$

$$\begin{aligned} &= \tfrac{1}{2}k(5k-1) + 5(k+1) - 3 \\ &= \tfrac{5}{2}k^2 + \tfrac{9}{2}k + 2 \\ &= \tfrac{1}{2}(5k^2 + 9k + 4) \\ &= \tfrac{1}{2}(k+1)(5k+4) \\ &= \tfrac{1}{2}(k+1)[5(k+1) - 1]. \end{aligned}$$

Thus, P_{k+1} is true and the proof is complete.

[7] (i) P_1 is true, since $1 \cdot 2^{1-1} = 1 + (1-1) \cdot 2^1 = 1$.

(ii) Assume P_k is true:

$1 + 2 \cdot 2 + 3 \cdot 2^2 + \cdots + k \cdot 2^{k-1} = 1 + (k-1) \cdot 2^k$. Hence

$1 + 2 \cdot 2 + 3 \cdot 2^2 + \cdots + k \cdot 2^{k-1} + (k+1) \cdot 2^k$

$$\begin{aligned} &= 1 + (k-1) \cdot 2^k + (k+1) \cdot 2^k \\ &= 1 + k \cdot 2^k - 2^k + k \cdot 2^k + 2^k \\ &= 1 + k \cdot 2^1 \cdot 2^k \\ &= 1 + [(k+1) - 1] \cdot 2^{k+1}. \end{aligned}$$

Thus, P_{k+1} is true and the proof is complete.

[9] (i) P_1 is true, since $(1)^1 = \dfrac{1(1+1)[2(1)+1]}{6} = 1.$

(ii) Assume P_k is true:

$1^2 + 2^2 + 3^2 + \cdots + k^2 = \dfrac{k(k+1)(2k+1)}{6}$. Hence

$$\begin{aligned} 1^2 + 2^2 + 3^2 + \cdots + k^2 + (k+1)^2 &= \frac{k(k+1)(2k+1)}{6} + (k+1)^2 \\ &= (k+1)\left[\frac{k(2k+1)}{6} + \frac{6(k+1)}{6}\right] \\ &= \frac{(k+1)(2k^2+7k+6)}{6} \\ &= \frac{(k+1)(k+2)(2k+3)}{6}. \end{aligned}$$

Thus, P_{k+1} is true and the proof is complete.

[11] (i) P_1 is true, since $\dfrac{1}{1(1+1)} = \dfrac{1}{1+1} = \dfrac{1}{2}$.

(ii) Assume P_k is true:

$\dfrac{1}{1\cdot 2} + \dfrac{1}{2\cdot 3} + \dfrac{1}{3\cdot 4} + \cdots + \dfrac{1}{k(k+1)} = \dfrac{k}{k+1}$. Hence

$$\begin{aligned} &\frac{1}{1\cdot 2} + \frac{1}{2\cdot 3} + \frac{1}{3\cdot 4} + \cdots + \frac{1}{k(k+1)} + \frac{1}{(k+1)(k+2)} \\ &\qquad = \frac{k}{k+1} + \frac{1}{(k+1)(k+2)} \\ &\qquad = \frac{k(k+2)+1}{(k+1)(k+2)} \\ &\qquad = \frac{k^2+2k+1}{(k+1)(k+2)} = \frac{k+1}{(k+1)+1}. \end{aligned}$$

Thus, P_{k+1} is true and the proof is complete.

[13] (i) P_1 is true, since $3^1 = \frac{3}{2}(3^1 - 1) = 3$.

(ii) Assume P_k is true:

$3 + 3^2 + 3^3 + \cdots + 3^k = \frac{3}{2}(3^k - 1)$. Hence

$$\begin{aligned} 3 + 3^2 + 3^3 + \cdots + 3^k + 3^{k+1} &= \tfrac{3}{2}(3^k - 1) + 3^{k+1} \\ &= \tfrac{3}{2}\cdot 3^k - \tfrac{3}{2} + 3\cdot 3^k = \tfrac{9}{2}\cdot 3^k - \tfrac{3}{2} \\ &= \tfrac{3}{2}(3\cdot 3^k - 1) = \tfrac{3}{2}(3^{k+1} - 1). \end{aligned}$$

Thus, P_{k+1} is true and the proof is complete.

[15] (i) P_1 is true, since $1 < 2^1$.

(ii) Assume P_k is true: $k < 2^k$. Now $k + 1 < k + k = 2(k)$ for $k > 1$.

From P_k, we see that $2(k) < 2(2^k) = 2^{k+1}$ and conclude that $k + 1 < 2^{k+1}$.

Thus, P_{k+1} is true and the proof is complete.

17 (i) P_1 is true, since $1 < \frac{1}{8}[2(1) + 1]^2 = \frac{9}{8}$.

(ii) Assume P_k is true: $1 + 2 + 3 + \cdots + k < \frac{1}{8}(2k + 1)^2$. Hence

$$\begin{aligned} 1 + 2 + 3 + \cdots + k + (k + 1) &< \tfrac{1}{8}(2k + 1)^2 + (k + 1) \\ &= \tfrac{1}{2}k^2 + \tfrac{3}{2}k + \tfrac{9}{8} = \tfrac{1}{8}(4k^2 + 12k + 9) \\ &= \tfrac{1}{8}(2k + 3)^2 = \tfrac{1}{8}[2(k + 1) + 1]^2. \end{aligned}$$

Thus, P_{k+1} is true and the proof is complete.

19 (i) For $n = 1$, $n^3 - n + 3 = 3$ and 3 is a factor of 3.

(ii) Assume 3 is a factor of $k^3 - k + 3$. The $(k + 1)$st term is

$$\begin{aligned} (k + 1)^3 - (k + 1) + 3 &= k^3 + 3k^2 + 2k + 3 \\ &= (k^3 - k + 3) + 3k^2 + 3k \\ &= (k^3 - k + 3) + 3(k^2 + k). \end{aligned}$$

By the induction hypothesis, 3 is a factor of $k^3 - k + 3$ and

3 is a factor of $3(k^2 + k)$, so 3 is a factor of the $(k + 1)$st term.

Thus, P_{k+1} is true and the proof is complete.

21 (i) For $n = 1$, $5^n - 1 = 4$ and 4 is a factor of 4.

(ii) Assume 4 is a factor of $5^k - 1$. The $(k + 1)$st term is

$5^{k+1} - 1 = 5 \cdot 5^k - 1 = 5 \cdot 5^k - 5 + 4 = 5(5^k - 1) + 4$.

By the induction hypothesis, 4 is a factor of 5^{k-1} and

4 is a factor of 4, so 4 is a factor of the $(k + 1)$st term.

Thus, P_{k+1} is true and the proof is complete.

Note: For Exercises 23-30, there are several ways to find j. Possibilities include solving the inequality, sketching the graphs of the functions that represent each side, and trial and error. Trial and error may be the easiest to use.

23 $n + 12 \leq n^2$

For j: $n^2 \geq n + 12 \Rightarrow n^2 - n - 12 \geq 0 \Rightarrow (n - 4)(n + 3) \geq 0 \Rightarrow n \geq 4$ $\{n>0\}$

(i) P_4 is true, since $4 + 12 \leq 4^2$.

(ii) Assume P_k is true: $k + 12 \leq k^2$. Hence

$(k + 1) + 12 = (k + 12) + 1 \leq (k^2) + 1 < k^2 + 2k + 1 = (k + 1)^2$.

Thus, P_{k+1} is true and the proof is complete.

25 $5 + \log_2 n \le n$

For j: By sketching $y = 5 + \log_2 x$ and $y = x$, we see that the solution for $x > 1$ must be larger than 5. See *Figure 25*. By trial and error, $j = 8$.

(i) P_8 is true, since $5 + \log_2 8 \le 8$.

(ii) Assume P_k is true: $5 + \log_2 k \le k$. Hence

$$\begin{aligned} 5 + \log_2(k+1) < 5 + \log_2(k+k) &= 5 + \log_2 2k \\ &= 5 + \log_2 2 + \log_2 k \\ &= (5 + \log_2 k) + 1 \le k + 1. \end{aligned}$$

Thus, P_{k+1} is true and the proof is complete.

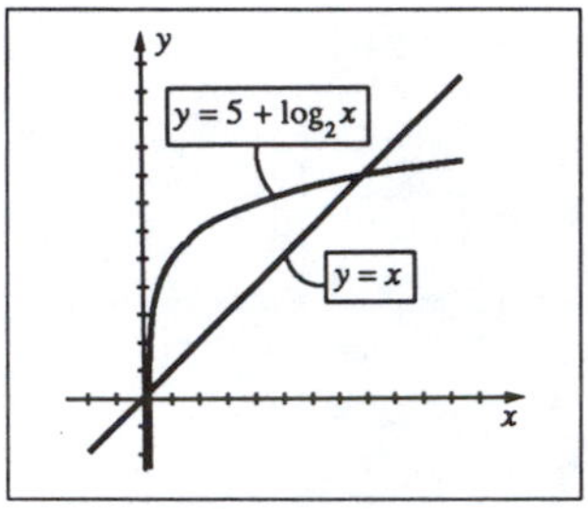

Figure 25

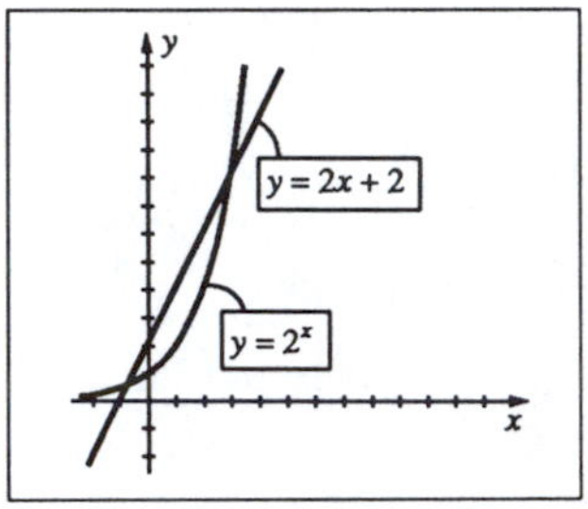

Figure 29

27 $2^n \le n!$

For j: Examining the pattern formed by letting $n = 1, 2, 3, 4$ leads us to the conclusion that $j = 4$.

(i) P_4 is true, since $2^4 \le 4!$.

(ii) Assume P_k is true: $2^k \le k!$. Hence

$2^{k+1} = 2 \cdot 2^k \le 2 \cdot k! < (k+1) \cdot k! = (k+1)!$.

Thus, P_{k+1} is true and the proof is complete.

29 $2n + 2 \le 2^n$

For j: By sketching $y = 2x + 2$ and $y = 2^x$, we see there is one positive solution. See *Figure 29*. By trial and error, $j = 3$.

(i) P_3 is true, since $2(3) + 2 \le 2^3$.

(ii) Assume P_k is true: $2k + 2 \le 2^k$. Hence

$2(k+1) + 2 = (2k+2) + 2 \le 2^k + 2^k = 2 \cdot 2^k = 2^{k+1}$.

Thus, P_{k+1} is true and the proof is complete.

31 (i) If $a > 1$ then $a^1 = a > 1$, so P_1 is true.

(ii) Assume P_k is true: $a^k > 1$.

Multiply both sides by a to obtain $a^{k+1} > a$, but since $a > 1$, we have $a^{k+1} > 1$.

Thus, P_{k+1} is true and the proof is complete.

[33] (i) For $n = 1$, $a - b$ is a factor of $a^1 - b^1$.

(ii) Assume $a - b$ is a factor of $a^k - b^k$. Following the hint for the $(k + 1)$st term,

$a^{k+1} - b^{k+1} = a^k \cdot a - b \cdot a^k + b \cdot a^k - b^k \cdot b = a^k(a - b) + (a^k - b^k)\,b.$

Since $(a - b)$ is a factor of $a^k(a - b)$ and since by the induction hypothesis, $a - b$ is a factor of $(a^k - b^k)$, it follows that $a - b$ is a factor of the $(k + 1)$st term.

Thus, P_{k+1} is true and the proof is complete.